Dear Educator,

Thank you for your interest in American Book Company's state-specific test preparation resources. Enclosed you will find the preview book and/or demonstration disk that you requested. We commend you for your interest in pursuing your students' success. Feel free to contact us with any questions about our books, software, or the ordering process.

Our Products Feature	**Your Students Will Improve**
Multiple-choice diagnostic tests	Confidence and mastery of subjects
Step-by-step instruction	Concept development
Frequent practice exercises	Critical thinking
Chapter reviews	Test-taking skills
Multiple-choice practice tests	Problem-solving skills

About American Book Company

American Book Company's writers and curriculum specialists have **over 100 years of combined teaching experience**, working with students from kindergarten through middle, high school, and adult education.

Our company specializes in **test preparation books and software** for high school graduation tests and exit exams. We currently offer test preparation materials for Alabama, Arizona, California, Florida, Georgia, Indiana, Louisiana, Maine, Maryland, Minnesota, Mississippi, Nevada, New Jersey, North Carolina, Ohio, Oklahoma, South Carolina, Tennessee, and Texas.

We also offer books and software for **middle school review and high school remediation**. The materials in the Basics Made Easy Series are aligned with the standards for the Iowa Test of Basic Skills and the Stanford 9 Achievement Test.

While some other book publishers offer general test preparation materials, our student workbooks and software are **specifically designed** to meet the unique requirements of **each state's exit exam** or graduation test. Whether the subject is language arts, math, reading, science, social studies, or writing, our books and software are designed to meet the standards published by the state agency responsible for the graduation test or exit exam. Our materials provide **no tricks or secret solutions** to passing standardized tests, just engaging instruction and practical exercises to help students master the concepts and skills they need.

While we cannot guarantee success, our products are designed to provide students with the concept and skill development they need for the graduation test or exit exam in their own state. We look forward to hearing from you soon.

Sincerely,

Joe Wood
Curriculum Specialist

PO Box 2638★Woodstock, GA 30188-1383★Phone: 1-888-264-5877★Fax: 1-866-827-3240
Web Site: www.americanbookcompany.com★E-mail: contact@americanbookcompany.com

SAT Mathematics

Test Preparation Guide

ERICA DAY

COLLEEN PINTOZZI

AMERICAN BOOK COMPANY

P. O. BOX 2638

WOODSTOCK, GEORGIA 30188-1383

TOLL FREE 1 (888) 264-5877   PHONE (770) 928-2834

TOLL FREE FAX 1 (866) 827-3240

WEB SITE: www.americanbookcompany.com

Acknowledgements

In preparing this book, we would like to acknowledge Mary Stoddard, Philip Jones, Marsha Torrens, and Eric Field for their contributions in editing and developing graphics for this book. We would also like to thank our many students whose needs and questions inspired us to write this text.

by American Book Company
P.O. Box 2638
Woodstock, GA 30188-1383

Printed in the United States of America
1/06

Contents

Acknowledgements ii

Preface xi

Diagnostic Test 1

Part 1 1

Part 2 3

Part 3 6

1 Numbers and Number Systems 11

1.1 Real Numbers 11

1.2 Integers 12

1.3 Even and Odd Integers 12

1.4 Prime Numbers 12

1.5 Digits 12

1.6 Absolute Value 13

1.7 Word Problems 14

1.8 Two-Step Problems 14

1.9 Order of Operations 15

1.10 Sets 17

1.11 Subsets 18

1.12 Intersection of Sets 19

1.13 Union of Sets 20

1.14 Reading Venn Diagrams 21

Chapter 1 Review 22

2 Fractions, Decimals, and Percents 24

2.1 Greatest Common Factor 24

2.2 Least Common Multiple 25

2.3 Fraction Review 25

2.4 Fraction Word Problems 26

2.5 Changing Fractions to Decimals 27

2.6 Changing Mixed Numbers to Decimals 28

2.7 Changing Decimals to Fractions 28

2.8 Changing Decimals with Whole Numbers to Mixed Numbers 29

2.9 Decimal Word Problems 29

2.10	Changing Percents to Decimals and Decimals to Percents	30
2.11	Changing Percents to Fractions and Fractions to Percents	31
2.12	Changing Percents to Mixed Numbers and Mixed Numbers to Percents	32
2.13	Changing to Percent Word Problems	33
2.14	Finding the Percent of the Total	34
2.15	Finding the Percent Increase or Decrease	35
2.16	Finding the Amount of a Discount	36
2.17	Finding the Discounted Sale Price	37
2.18	Sales Tax	38
	Chapter 2 Review	39
3	**Exponents and Roots**	**42**
3.1	Understanding Exponents	42
3.2	Multiplication with Exponents	43
3.3	Division with Exponents	44
3.4	Square Root	45
3.5	Simplifying Square Roots	45
3.6	Adding and Subtracting Roots	46
3.7	Multiplying Roots	47
3.8	Dividing Roots	48
	Chapter 3 Review	49
4	**Introduction to Algebra**	**50**
4.1	Algebra Vocabulary	50
4.2	Substituting Numbers for Variables	51
4.3	Understanding Algebra Word Problems	52
4.4	Setting Up Algebra Word Problems	54
4.5	Changing Algebra Word Problems to Algebraic Equations	55
4.6	Substituting Numbers in Formulas	56
4.7	Properties of Addition and Multiplication	57
	Chapter 4 Review	58
5	**Introduction to Graphing**	**60**
5.1	Graphing on a Number Line	60
5.2	Graphing Fractional Values	60
5.3	Recognizing Improper Fractions, Decimals, and Square Root Values on a Number Line	62
5.4	Plotting Points on a Vertical Number Line	64

5.5	Cartesian Coordinates	65
5.6	Identifying Ordered Pairs	66
	Chapter 5 Review	68
6	**Solving One-Step Equations and Inequalities**	**69**
6.1	One-Step Algebra Problems with Addition and Subtraction	69
6.2	One-Step Algebra Problems with Multiplication and Division	70
6.3	Multiplying and Dividing with Negative Numbers	72
6.4	Variables with a Coefficient of Negative One	73
6.5	Graphing Inequalities	74
6.6	Solving Inequalities by Addition and Subtraction	75
6.7	Solving Inequalities by Multiplication and Division	76
	Chapter 6 Review	77
7	**Solving Multi-Step Equations and Inequalities**	**79**
7.1	Two-Step Algebra Problems	79
7.2	Two-Step Algebra Problems with Fractions	80
7.3	More Two-Step Algebra Problems with Fractions	81
7.4	Combining Like Terms	82
7.5	Solving Equations with Like Terms	82
7.6	Removing Parentheses	84
7.7	Multi-Step Algebra Problems	85
7.8	Solving Radical Equations	87
7.9	Multi-Step Inequalities	88
7.10	Solving Equations and Inequalities with Absolute Values	90
7.11	More Solving Equations and Inequalities with Absolute Values	91
7.12	Inequality Word Problems	93
	Chapter 7 Review	94
8	**Rates, Ratios, and Proportions**	**95**
8.1	Time of Travel	95
8.2	Rate	96
8.3	More Rates	97
8.4	Distance	98
8.5	Ratio Problems	99
8.6	Writing Ratios Using Variables	100
8.7	Solving Proportions	101
8.8	Ratio and Proportion Word Problems	102

8.9	Proportional Reasoning	103
8.10	Direct and Indirect Variation	104
	Chapter 8 Review	107
9	**Polynomials**	**108**
9.1	Adding and Subtracting Monomials	108
9.2	Adding Polynomials	109
9.3	Subtracting Polynomials	110
9.4	Multiplying Monomials	111
9.5	Multiplying Monomials by Polynomials	112
9.6	Dividing Polynomials by Monomials	113
9.7	Removing Parentheses and Simplifying	114
9.8	Multiplying Two Binomials	115
9.9	Simplifying Expressions with Exponents	116
	Chapter 9 Review	117
10	**Factoring**	**118**
10.1	Factor By Grouping	121
10.2	Factoring Trinomials	122
10.3	More Factoring Trinomials	124
10.4	Factoring More Trinomials	125
10.5	Factoring Trinomials with Two Variables	127
10.6	Factoring the Difference of Two Squares	128
10.7	Simplifying Algebraic Ratios	130
	Chapter 10 Review	131
11	**Solving Quadratic Equations**	**132**
11.1	Solving the Difference of Two Squares	134
11.2	Solving Perfect Squares	136
11.3	Completing the Square	137
11.4	Using the Quadratic Formula	138
	Chapter 11 Review	139
12	**Graphing and Writing Equations and Inequalities**	**140**
12.1	Graphing Linear Equations	140
12.2	Graphing Horizontal and Vertical Lines	142
12.3	Finding the Distance Between Two Points	143
12.4	Finding the Midpoint of a Line Segment	144

12.5 Finding the Intercepts of a Line 145
12.6 Understanding Slope 146
12.7 Slope-Intercept Form of a Line 148
12.8 Verify That a Point Lies on a Line 149
12.9 Graphing a Line Knowing a Point and Slope 150
12.10 Finding the Equation of a Line Using Two Points or a Point and Slope 151
12.11 Graphing Inequalities 152
Chapter 12 Review 155

13 Applications of Graphs **157**
13.1 Changing the Slope or Y-Intercept of a Line 157
13.2 Equations of Perpendicular Lines 159
13.3 Writing an Equation From Data 161
13.4 Graphing Linear Data 162
13.5 Identifying Graphs of Linear Equations 164
13.6 Graphing Non-Linear Equations 165
13.7 Finding the Vertex of a Quadratic Equation 166
13.8 Identifying Graphs of Real-World Situations 167
Chapter 13 Review 170

14 Systems of Equations and Systems of Inequalities **172**
14.1 Finding Common Solutions for Intersecting Lines 174
14.2 Solving Systems of Equations by Substitution 175
14.3 Solving Systems of Equations by Adding or Subtracting 176
14.4 Graphing Systems of Inequalities 178
14.5 Solving Word Problems with Systems of Equations 179
14.6 Consecutive Integer Problems 180
Chapter 14 Review 182

15 Relations and Functions **184**
15.1 Relations 184
15.2 Determining Domain and Range From Graphs 186
15.3 Domain and Range of Quadratic Equations 188
15.4 Functions 191
15.5 Function Notation 192
15.6 Recognizing Functions 193
15.7 Qualitative Behavior of Graphs 196
15.8 Relations That Can Be Represented by Functions 198

15.9	Exponential Growth and Decay	200
15.10	Piecewise Functions	202
	Chapter 15 Review	204
16 Series, Sequences, and Algorithms		**205**
16.1	Number Patterns	205
16.2	Geometric Patterns	206
16.3	Limits of Series and Sequences	208
16.4	Summing Arithmetic and Geometric Series	210
16.5	Algorithms	212
16.6	More Algorithms	213
16.7	Inductive Reasoning and Patterns	214
16.8	Mathematical Reasoning/Logic	218
16.9	Deductive and Inductive Arguments	219
	Chapter 16 Review	221
17 Statistics		**223**
17.1	Range	223
17.2	Mean	224
17.3	Finding Data Missing From the Mean	225
17.4	Median	226
17.5	Mode	227
17.6	Applying Measures of Central Tendency	228
17.7	Stem-and-Leaf Plots	229
17.8	More Stem-and-Leaf Plots	230
17.9	Quartiles and Extremes	231
17.10	Box-and-Whisker Plots	232
17.11	Scatter Plots	233
17.12	The Line of Best Fit	235
	Chapter 17 Review	236
18 Data Interpretation		**237**
18.1	Tally Charts and Frequency Tables	237
18.2	Histograms	238
18.3	Reading Tables	239
18.4	Bar Graphs	240
18.5	Line Graphs	241
18.6	Circle Graphs	242

	Chapter 18 Review	243
19 Probability and Counting		**245**
19.1	Probability	245
19.2	Independent and Dependent Events	247
19.3	Tree Diagrams	249
19.4	Probability Distributions	251
19.5	Geometric Probability	253
19.6	Permutations	256
19.7	More Permutations	258
19.8	Combinations	259
19.9	More Combinations	260
	Chapter 19 Review	261
20 Angles		**264**
20.1	Types of Angles	265
20.2	Measuring Angles	266
20.3	Central Angles	267
20.4	Adjacent Angles	269
20.5	Vertical Angles	270
20.6	Complementary and Supplementary Angles	271
20.7	Corresponding, Alternate Interior, and Alternate Exterior Angles	272
20.8	Sum of Interior Angles of a Polygon	273
20.9	Congruent Figures	274
20.10	Similar and Congruent	276
	Chapter 20 Review	277
21 Triangles		**279**
21.1	Types of Triangles	279
21.2	Interior Angles of a Triangle	279
21.3	Similar Triangles	280
21.4	Side and Angle Relationships	282
21.5	Pythagorean Theorem	283
21.6	Finding the Missing Leg of a Right Triangle	284
21.7	Applications of the Pythagorean Theorem	285
21.8	Special Right Triangles	287
21.9	Introduction to Trigonometric Ratios	289
	Chapter 21 Review	294

22 Plane and Solid Geometry **296**
22.1 Lines and Line Segments 296
22.2 Types of Polygons 296
22.3 Perimeter 297
22.4 Area of Squares and Rectangles 298
22.5 Area of Triangles 299
22.6 Parts of a Circle 300
22.7 Circumference 301
22.8 Area of a Circle 302
22.9 Two-Step Area Problems 303
22.10 Geometric Relationships of Plane Figures 305
22.11 Solids 307
22.12 Understanding Volume 307
22.13 Volume of Rectangular Prisms 308
22.14 Volume of Cubes 309
22.15 Volume of Cylinders 310
Chapter 22 Review 311

23 Transformations **313**
23.1 Drawing Geometric Figures on a Cartesian Coordinate Plane 313
23.2 Reflections 316
23.3 Translations 319
23.4 Rotations 321
23.5 Transformation Practice 322
23.6 Dilations 323
Chapter 23 Review 325

Practice Test 1 **328**
Part 1 328
Part 2 330
Part 3 333

Practice Test 2 **335**
Part 1 335
Part 2 337
Part 3 341

Index **344**

Preface

SAT Mathematics Test Preparation Guide will help you review and learn important concepts and skills related to high school mathematics. To help identify which areas are of greater challenge for you, first take the diagnostic test, then complete the evaluation chart with your instructor in order to help you identify the chapters which require your careful attention. When you have finished your review of all of the material your teacher assigns, take the practice tests to evaluate your understanding of the material presented in this book. **The materials in this book are based on the standards in mathematics published by the College Board.**

This book contains several sections. These sections are as follows: 1) A Diagnostic Test; 2) Chapters that teach the concepts and skills for the SAT Mathematics Test; 3) Two Practice Tests. Answers to the tests and exercises are in a separate manual.

ABOUT THE AUTHORS

Erica Day has a Bachelor of Science Degree in Mathematics and working on a Master of Science Degree in Mathematics. She graduated with high honors from Kennesaw State University in Kennesaw, Georgia. She has also tutored all levels of mathematics, ranging from high school algebra and geometry to university-level statistics, calculus, and linear algebra. She is currently writing and editing mathematics books for American Book Company, where she has coauthored numerous books, such as ***Passing the Georgia Algebra I End of Course***, ***Passing the Georgia High School Graduation Test in Mathematics***, ***Passing the Arizona AIMS in Mathematics***, and ***Passing the New Jersey HSPA in Mathematics***, to help students pass graduation and end of course exams.

Colleen Pintozzi has taught mathematics at the middle school, junior high, senior high, and adult level for 22 years. She holds a B.S. degree from Wright State University in Dayton, Ohio and has done graduate work at Wright State University, Duke University, and the University of North Carolina at Chapel Hill. She is the author of many mathematics books including such best-sellers as ***Basics Made Easy: Mathematics Review***, ***Passing the New Alabama Graduation Exam in Mathematics***, ***Passing the Louisiana LEAP 21 GEE***, ***Passing the Indiana ISTEP+ GQE in Mathematics***, ***Passing the Minnesota Basic Standards Test in Mathematics***, and ***Passing the Nevada High School Proficiency Exam in Mathematics.***

New SAT Mathematics
Reference Sheet

Notes

1. The use of a calculator is permitted.
2. All numbers used are real numbers.
3. Figures that accompany problems in this test are intended to provide information useful in solving the problems.
   They are drawn as accurately as possible EXCEPT when it is stated in a specific problem that the figure is not drawn to scale. All figures lie in a plane unless otherwise indicated.
4. Unless otherwise specified, the domain of any function f is assumed to be the set of all real numbers x for which $f(x)$ is a real number.

Reference Information

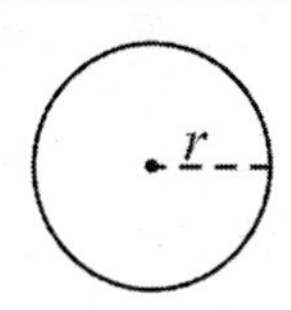

$A = \pi r^2$
$C = 2\pi r^2$

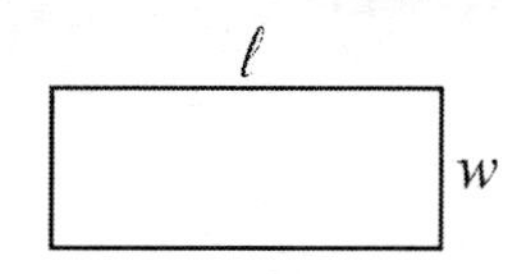

$A = \ell w$

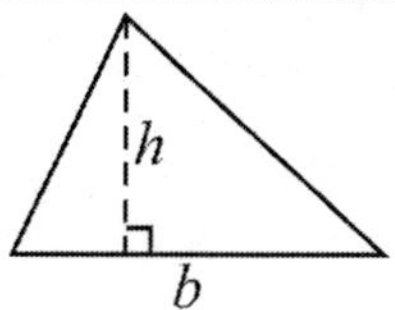

$A = \frac{1}{2}bh$

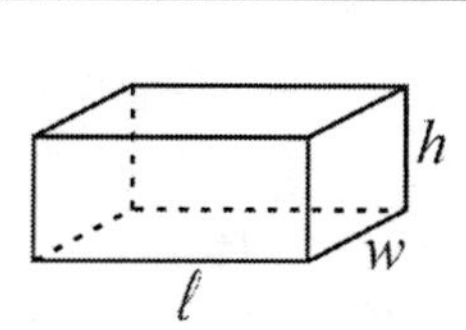

$V = \ell wh$

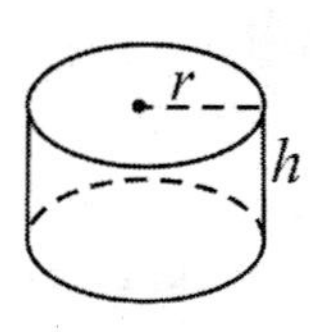

$V = \pi r^2 h$

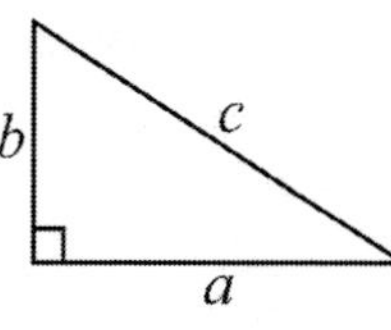

$c^2 = a^2 + b^2$

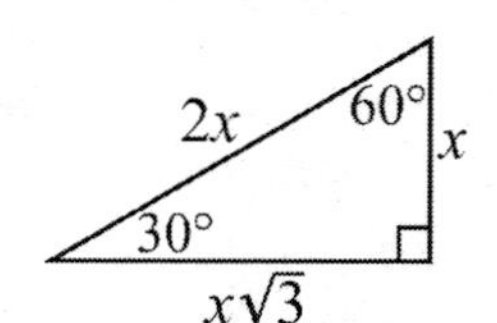

s 45° $s\sqrt{2}$ 45° s

Special Right Triangles

The number of degrees of arc in a circle is 360.

The sum of the measures in a degrees of the angles of a triangle is 180.

Diagnostic Test

Part 1

Questions 1–8 of part 1 of this test are multiple choice, and questions 9–18 are student-produced response questions. For questions 9–18, use the grids provided by your instructor. You have 25 minutes to complete this part of the test.

Note to instructor: A blank sheet of grids is located at the end of the answer key.

1. In the figure below, $x \| y$. If $a = 115°$, what is $b + g$?

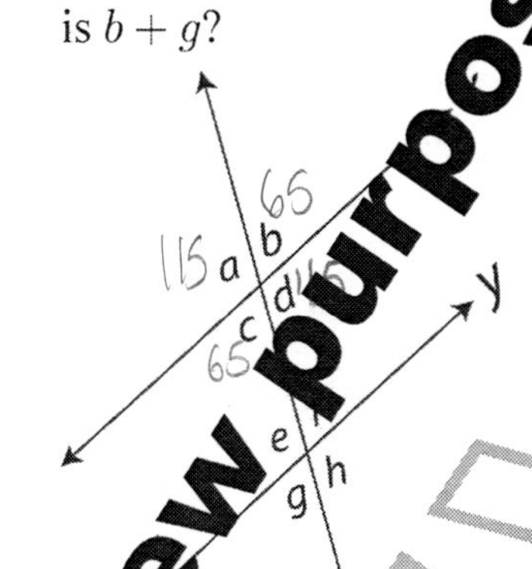

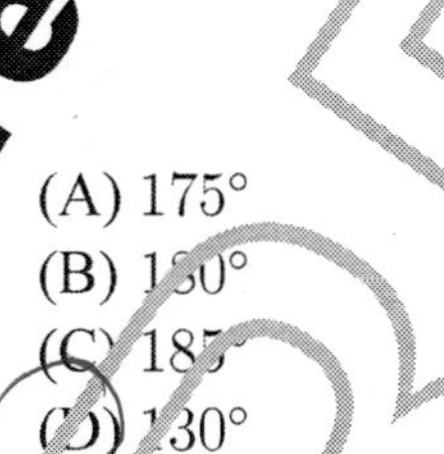

(A) 175°
(B) 180°
(C) 185°
(D) 130°
(E) 135°

2. For all numbers a and b, let $a * b$ be defined as $a * b = 2a^2 + 3ab - b^2$. What is the value of $2 * (1 * 4)$?

(A) -12
(B) -8
(C) 0
(D) 16
(E) 24

3. If x is an even integer and y is an odd integer, which of the following is an even integer?

(A) $3x + y$
(B) $x + 3y$
(C) $3(x + y)$
(D) $3xy$
(E) $x - 3y$

4. Which of the following could be the equation of the graph below?

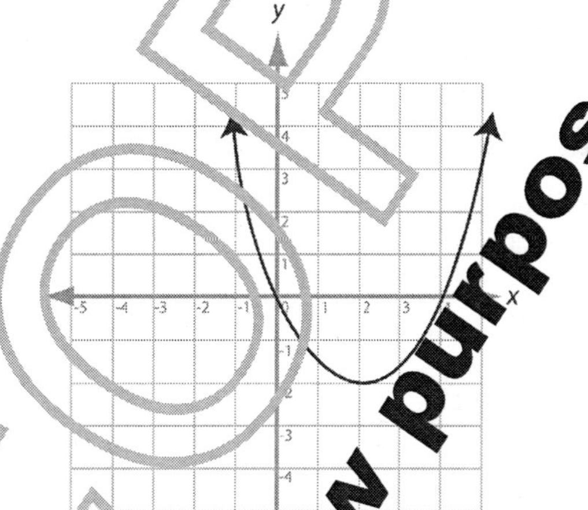

(A) $f(x) = (x + 2)^2 - 2$
(B) $f(x) = (x - 2)^2 + 2$
(C) $f(x) = (x + 2)^2 + 2$
(D) $f(x) = (x - 2)^2 - 2$
(E) $f(x) = 2x^2 - 2$

5. What is the average (arithmetic mean) of $3x - 7$, $8x + 3$, $-6x + 5$, and $3x + 1$?
   (A) $x + \frac{1}{2}$
   (B) $x + 4$
   (C) $2x + \frac{1}{2}$
   (D) $2x + 4$
   (E) $x + 4$

6. If $6(x - y) + 2 = 26$, then $x - y = ?$
   (A) 2
   (B) 3
   (C) 4
   (D) 5
   (E) 6

7. If $x + y < 6$ and $x - y > 8$, which of the following pairs are the values of x and y?
   (A) $(2, -3)$
   (B) $(2, $ [illegible]$)$
   (C) [illegible]
   (D) ([illegible]$, -5)$
   (E) $(4, 6)$

8. Below is the graph of $y = f(x)$. If so, for how many values of x does $f(x) = 0$?

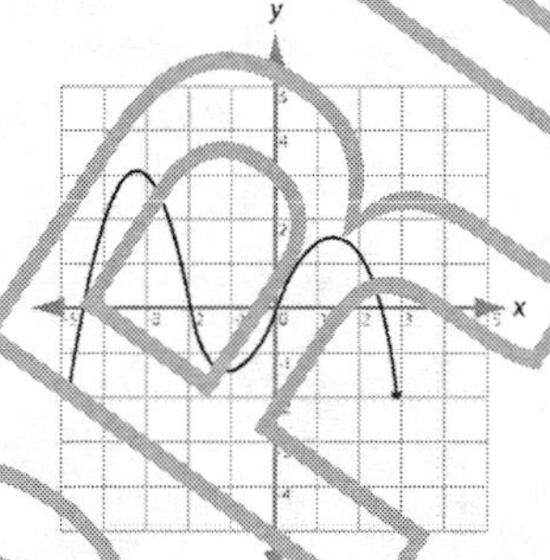

   (A) None
   (B) 1
   (C) 2
   (D) 3
   (E) 4

9. If the area of a triangle is 48 and its base is 12, what is the length of the altitude to that base?

10. Jack has 2 hip hop, 3 pop, and 1 country CD in his car. If he pulls out 2 CDs out without looking, what is the probability that both are pop?

11. A boat travels 30 miles upstream in 4 hours. The trip downstream takes 2.5 hours. Find the rate of the boat in still water.

12. 12. $\triangle ABC$ and $\triangle XYZ$ are similar. Find x if $z = 3$, $a = 6$, and $c = 4.5$.

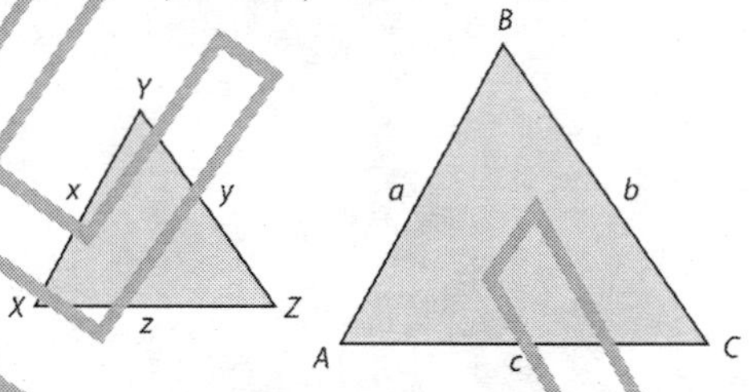

13. On a 180-yard hole, a golfer hits the ball 175 yards, 3° off-line. How far is the ball from the cup?

14. Ryan is constructing a cylinder that can hold 90.8 cubic centimeters. If the diameter of the cylinder is [illegible].4 centimeters, what should the height measure? (Round to the nearest tenth.)

15. Find the radius of a sphere whose surface area equals 91.6 in^2. Round to the nearest tenth.

16. A baker earns \$0.15 profit per glazed doughnut g, and \$0.40 profit per candy doughnut j. If a customer wants to buy no more than a dozen doughnuts and wants to try at least one of each kind, what is the maximum profit the baker can earn?

17. Find the length of radius of the base of a cylinder [illegible] prism if the height is 9 cm and the volume is 1017.36 cm^3. Round to the nearest whole number.

18. What is the next number in this sequence? 0.03, 0.12, 0.48, 1.92, _____

Part 2

Questions 1–20 of part 2 of this test are multiple choice. You have 25 minutes to complete this part of the test.

1. If $(8^n)^3 = 8^{21}$, what is the value of n?

(A) 7
(B) 10
(C) 18
(D) 24
(E) 63

2. What is the equation of a line parallel to the y-axis and four units to the left of the y-axis in the xy plane?

(A) $x = 4$
(B) $x = 0$
(C) $x = -4$
(D) $y =$ [illegible]
(E) $y = -4$

3. In the figure below, three lines intersect at a point. If $a = 55°$ and [illegible] $= 65°$, what is the value of e?

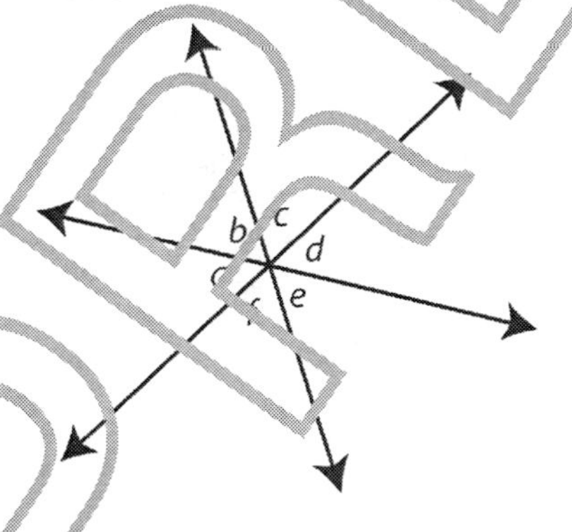

(A) 50°
(B) 55°
(C) 60°
(D) 65°
(E) 70°

4. If both the length and width of the rectangle are doubled, what will be the area of the resulting rectangle in square centimeters?

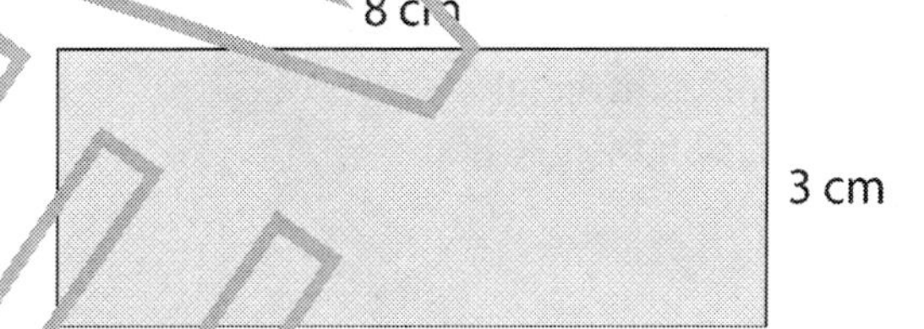

(A) 22
(B) 24
(C) 44
(D) 46
(E) 96

5. All of the following are factors of 153 except

(A) 3
(B) 6
(C) 9
(D) 17
(E) 51

6. An automobile manufacturer produces two kinds of cars, the x and the y. The company must always produce twice as many x as y and at least 200 cars but no more than 1,000 cars per day. Which of the following represents this situation?

(A) $200 \leq 2y \leq 1,000$
(B) $200 \leq 2x + y \leq 1,000$
(C) $200 \leq x + 2y \leq 1,000$
(D) $200 \leq x + y \leq 1,000$
(E) None of the Above

7. What is the value of y so that the mean of $\{3, 5, 7, 8, y\}$ is 9?

(A) 27
(B) 26
(C) 26
(D) 22
(E) 21

8. If y varies inversely with x. $y = 14$ when $x = 3$, what is x when $y = 84$?

(A) 28
(B) 6
(C) 2
(D) $\frac{1}{2}$
(E) 24

9. 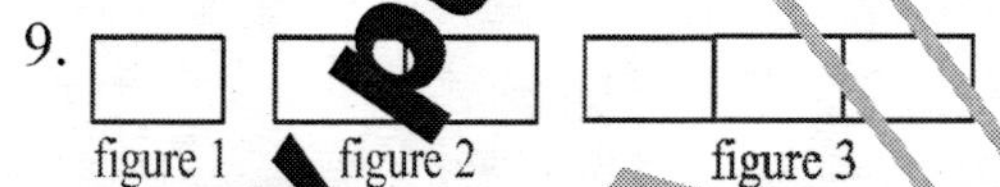

The figures above are formed by placing rectangles side by side. The rectangle is 1 unit wide and 2 units long.
Look at the chart below.

Figure	Perimeter
1	6
2	10
3	14
4	
5	

Choose the correct perimeters for figures 4 and 5, and the mathematical sentence for the perimeter of the nth figure.

(A) Figure 4, P = 18
Figure 5, P = 22
$P = 4n + 1$
(B) Figure 4, P = 18
Figure 5, P = 22
$P = 4n + 2$
(C) Figure 4, P = 17
Figure 5, P = 21
$P = 2n + 1$
(D) Figure 4, P = 20
Figure 5, P = 24
$P = n + 5$

10. Find the area of the shaded region.

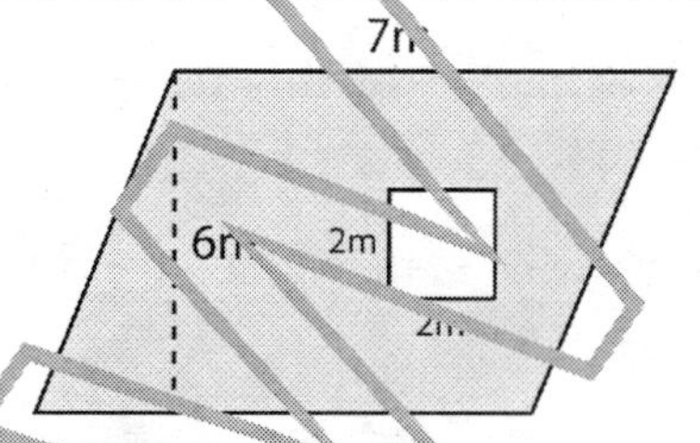

(A) 21 m^2
(B) 36 m^2
(C) 38 m^2
(D) 40 m^2
(E) 42 m^2

11. In $\triangle STU, m\angle S = 9x - 15, m\angle T = 5x + 3,$ and $m\angle U = 2x + 15$. Which inequality shows the relationship between the lengths of the sides of the triangle?

(A) $TU > SU > ST$
(B) $TS > TU > SU$
(C) $TS > SU > TU$
(D) $TU > ST > SU$
(E) $ST > TU > SU$

12. Find the next two terms of the geometric sequence $64, -32, 16, -8, \ldots$

(A) $-4, -2$
(B) $4, -2$
(C) $-4, 2$
(D) $-16, -32$
(E) $16, -32$

13. Jessica is trying to determine the height of a mast on a sailboat. The mast casts a shadow 10 feet long, while Jessica's shadow is 35 inches long. If Jessica is 5 feet 3 inches tall, what is the height of the mast?

(A) 34 feet
(B) 29 feet
(C) 24 feet
(D) [illegible] feet
(E) 18 $feet$

14. Find the slope of a line that passes through $\left(\frac{1}{3}, \frac{1}{x}\right)$ and $\left(\frac{3}{x}, \frac{1}{2}\right)$.

(A) $\frac{3(2-x)}{2(x-9)}$

(B) $\frac{3(2-x)}{2(x-3)}$

(C) $\frac{2(x-2)}{3(x-3)}$

(D) $-\frac{3(2-x)}{2(x-3)}$

(E) $-\frac{3(x-2)}{2(x-9)}$

15. Simplify: $\frac{x^2 \text{[illegible]} 2x + 15}{x^2 + x - 6}$, if $x \neq 2$ or $x \neq 3$.

(A) $\frac{x - \text{[illegible]}}{x + \text{[illegible]}}$

(B) $\frac{\text{[illegible]} 5}{x - 2}$

(C) $\frac{x+5}{x-2}$

(D) $\frac{x+5}{x+2}$

(E) Cannot be simplified further

16. What is the measure of $\overline{BN}$ if $BC = 18$ and $\overline{MN}$ is a median of trapezoid $ABCD$?

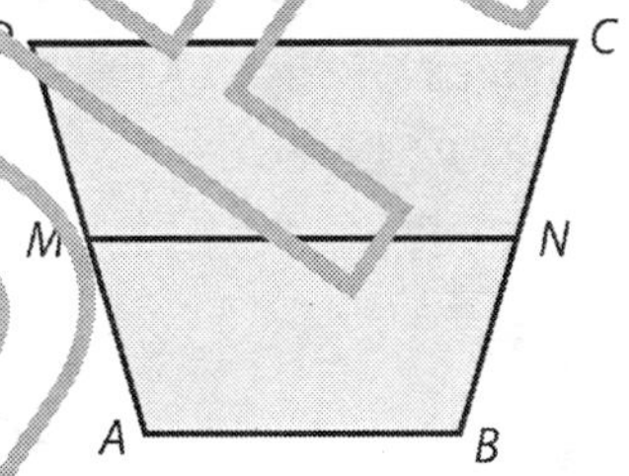

(A) 3
(B) 6
(C) 9
(D) 12
(E) 18

17. Andrea has 10 more jellybeans than her friend Chelsea, but she has half as many as Rebecca. Which expression below best describes Rebecca's jelly beans?

(A) $R = 2C + 20$
(B) $R = C + 10$
(C) $R = A + \frac{1}{2}C$
(D) $R = 2A + 10$
(E) $R = \frac{1}{2}C + 5$

18. In the inequality $3 + \sqrt{x + 11} > 7$, what must x be greater than or equal to?

(A) 0
(B) [illegible]
(C) 5
(D) 7
(E) 10

19. Determine the ratio of the volumes for the similar prisms.

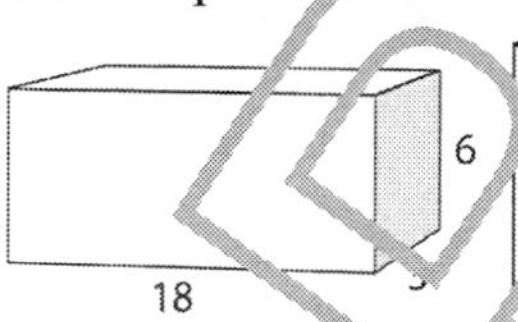

(A) $\frac{27}{343}$
(B) $\frac{9}{49}$
(C) $\frac{9}{21}$
(D) $\frac{1}{21}$
(E) $\frac{1}{3}$

20. An aquarium is 24 inches long, 12 inches wide, and 16 inches high. If a larger, similar aquarium holds 9,[illegible]000 cubic inches of water, what are its dimensions?

(A) 42 in × [illegible] in × 36 in
(B) 37 in × 18 in × 25 in
(C) 36 in × 18 in × 24 in
(D) [illegible] in × 21 in × 31 in
(E) [illegible] in × 15 in × 20 in

Part 3

Questions 1–16 of part 3 of this test are multiple choice. You have 20 minutes to complete this part of the test.

1. If $3,200,000 = 3.2 \times 10^k$, then $k =$

(A) 4
(B) 5
(C) 6
(D) -6
(E) -5

2. Using the figure below, $x + y + z =$

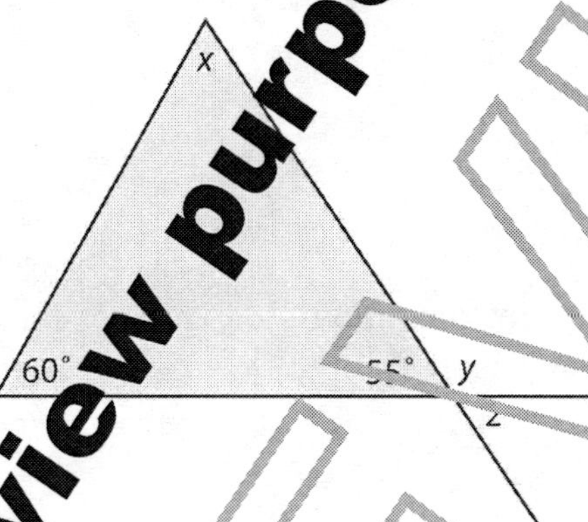

(A) 175°
(B) 180°
(C) 240°
(D) 245°
(E) 250°

3. Michelle wants to buy a t-shirt for $39, jeans for $41, and a pair of boots for $45. There is a 7% sales tax. If she gave the sales clerk $150, how much change would she get back?

(A) $15.18
(B) $16.25
(C) $24.93
(D) $25.00
(E) $33.75

4. Lauren had a box of 48 candy bars to sell for a club fundraiser. She sold half of the bars on her own, and her father sold half of the remaining bars at work. If no other bars were sold, what fraction of Lauren's original bars were sold?

(A) $\frac{1}{3}$
(B) $\frac{1}{4}$
(C) $\frac{1}{2}$
(D) $\frac{2}{3}$
(E) $\frac{3}{4}$

5. Write an equation of a line that has an undefined slope and passes through the point $(3, -4)$.

(A) $y = 3$
(B) $x = 3$
(C) $y = -4$
(D) $x = -4$
(E) None of the Above

6. Using a standard deck of cards, what is the probability of choosing a red 7?

(A) $\frac{2}{13}$
(B) $\frac{3}{26}$
(C) $\frac{1}{13}$
(D) $\frac{1}{[illegible]}$
(E) $\frac{1}{2}$

7. A phone company has a monthly fee of \$7.95 and charges a rate of \$0.05 per minute. Another phone company has a monthly fee of \$9.95 and charges a rate of \$0.03 per minute. At how many minutes would the two companies have equal charges?

(A) They will never have the same charges.
(B) 50
(C) 100
(D) 150
(E) 200

8. Suppose one runner can run 8 laps in 12 minutes, and another runner can run 8 laps in 10 minutes. If they start at the same time, after how many laps by the fast runner will he lap the slow runner?

(A) 5
(B) 6
(C) 7
(D) 7.5
(E) 8

9. Suppose the perimeter of a triangle is given by $P = 13x + 5y$, and two of its sides have lengths of $2x + y$ and $8x + 4y$. What is the length of the third side?

(A) $23x + 10y$
(B) $23x + 5y$
(C) $6x + 3y$
(D) $3x + y$
(E) $3x$

10. In $\triangle MNP$, $\overline{MN} \cong \overline{MP}$, $m\angle P$ is 10 more than 3 times a number, $m\angle N$ is 8 less than 5 times the same number. Find $m\angle M$.

(A) $37°$
(B) $48°$
(C) $59°$
(D) $106°$
(E) $116°$

11. The dart board is made up of concentric circles. The center circle has a diameter of 4 inches. Each successive ring has a radius 2 inches greater than the ring before. Find the probability of scoring 80 points.

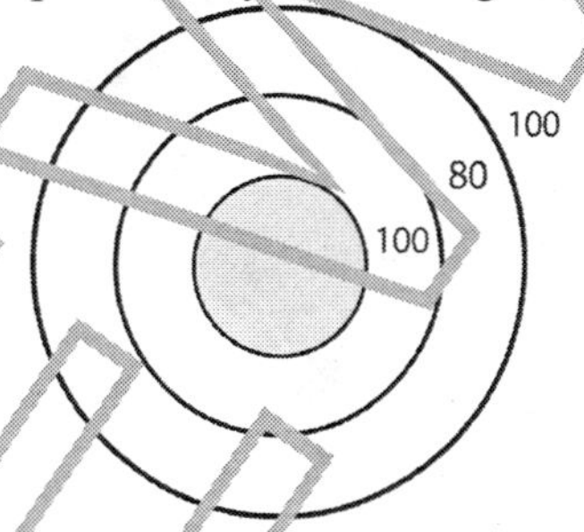

(A) $\frac{1}{3}$
(B) $\frac{1}{2}$
(C) $\frac{5}{16}$
(D) $\frac{5}{9}$
(E) $\frac{7}{16}$

12. On a floor plan, 1 inch represents 20 feet. If the dimensions of a room on the plan are $\frac{7}{8}$ inch by $1\frac{1}{4}$ inches, what are the actual dimensions of the room?

(A) 18 ft by 21.5 ft
(B) 17.5 ft by 25 ft
(C) 17 ft by 21.5 ft
(D) 16 ft by 25 ft
(E) 15.5 ft by 21.5 ft

13. The function $f(x) = \dfrac{x-3}{x-3}$ has which of the following properties?

(A) It has a hole at $(3, 0)$.
(B) It has a hole at $(3, 1)$.
(C) $f(x) = 0$ for all values of x.
(D) $f(x) = 1$ for all values of x.
(E) None of the Above

14. Solve $|x+5|-3=-1$.

(A) $(-7,-3)$
(B) $(-5,2)$
(C) $(3,7)$
(D) $(2,5)$
(E) $(-2,-5)$

15. Find the vertex of the graph of the function $f(x)=3x^2-6x+1$.

(A) $(1,4)$
(B) $(-1,4)$
(C) $(1,-4)$
(D) $(-1,-2)$
(E) $(1,-2)$

16. Simplify $5\sqrt{18}-2\sqrt{75}+6\sqrt{27}$.

(A) $15\sqrt{2}+8\sqrt{3}$
(B) $23\sqrt{5}$
(C) $10\sqrt{3}-15\sqrt{2}$
(D) $54-10\sqrt{3}$
(E) Cannot be simplified

Evaluation Chart for the Diagnostic Mathematics Test

Directions: On the following chart on this page and the next, circle the question numbers that you answered incorrectly. Then turn to the appropriate topics (listed by chapters), read the explanations, and complete the exercises. Review the other chapters as needed. Finally, complete ***SAT Mathematics Test Preparation Guide*** Practice Tests for further review.

		Questions Part 1	Questions Part 2	Questions Part 3	Pages
Chapter 1:	Numbers and Number Systems	3	5		11–23
Chapter 2:	Fractions, Decimals, and Percents	15		3, 4	24–41
Chapter 3:	Exponents and Roots		1	1, 6	42–49
Chapter 4:	Introduction to Algebra		17		50–59
Chapter 5:	Introduction to Graphing				60–68
Chapter 6:	Solving One-Step Equations and Inequalities				69–78
Chapter 7:	Solving Multi-Step Equations and Inequalities	6, 7	6, 18	14	79–94
Chapter 8:	Rates, Ratios, and Proportions	11	8, 13	8, 12	95–107
Chapter 9:	Polynomials				108–117
Chapter 10:	Factoring		15		118–131
Chapter 11:	Solving Quadratic Equations				132–139
Chapter 12:	Graphing and Writing Equations and Inequalities		2, 14	5	140–156
Chapter 13:	Applications of Graphs	4		13, 15	157–171
Chapter 14:	Systems of Equations and Systems of Inequalities			7	172–183

		Questions Part 1	Questions Part 2	Questions Part 3	Pages
Chapter 15:	Relations and Functions	8			184–204
Chapter 16:	Series, Sequence, and Algorithms	2, 18	9, 12		205–222
Chapter 17:	Statistics	5	7		223–236
Chapter 18:	Data Interpretation				237–244
Chapter 19:	Probability and Counting	10		6, 11	245–263
Chapter 20:	Angles	1	3	2, 10	264–278
Chapter 21:	Triangles	9, 12, 13	11, 16		279–295
Chapter 22:	Plane and Solid Geometry	14, 15, 17	4, 10, 19, 20	9	296–312
Chapter 23:	Transformations				313–326

Chapter 1
Numbers and Number Systems

Real Numbers

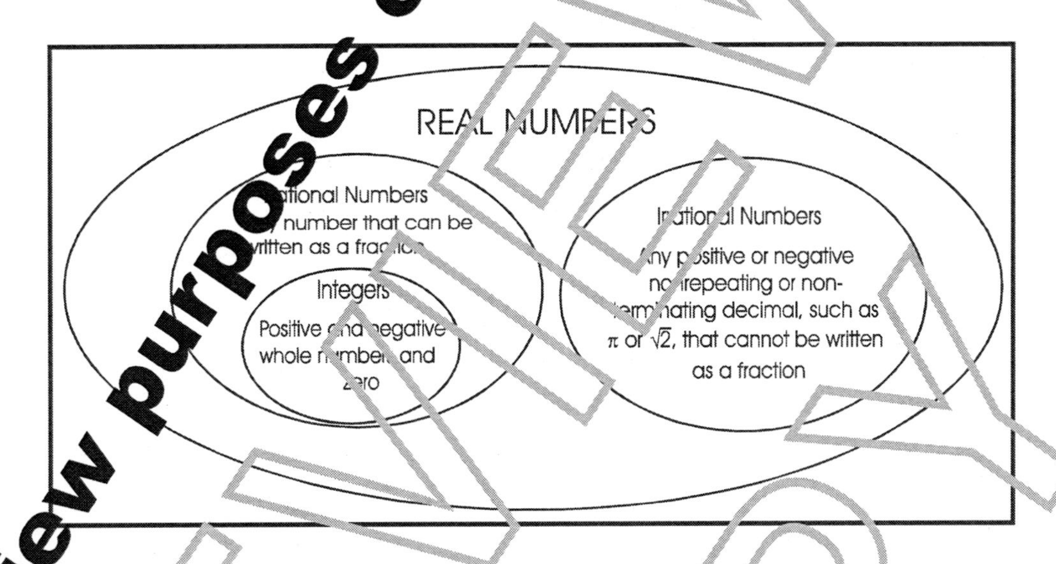

Real numbers include all positive and negative numbers and zero. Included in the set of real numbers are positive and negative fractions, decimals, and rational and irrational numbers.

Use the diagram above and your calculator to answer the following questions.

1. Using your calculator, find the square root of 7. Does it repeat? Does it end? Is it rational or an irrational number?
2. Find $\sqrt{25}$. Is it rational or irrational? Is it an integer?
3. Is an integer an irrational number?
4. Is an integer a real number?
5. Is $\frac{[illegible]}{8}$ a real number? Is it rational or irrational?

Identify the following numbers as rational (R) or (I).

6. 5π
7. $\sqrt{8}$
8. $\frac{1}{3}$
9. -7.2
10. $-\frac{3}{4}$
11. $\frac{\sqrt{2}}{2}$
12. $9+\pi$
13. 1.0004
14. $-\frac{4}{5}$
15. $1.1\overline{8}$
16. $\sqrt{8}$[illegible]
17. $\frac{[illegible]}{4}$
18. $-\sqrt{36}$
19. $17\frac{1}{2}$
20. $-\frac{5}{3}$

1.2 Integers

In elementary school, you learned to use whole numbers.

$$\textbf{Whole numbers} = \{0, 1, 2, 3, 4, 5, ...\}$$

For most things in life, whole numbers are all we need to use. However, when a checking account falls below zero or the temperature falls below zero, we need a way to express that. Mathematicians have decided that a negative sign, which looks exactly like a subtraction sign, would be used in front of a number to show that the number is below zero. All the negative whole numbers and positive whole numbers plus zero make up the set of integers.

$$\textbf{Integers} = \{..., -4, -3, -2, -1, 0, 1, 2, 3, 4, ...\}$$

1.3 Even and Odd Integers

Integers are all the negative and positive whole numbers. Zero is neither positive or negative.

$$\textbf{Integers} = \{..., -4, -3, -2, -1, 1, 2, 3, 4, ...\}$$

Even integers are all the integers that are evenly divisible by 2. This means an integer is even if it can be divided by 2 without any remainders.

$$\textbf{Even Integers} = \{..., -6, -4, -2, 0, 2, 4, 6, ...\}$$

Odd integers are all the integers that are not evenly divisible by 2.

$$\textbf{Odd Integers} = \{..., -7, -5, -3, -1, 1, 3, 5, 7, ...\}$$

1.4 Prime Numbers

A **prime number**, also called a prime, is a number that can only be divided by itself and 1. All prime numbers are positive and odd.

$$\textbf{Prime Numbers} = \{1, 3, 5, 7, 11, 13, ...\}$$

1.5 Digits

Digits are specific units. The only numbers that are digits are 0, 1, 2, 3, 4, 5, 6, 7, 8, 9. For example, 3 is a digit in the number 1, 368. There are four specific digits in this number.

1.6 Absolute Value

The absolute value of a number is the distance the number is from zero on the number line.

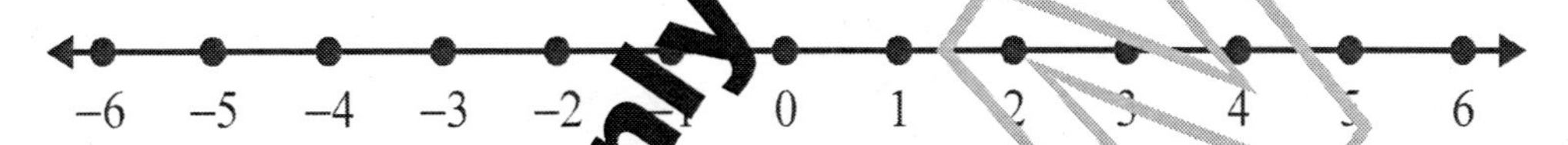

The absolute value of 6 is written $|6|$. $|6| = 6$
The absolute value of -6 is written $|-6|$. $|-6| = 6$

Both 6 and -6 are the same distance, 6 spaces, from zero so their absolute value is the same: 6.

Examples:

$|-4| = 4$ $-|-4| = -4$ $|-9| + 5 = 9 + 5 = 14$

$|9| - |8| = 9 - 8 = 1$ $|6| - |-6| = 6 - 6 = 0$ $|-5| + |-2| = 5 + 2 = 7$

Simplify the following absolute value problems.

1. $|9| =$ ______
2. $-|5| =$ ______
3. $|-25| =$ ______
4. $-|-12| =$ ______
5. $-|64| =$ ______
6. $|-2| =$ ______
7. $-|-3| =$ ______
8. $|-4| - |3| =$ ______
9. $|-8| - |-4| =$ ______
10. $|5| + |-4| =$ ______
11. $|-2| + |6| =$ ______
12. $|10| + |8| =$ ______
13. $|-2| + |4| =$ ______
14. $|-3| + |-4| =$ ______
15. $|7| - |-5| =$ ______

1.7 Word Problems

1. If Jacob averages 15 points per basketball game, how many points will he score in a season with 12 games?

2. A cashier can ring up 12 items per minute. How long will it take the cashier to reach a total for a customer with 72 items?

3. Mrs. Randolph has 26 students in 1st period, 32 students in 2nd period, 27 students in 3rd period, and 30 students in 4th period. What is the total number of students Mrs. Randolph teaches?

4. When Blake started on his trip, his odometer read 109, 875. At the end of his trip it read 110, 480. How many miles did he travel?

5. A school cafeteria has 52 tables. If each table seats 14 people, how many people can be seated in the cafeteria?

6. Leadville, Colorado is 14, 286 feet above sea level. Denver, Colorado is 5, 280 feet above sea level. What is the difference in elevation between these two cities?

7. The local bakery made 288 doughnuts on Friday morning. How many dozen doughnuts did they make?

8. Mattie ate 14 chocolate-covered raisins. Her big brother ate 5 times as many. How many chocolate-covered raisins did her brother eat?

1.8 Two-Step Problems

1. There are 25 miniature chocolate bars in a bag. There are 20 bags in a carton. Damon needs to order 10, 000 miniature chocolate bars. How many cartons will he need to order?

2. LeAnn needs 2, 400 boxes for her business. The boxes she needs come in bundles of 50 that weigh 45 pounds per bundle. What will be the total weight of the 2, 400 boxes she needs?

3. Seth uses 20 nails to make a birdhouse. He wants to make 60 birdhouses to sell at the county fair. There are 30 nails in a box. How many boxes will he need?

4. There are 12 computer disks in a box. There are 10 boxes in a carton. John ordered 16 cartons. How many disks is he getting?

5. The Do-Nut Factory packs 13 doughnuts in each baker's dozen box. They also sell cartons of doughnuts which have 6 baker's dozen boxes. Duncan needs to feed 780 people. Assuming each person eats only 1 doughnut, how many cartons will he need to buy from the doughnut factory?

6. Brittany has 2 dogs, a Saint Bernard and a Golden Retriever. The Saint Bernard eats twice as much as the Golden Retriever. The retriever eats 5 pounds of food in 6 days. How many pounds of food do the two dogs eat in 30 days?

7. Each of the 4 engines on a jet uses 500 gallons of fuel per hour. How many gallons of fuel are needed for a 5-hour flight with enough fuel for an additional 2 hours as a safety precaution?

8. The Farmer's Dairy has 1, 620 pounds of butter to package. They are packaging the butter in five-pound tubs to distribute to restaurants. If they put 12 tubs in a case, how many cases of butter can they fill?

1.9 Order of Operations

In long math problems with $+$, $-$, $\times$, $\div$, $()$, and exponents in them, you have to know what to do first. Without following the same rules, you could get different answers. If you will memorize the silly sentence, Please Excuse My Dear Aunt Sally, you can memorize the order you must follow.

Please — **"P"** stands for parentheses. You must get rid of parentheses first.
Examples: $3\,(1+4) = 3\,(5) = 15$
$6\,(10-6) = 6\,(4) = 24$

Excuse — **"E"** stands for exponents. You must eliminate exponents next.
Example: $4^2 = 4 \times 4 = 16$

My Dear — **"M"** stands for multiply. **"D"** stands for divide. Start on the left of the equation and perform all multiplications and divisions in the order in which they appear.

Aunt Sally — **"A"** stands for add. **"S"** stands for subtract. Start on the left and perform all additions and subtractions in the order they appear.

Example 1: $12 \div 2\,(6-3) + 3^2 - 1$		
Please	Eliminate **parentheses**. $6 - 3 = 3$ so now we have	$12 \div 2 \times 3 + 3^2 - 1$
Excuse	Eliminate **exponents**. $3^2 = 9$ so now we have	$12 \div 2 \times 3 + 9 - 1$
My Dear	**Multiply** and **divide** next in order from left to right	$12 \div 2 = 6$ then $6 \times 3 = 18$
Aunt Sally	Last, we **add** and **subtract** in order from left to right	$18 + 9 - 1 = 26$

Simplify the following problems.

1. $6 + 9 \times 2 - 4$
2. $3\,(4 + 2) - 6^2$
3. $3\,(6 - 3) - 2^3$
4. $49 \div 7 - 3 \times 3$
5. $10 \times 4 - (7 - 2)$
6. $2 \times 3 \div 6 \times 4$
7. $4^3 \div 8\,(4 + 2)$
8. $7 + 8\,(14 - 6) \div 4$
9. $(2 + 8 - 12) \times 4$
10. $4\,(8 - 13) \times 4$
11. $8 + 4^2 \times 2 - 6$
12. $3^2\,(4 + 6) + 3$
13. $(12 - 6) + 27 \div 3^2$
14. $82^0 - 1 + 4 \div 2^2$
15. $1 + (2 - 3) + 8$
16. $12 - 4\,(7 - 2)$
17. $18 \div (6 + 3) + 12$
18. $10^2 + 3^3 - 2 \times 3$
19. $4^2 + (7 + 2) \div 3$
20. $7 \times 4 - 9 \div 3$

When a problem has a fraction bar, simplify the top of the fraction (numerator) and the bottom of the fraction (denominator) separately using the rules for order of operations. You treat the top and bottom as if they were separate problems. Then reduce the fraction to lowest terms.

Example 2: $\dfrac{2(4-3)-6}{5^2+3(2+1)}$

Please	Eliminate **parentheses**. $(4-3)=1$ and $(2+1)=3$	$\dfrac{2\times 1-6}{5^2+3\times 3}$
Excuse	Eliminate **exponents**. $5^2=25$	$\dfrac{2\times 1-6}{25+3\times 3}$
My **D**ear	**Multiply** and **divide** in the numerator and denominator separately. $3\times 3=9$ and $2\times 1=2$	$\dfrac{2-6}{25+9}$
Aunt **S**ally	**Add** and **subtract** in the numerator and denominator separately. $2-6=-4$ and $25+9=34$	$\dfrac{-4}{34}$

Now reduce the fraction to lowest terms. $\dfrac{-4}{34}=\dfrac{-2}{17}$

Simplify the following problems.

1. $\dfrac{2^2+4}{5+3(8+1)}$
2. $\dfrac{8^2-(4+11)}{4^2-3^2}$
3. $\dfrac{5-2(4-3)}{2(1-8)}$
4. $\dfrac{10+(2-4)}{4(2+6)-2^2}$
5. $\dfrac{3^3-8(1+2)}{-10-(3+8)}$
6. $\dfrac{(9-3)+3^2}{-5-2(4+1)}$
7. $\dfrac{16-3(10-6)}{(13+15)-5^2}$
8. $\dfrac{(2-5)-11}{12-2(3+1)}$
9. $\dfrac{7+(8-16)}{6^2-5^2}$
10. $\dfrac{16-(12-3)}{8(2+3)-5}$
11. $\dfrac{-3(9-7)}{7+9-2^3}$
12. $\dfrac{4-(2+7)}{13+(6-9)}$
13. $\dfrac{5(3-8)-2^2}{7-3(6+1)}$
14. $\dfrac{(\text{[illegible]}-8)+5}{3^2-(5+9)}$
15. $\dfrac{6^2-4(7+3)}{8+(9-3)}$

1.10 Sets

A **set** contains an object or objects that are clearly identified. The objects in a set are called **elements** or **members**. A set can also be **empty**. The symbol $\emptyset$ is used to denote the empty set.

Members of a set can be described in two ways. One way is to give a **rule** of description that clearly defines the members. Another way is to make a **roster** of each member of the set, mentioning each member only once in any order.

Rule		Roster
{the letters in the word "school"}	=	{s, c, h, o, l}

A **complement** of a set is everything that does not belong to the set. If there exists a set A, then the complement is denoted by A' or $\overline{A}$. Take the example above. Let $A =$ {the letters in the word "school"}. The complement is the set that contains every other letter in the alphabet that is not in the word "school". $A' =$ {a, b, d, e, f, g, i, j, k, m, n, p, q, r, t, u, v, w, x, y, z}.

The symbol $\in$ is used to show that an object is the member of a set. For example, $\mathbf{3} \in \{\mathbf{1}, \mathbf{2}, \mathbf{3}, \mathbf{4}\}$.
The symbol $\notin$ means "is not a member of". For example, $\mathbf{8} \notin \{\mathbf{1}, \mathbf{2}, \mathbf{3}, \mathbf{4}\}$.
The symbol $\neq$ means "is not equal to".
For example, **{the cities in Texas} $\neq$ {New York, Philadelphia, Nashville}**

Read each of the statements below and tell whether they are true or false.

1. {the days of the week} = {January, March, December}
2. {The first five letters of the alphabet} = {a, b, c, d, e}
3. 5 $\notin$ {all odd numbers}
4. Friday $\in$ {the days of the week}
5. {the last three letters of the alphabet} $\neq$ {x, y, z}
6. $t \notin$ {the letters in the word "yellow"}
7. the letters in the word "funny" $\in$ {the letters of the alphabet}
8. {The letters in the word "Alabama"} = {a, b, l, m}
9. {living unicorns} $\neq \emptyset$
10. {the letters in the word "horse"} $\neq$ {the letters in the word "shore"}

Identify each set by making a roster (see above). If a set has no members, use $\emptyset$.

11. {the letters in the word "hat" that are also in the word "thin"}
12. {the letters in the word "Mississippi"}
13. {the provinces of Canada that border the state of Texas}
14. {the letters in the word "kitchen" and also in the word "dinner"}
15. {the days of the week that have the letter "n"}
16. {the letters in the word "June" that are also in the word "April"}
17. {the letters in the word "instruments" that are not in the word "telescope"}
18. {the digits in the number "19,582" that are also in the number "56,871"}

1.11 Subsets

If every member of set A is also a member of set B, then set A is a **subset** of set B.

The symbol $\subseteq$ means "is a subset of."

For example, every member of the set $\{1, 2, 3, 4\}$ is a member of the set $\{1, 2, 3, 4, 5, 6, 7\}$. Therefore, $\{1, 2, 3, 4\} \subseteq \{1, 2, 3, 4, 5, 6, 7\}$. The relationship is pictured in the following diagram:

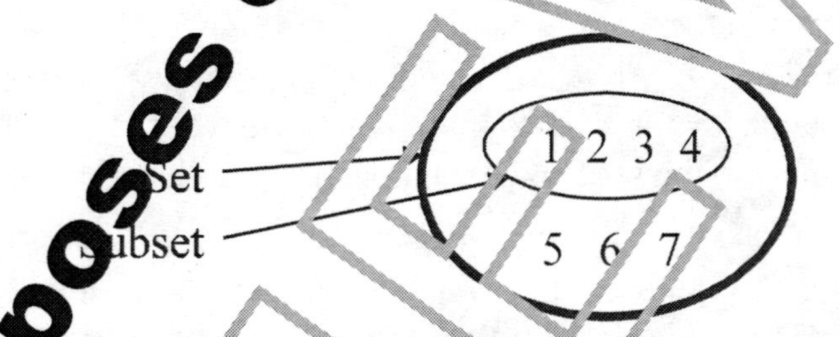

The symbol $\nsubseteq$ means "is **not** a subset of." For example, not every member of the set {Ann, Sue} is a member of the set {Ann, John, Cindy}. Therefore, {Ann, Sue} $\nsubseteq$ {Ann, John, Cindy}.

Read each of the statements below and tell whether they are true or false.

1. {a, b, c} $\nsubseteq$ {a, b, c, d}
2. {1, 3, 5, 7} $\subseteq$ {1, 2, 4, 5, 6, 7}
3. {a, e, i, o, u} $\subseteq$ {the vowels of the alphabet}
4. {dogs} $\subseteq$ {poodles, bull dogs, collies}
5. {fruit} $\nsubseteq$ {apples, grapes, bananas}
6. {10, 20, 30, 40} $\subseteq$ {20, 30, 40, 50}
7. {English, Spanish, French} $\subseteq$ {languages}
8. {duck, swan, penguin} $\nsubseteq$ {mammals}
9. {1, 2, 5, 10} $\subseteq$ {whole numbers}
10. {Atlanta} $\nsubseteq$ {U.S. state capitals}

The set {Pam} has two subsets: {Pam} and $\emptyset$. The set {Emily, Brad} has 4 subsets: {Emily}, {Brad}, {Emily, Brad} and $\emptyset$. The $\emptyset$ is a subset of any set.

For each of the following sets, list all of the possible subsets.

1. {a}
2. {1, 2}
3. {r, s, t}
4. {1, 3, 5, 7}
5. {Joe, Ed}

1.12 Intersection of Sets

To find the intersection of two sets, you need to identify the members that the two sets share in common. The symbol for intersection is ∩. **A Venn diagram** shows how sets intersect.

Roster

$\{2, 4, 6, 8\} \cap \{4, 6, 8, 10\} = \{4, 6, 8\}$

The shaded area is the intersection of the two sets. It shows which numbers both sets have in common.

Venn Diagram

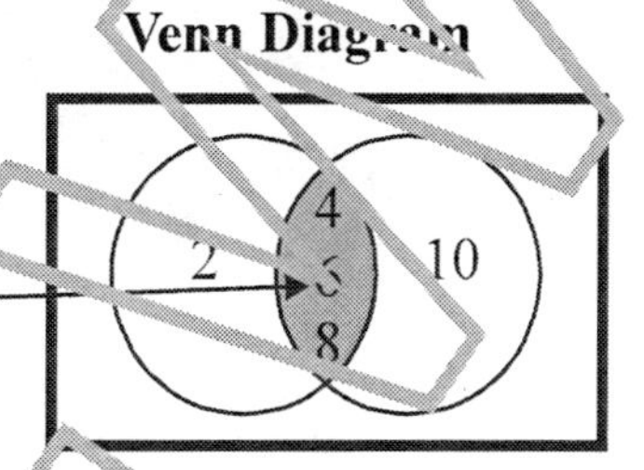

Find the intersection of the following sets.

1. {Ben, Jan, Dan, Tom} ∩ {Dan, Mike, Kate, Jan} =
2. {pink, purple, yellow} ∩ {purple, green, blue} =
3. {2, 4, 6, 8, 10} ∩ {1, 2, 3, 4, 5, 6, 7, 8, 9, 10} =
4. {a, e, i, o, u} ∩ {a, b, c, d, e, f, g, h, i, j, k} =
5. {pine, oak, walnut, maple} ∩ {maple, poplar} =
6. {100, [illegible], 95, 78, 62} ∩ {57, 82, 95, 98, 99} =
7. {orange, kiwi, coconut, pineapple} ∩ {pear, apple, orange} =

Look at the Venn diagram at the right to answer the questions below. Show your answers in roster form. Do the problem in parentheses first.

8. $A \cap B =$
9. $(A \cap B) \cap C =$
10. $A \cap C =$
11. $B \cap C =$
12. $(A \cap C) \cap B =$

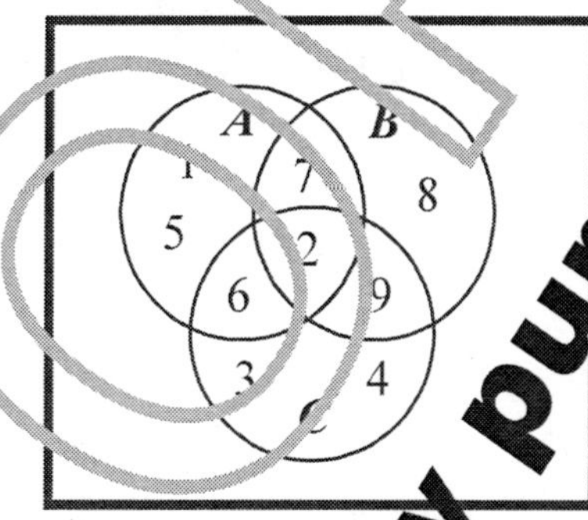

13. blue ∩ green =
14. (purple ∩ blue) ∩ green =
15. blue ∩ purple =
16. (blue ∩ green) ∩ purple =
17. green ∩ purple =

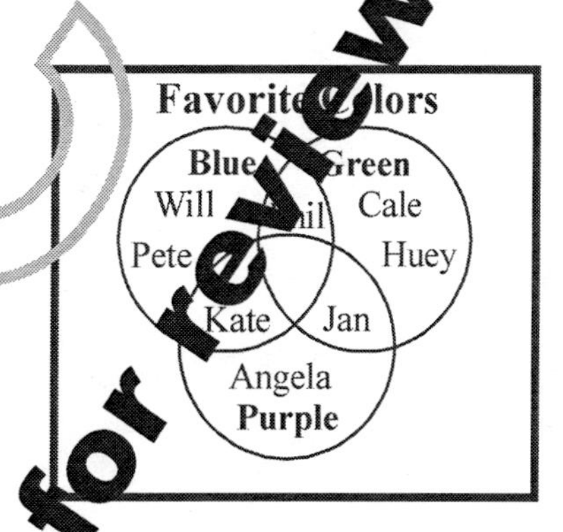

1.13 Union of Sets

The union of two sets means to put the members of two sets together into one set without repeating any members. The symbol for union is ∪.

$$\{1, 2, 3, 4\} \cup \{3, 4, 5, 6\} = \{1, 2, 3, 4, 5, 6\}$$

The union of theses two sets is the shaded area in the Venn diagram below.

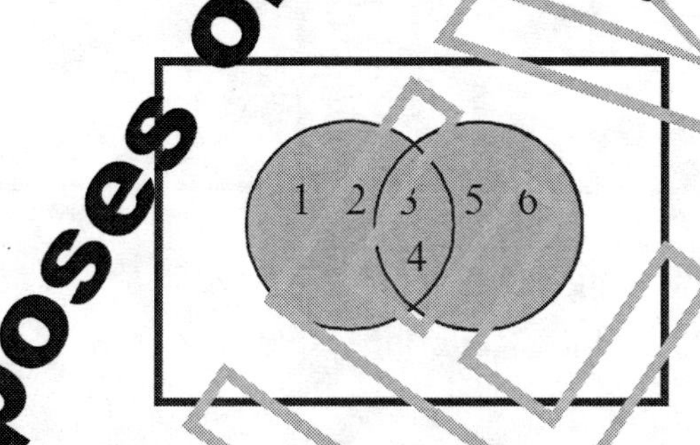

Find the union of the sets below.

1. {apples, pears, oranges} ∪ {pears, bananas, apples} =
2. {5, 10, 15, 20, 25} ∪ {10, 20, 30, 40} =
3. {Ted, Steve, Kevin, Michael} ∪ {Kevin, George, Kenny} =
4. {raisins, prunes, apricots} ∪ {peanuts, almonds, coconut} =
5. {sales, marketing, accounting} ∪ {receiving, shipping, sales} =
6. {beef, pork, chicken} ∪{chicken, tuna, shark} =

Refer to the following Venn diagrams to answer the questions below. Identify each of the following sets by roster.

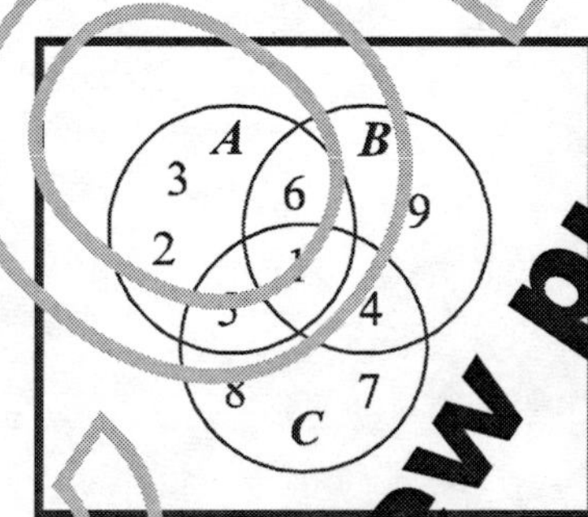

7. $A \cup C =$

8. $C \cup B =$

9. $B \cup A =$

10. $A \cup B \cup C =$

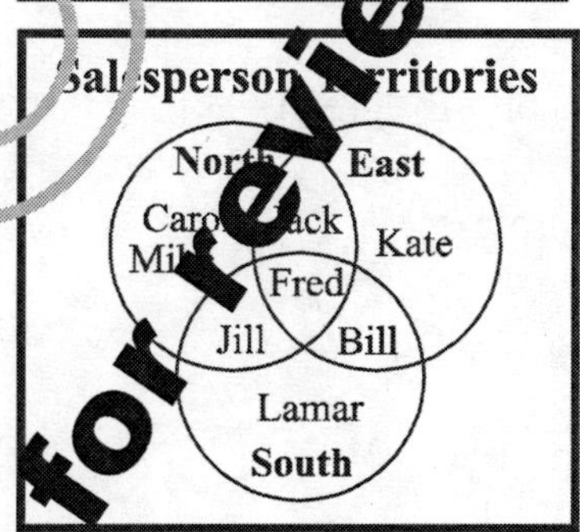

11. North ∪ East =

12. North ∪ South =

13. East ∪ South =

14. North ∪ East ∪ South =

1.14 Reading Venn Diagrams

Venn diagrams are a visual way to see two or more variables. They show whether or not the variables intersect. A Venn diagram also shows the union of the events.

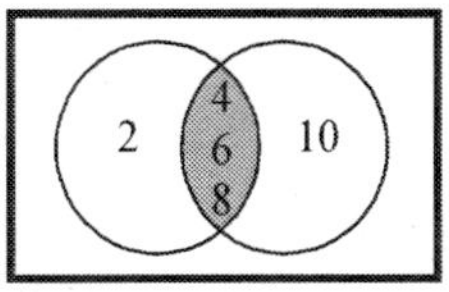

$\{2,4,6,8\} \cap \{4,6,8,10\} = \{4,6,8\}$
The shaded area is the intersection of the two sets.

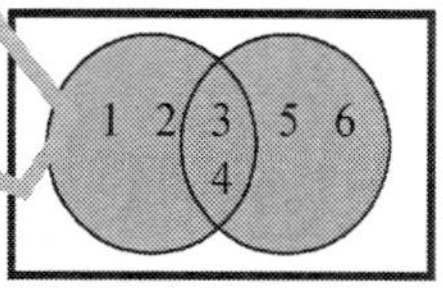

$\{1,2,3,4\} \cup \{3,4,5,6\} = \{1,2,3,4,5,6\}$
The shaded area is the union of the two sets.

Example 3: Below is a Venn diagram of how many students play football and baseball. Find the intersection and the union of the two events below.

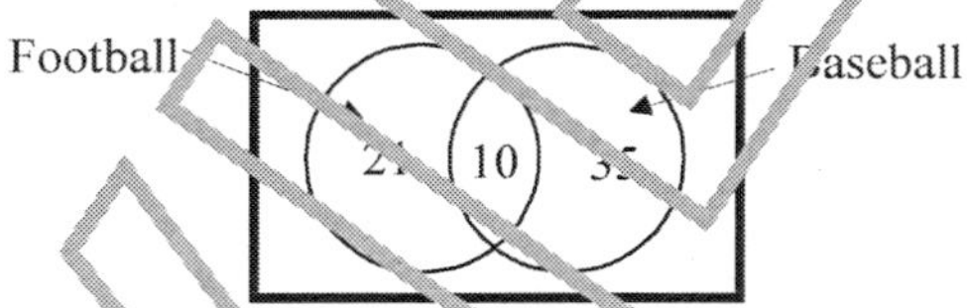

Step 1: First, you must figure out how many student play each sport by interpreting the diagram. Ten students play both sports, since they are counted on the football and baseball side. **The intersection is 10 students.**

Step 2: To find the union, you must add all the players together. The union is $21 + 35 + 10 = 66$ **students**.

Look at the Venn diagram to answer the questions below.

1. Dogs $\cap$ Cats
2. Dogs $\cup$ Birds
3. Cats $\cap$ Birds
4. Dogs $\cap$ Cats $\cap$ Birds
5. Dogs $\cup$ Cats
6. Dogs $\cup$ Cats $\cup$ Birds
7. Dogs $\cap$ Birds
8. Cats $\cup$ Birds
9. Dogs $\cup$ (Cats $\cup$ Birds)
10. (Dogs $\cap$ Cats) $\cap$ Birds

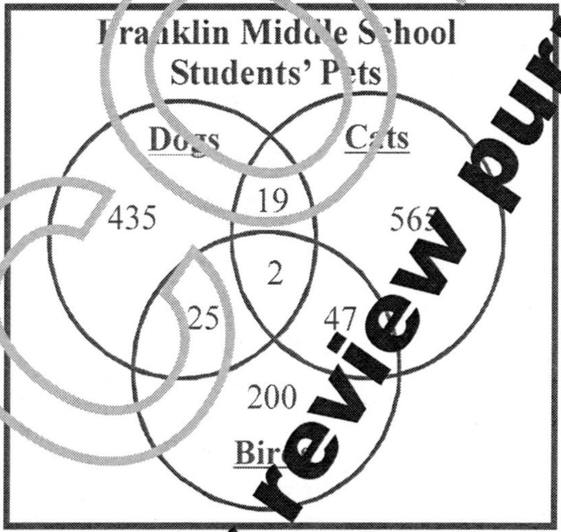

Chapter 1 Review

Solve the following absolute value problems.

1. $|4|$ 2. $|-6|$ 3. $|-3| + |7|$ 4. $|-2| - |-6|$ 5. $|8| - |-5|$

Solve the following word problems.

6. Concession stand sales for a football game totaled \$1,563. The actual cost for the food and beverages was \$395. How much profit did the concession stand make?

7. An orange grove worker can harvest 480 oranges per hour by hand. How many oranges can the worker harvest in an 8-hour day?

8. Tom has 155 head of cattle. Each eats 8 pounds of grain per day. How many pounds of grain does Tom need to feed his cattle for 10 days?

9. When you grind 3 cups of grain, you get 5 cups of flour. How many cups of grain must you grind to get 40 cups of flour?

10. The Beta Club is raising money by selling boxes of candy. It sold 152 boxes on Monday, 236 boxes on Tuesday, 107 boxes on Wednesday, and 93 boxes on Thursday. How many total boxes did the Beta Club sell?

Read each of the statements below and tell whether they are true or false.

11. {odd whole numbers} ⊆ {all whole numbers}
12. {yearly seasons} ≠ {spring, summer, fall, winter}
13. {Monday, Tuesday, Wednesday} ⊆ {days of the week}
14. United States of America ∉ {countries in North America}
15. {green, purple} ⊈ {primary colors}
16. {plants with red flowers} = ∅
17. {letters in the word "subsets"} = {b, u, s, e, t}
18. Milky Way ∉ {galaxies in the universe}
19. {Houston, Dallas } ⊆ {cities in Texas }
20. George Washington ∈ {former presidents of the United States}

Complete the following statements.

21. $\{3, 6, 9, 12, 15\} \cap \{0, 5, 10, 15, 20\} =$
22. {Felix, Mark, Kate} ∪ {Mark, Carol, Jack} =
23. {letters in "perfect"} ∩ {letters in "profit"} =
24. {Rome, London, Paris} ∩ {Italy, England, France} =
25. {black, white, gray} ∪ {red, white, blue} =
26. $\{1, 2, 3, 4, 5, 6\} \cup \{2, 4, 6, 8, 10, 12\} =$

Use the following Venn diagram to answer questions 1 and 2.

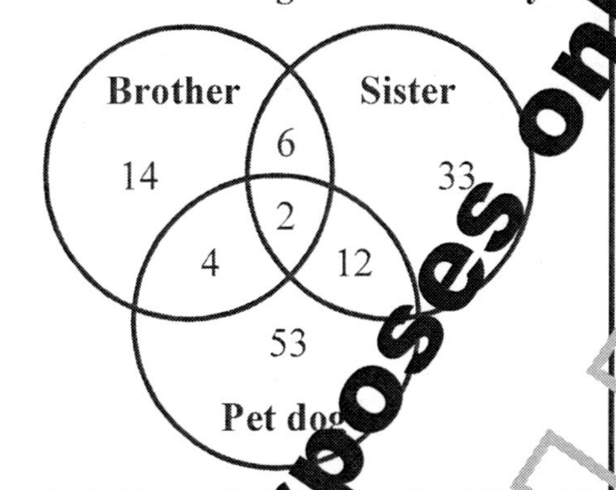

27. In the Venn diagram above, how many members are in the following set? {sister ∪ pet dog}
    (A) 12
    (B) 14
    (C) 100
    (D) 110

28. In the Venn diagram above, how many members are in the following set? {brother} ∩ {sister}
    (A) 6
    (B) 8
    (C) 53
    (D) 55

29. Which of the sets below does not contain {a, e, o} as a subset?
    (A) {letters of the alphabet}
    (B) {vowels in the alphabet}
    (C) {letters in the word "predator"}
    (D) {the first 5 letters of the alphabet}

30. Which of the following statements is true?
    (A) $\{1, 2, 4, 5\} \cup \emptyset = \emptyset$
    (B) {c} ∉ {c, a, t}
    (C) {f, g} ⊆ {f, g, h, i}
    (D) $\{k, l, m\} \cup \emptyset = \emptyset$

31. Which of the following statements is false?
    (A) {t, o} ⊆ {letters in the word "today"}
    (B) $8 \notin \{0, 2, 4, 6, 8\}$
    (C) $\emptyset = \{\ \}$
    (D) {a, b, c} ∩ {b, c, d, e} = {b, c}

32.

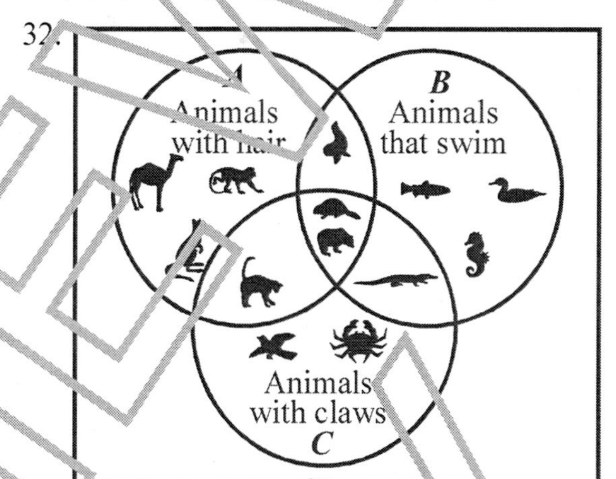

According to the above diagram, which one of the following statements is false?
    (A) $A \cap B = \{\ ,\ ,\ \}$
    (B) $\in B$
    (C) $\{\ ,\ ,\ \} \subseteq A$
    (D) $B \cup C \neq \emptyset$

33. Which of the following sets equals ∅?
    (A) {all whole numbers}
    (B) {all letters in the alphabet}
    (C) {all fish with feathers}
    (D) {all flowering plants}

34. {Karen, John, Sue} ∩ {John, Perry, Kay} =
    (A) ∅
    (B) {John}
    (C) {Karen, Sue, Perry, Kay}
    (D) {Karen, John, Sue, Perry, Kay}

Chapter 2
Fractions, Decimals, and Percents

2.1 Greatest Common Factor

To reduce fractions to their simplest form, you must be able to find the greatest common factor.

Example 1: Find the greatest common factor (GCF) of 16 and 24.

To find the **greatest common factor (GCF)** of two numbers, first list the factors of each number. The **factors** are all the numbers that will divide evenly into the numbers that you want to find the factors for.

The factors of 16 are: 1, 2, 4, 8, and 16

The factors of 24 are: 1, 2, 3, 4, 6, 8, 12, and 24

What is the **largest** number they both have in common? **8**

8 is the **greatest** (largest number) **common factor**.

Find all the factors and the greatest common factor (GCF) of each pair of numbers below.

	Pairs	Factors	GCF		Pairs	Factors	GCF
1.	10			10.	6		
	15				42		
2.	12			11.	14		
	16				63		
3.	18			12.	9		
	36				51		
4.	27			13.	18		
	45				45		
5.	32			14.	12		
	40				20		
6.	16			15.	16		
	48				40		
7.	14			16.	10		
	42				45		
8.	4			17.	18		
	26				30		
9.	8			18.	15		
	28				25		

2.2 Least Common Multiple

To find the **least common multiple (LCM)**, of two numbers, first list the multiples of each number. The multiples of a number are 1 times the number, 2 times the number, 3 times the number, and so on.

The multiples of 6 are: 6, 12, 18, 24, 30, ...

The multiples of 10 are: 10, 20, 30, 40, 50, ...

What is the smallest multiple they both have in common? 30

30 is the **least** (smallest number) **common multiple** of 6 and 10.

Find the least common multiple (LCM) of each pair of numbers below.

	Pairs	Multiples	LCM		Pairs	Multiples	LCM
1.	6	6, 12, 18, 24, 30	30	10.	6		
	15	15, 30			7		
2.	12			11.	4		
	16				18		
3.	18			12.	7		
	[illegible]				5		
4.	7			13.	30		
	3				45		
5.	12			14.	3		
	8				8		
6.	6			15.	12		
	8				9		
7.	4			16.	5		
	14				45		
8.	9			17.	3		
	6				5		
9.	2			18.	4		
	15				22		

2.3 Fraction Review

Change to mixed numbers or whole numbers.

1. $\frac{11}{2} =$ 3. $\frac{9}{4} =$ 5. $\frac{16}{4} =$ 7. $\frac{13}{4} =$ 9. $\frac{30}{10} =$ 11. $\frac{10}{4} =$

2. $\frac{8}{3} =$ 4. $\frac{20}{9} =$ 6. $\frac{21}{5} =$ 8. $\frac{42}{7} =$ 10. $\frac{18}{7} =$ 12. $\frac{11}{3} =$

Change to an improper fraction.

13. $2\frac{1}{8} =$	15. $4\frac{2}{3} =$	17. $9\frac{2}{5} =$	19. $6\frac{2}{7} =$	21. $2\frac{[illegible]}{8} =$	23. $8\frac{1}{9} =$
14. $8\frac{1}{4} =$	16. $9\frac{1}{3} =$	18. $6\frac{3}{4} =$	20. $2\frac{3}{4} =$	22. $9\frac{1}{[illegible]} =$	24. $7\frac{2}{7} =$

Reduce to lowest terms.

25. $\frac{2}{8} =$	27. $\frac{5}{15} =$	29. $\frac{10}{12} =$	31. $\frac{6}{18} =$	33. $\frac{4}{8} =$	35. $\frac{6}{14} =$
26. $\frac{10}{15} =$	28. $\frac{6}{24} =$	30. $\frac{16}{20} =$	32. $\frac{8}{24} =$	34. $\frac{8}{14} =$	36. $\frac{7}{28} =$

Perform the following operations. Reduce each answer to lowest terms.

37. $5\frac{1}{2} + 3\frac{5}{8}$	43. $6\frac{1}{4} - 2\frac{[illegible]}{5}$	49. $\frac{3}{4} \times \frac{7}{9}$	55. $28 \div 2\frac{2}{3}$
38. $7\frac{3}{8} + 4\frac{1}{2}$	44. $9\frac{1}{8} - 7\frac{3}{[illegible]}$	50. $\frac{1}{3} \times 16\frac{1}{2}$	56. $1\frac{1}{3} \div 2$
39. $9\frac{3}{4} + 4\frac{1}{2}$	45. $5\frac{1}{4} - 2\frac{2}{3}$	51. $10\frac{1}{2} \times 4\frac{2}{3}$	57. $1\frac{1}{[illegible]} \div 2\frac{[illegible]}{9}$
40. $2\frac{2}{3} + 2\frac{1}{4}$	46. $8\frac{1}{8} - 2\frac{[illegible]}{10}$	52. $4\frac{1}{6} \times 2\frac{4}{5}$	58. $15 \div 1\frac{1}{3}$
41. $4\frac{5}{6} +$ [illegible]	47. $\frac{3}{4} \times 100$	53. $9 \times 7\frac{2}{3}$	59. $8\frac{1}{4} \div \frac{1}{2}$
42. $7 -$ [illegible]	48. $\frac{2}{5} \times 80$	54. $2 \times 1\frac{7}{8}$	60. $20 \div 2\frac{2}{5}$

2.4 Fraction Word Problems

Solve and reduce answers to lowest terms.

1. Sara buys candy by the pound during the summer. During the first week of summer she buys $1\frac{1}{3}$ pounds of candy, during the second she buys $\frac{3}{4}$ of a pound, and during the third she buys $\frac{4}{5}$ pound. How many pounds did she buy during the first three weeks of summer?
2. Beth has a bread machine that makes a loaf of bread that weighs $1\frac{1}{2}$ pounds. If she makes a loaf of bread for each of her three sisters, how many pounds of bread will she make?
3. Rick chews on a piece of bubble gum for 120 minutes. About every $1\frac{1}{4}$ minutes, he blows a bubble. How many bubbles does Rick make?
4. Juan was competing in a 1000-meter race, but he had to pull out of the race after running $\frac{3}{4}$ of it. How many meters did he run?
5. Tad needs to measure where the free-throw line should be in front of his basketball goal. He knows his feet are $1\frac{1}{8}$ feet long and the free-throw line should be 15 feet from the backboard. How many toe-to-toe steps does Tad need to take to mark off 15 feet?
6. Mary gives her puppy a bath and uses $5\frac{1}{2}$ gallons of water. She throws away $3\frac{2}{3}$ gallons of the water. How much water does she have left?

2.5 Changing Fractions to Decimals

Example 2: Change $\frac{1}{8}$ to a decimal.

Step 1: To change a fraction to a decimal, simply divide the top number by the bottom number.

$$8\overline{)1}$$

Step 2: Add a decimal point and a 0 after the 1 and divide.

$$\begin{array}{r} 0.1 \\ 8\overline{)1.0} \\ -8 \\ \hline 2 \end{array}$$

Step 3: Continue adding 0's and dividing until there is no remainder.

$$\begin{array}{r} 0.125 \\ 8\overline{)1.000} \\ -8 \\ \hline 20 \\ -16 \\ \hline 40 \\ -40 \\ \hline 0 \end{array}$$

In some problems the number after the decimal point begins to repeat. Take, for example, the fraction $\frac{4}{11}$. $4 \div 11 = 0.363636$ and the 36 keeps repeating forever. To show that 36 repeats, simply write a bar above the numbers that repeat, $0.\overline{36}$.

Change the following fractions to decimals.

1. $\frac{4}{5}$	5. $\frac{1}{10}$	9. $\frac{3}{5}$	13. $\frac{7}{9}$	17. $\frac{3}{16}$
2. $\frac{2}{3}$	6. $\frac{5}{8}$	10. $\frac{7}{10}$	14. $\frac{9}{10}$	18. $\frac{3}{4}$
3. $\frac{1}{2}$	7. $\frac{5}{6}$	11. $\frac{4}{11}$	15. $\frac{1}{4}$	19. $\frac{8}{9}$
4. $\frac{5}{9}$	8. $\frac{1}{6}$	12. $\frac{1}{9}$	16. $\frac{3}{8}$	20. $\frac{5}{12}$

2.6 Changing Mixed Numbers to Decimals

If there is a whole number with a fraction, write the whole number to the left of the decimal point. Then change the fraction to a decimal.

Examples: $4\frac{1}{10} = 4.1$ $\quad 16\frac{2}{3} = 16.\overline{6}$ $\quad 12\frac{7}{8} = 12.875$

Change the following mixed numbers to decimals.

1. $5\frac{2}{3}$	5. $30\frac{1}{3}$	9. $6\frac{4}{5}$	13. $7\frac{1}{4}$	17. $10\frac{1}{10}$
2. $8\frac{5}{11}$	6. [illegible]	10. $13\frac{1}{2}$	14. $12\frac{1}{3}$	18. $20\frac{2}{5}$
3. $15\frac{3}{5}$	7. $1\frac{7}{8}$	11. $12\frac{4}{5}$	15. $1\frac{5}{8}$	19. $4\frac{9}{10}$
4. $13\frac{2}{3}$	8. $4\frac{9}{100}$	12. $11\frac{5}{8}$	16. $2\frac{3}{4}$	20. $5\frac{4}{11}$

2.7 Changing Decimals to Fractions

Example 3: Change 0.25 to a fraction.

Step 1: Copy the decimal without the point. This will be the top number of the fraction. $\frac{25}{\ }$

Step 2: The bottom number is a 1 with as many 0's after it as there are digits in the top number. $\frac{25 \leftarrow \text{Two digits}}{100 \leftarrow \text{Two 0's}}$

Step 3: You then need to reduce the fraction. $\frac{25}{100} = \frac{1}{4}$

Examples: $0.2 = \frac{2}{10} = \frac{1}{5}$ $\quad 0.65 = \frac{65}{100} = \frac{13}{20}$ $\quad 0.125 = \frac{125}{1000} = \frac{1}{8}$

Change the following decimals to fractions.

1. 0.55	5. 0.75	9. 0.71	13. 0.35
2. 0.6	6. 0.82	10. 0.64	14. 0.96
3. 0.12	7. 0.3	11. 0.56	15. 0.125
4. 0.9	8. 0.42	12. 0.[illegible]4	16. 0.375

2.8 Changing Decimals with Whole Numbers to Mixed Numbers

Example 4: Change 14.28 to a mixed number.

Step 1: Copy the portion of the number that is whole. 14

Step 2: Change .28 to a fraction. $14\frac{28}{100}$

Step 3: Reduce the fraction. $14\frac{28}{100} = 14\frac{7}{25}$

Change the following decimals to mixed numbers.

1. 7.125	5. 16.95	9. 5.7	13. 13.9
2. 99.5	6. 3.625	10. 45.425	14. 32.65
3. 2.13	7. 4.42	11. 15.8	15. 17.25
4. 5.1	8. 15.34	12. 8.16	16. 9.82

2.9 Decimal Word Problems

1. Mi[illegible] can have his bike fixed for $19.99, or he can buy the new part for his bike and replace it himself for $8.79. How much would he save by fixing his bike himself?
2. Megan buys 5 boxes of cookies for $3.75 each. How much does she spend?
3. Will subscribes to a monthly sports magazine. His one-year subscription costs $29.97. If he pays for the subscription in 3 equal installments, how much is each payment?
4. Pat purchases 2.5 pounds of jelly beans at $0.95 per pound. What is the total cost of the jelly beans?
5. The White family took $650 cash with them on vacation. At the end of their vacation, they had $4.67 left. How much cash did they spend on vacation?
6. Acer Middle School spends $1,443.20 on 55 math books. How much does each book cost?
7. The Junior Beta Club needs to raise $1,513.75 to go to a national convention. If they decide to sell candy bars at $1.25 each, how many must they sell to meet their goal?
8. Fleta owns a candy store. On Monday, she sold 6.5 pounds of chocolate, 8.34 pounds of jelly beans, 4.9 pounds of sour snaps, and 5.64 pounds of yogurt-covered raisins. How many pounds of candy did she sell in total?
9. Randal purchased a rare coin collection for $1,803.95. He sold it at auction for $2,700. How much money did he make on the coins?
10. A leather jacket that normally sells for $259.99 is on sale now for $197.88. How much can you save if you buy it now?

2.10 Changing Percents to Decimals and Decimals to Percents

To change a **percent** to a **decimal**, move the **decimal** point two places to the left, and drop the **percent** sign. If there is no decimal point shown, it is understood to be after the number and before the percent sign. Sometimes you will need to add a "0". (See 5% below.)

Example 5: $14\% = 0.14$ $5\% = 0.05$ $100\% = 1$ $103\% = 1.03$

↗ **(decimal point)**

Change the following percents to decimal numbers.

1. 18%
2. 23%
3. 9%
4. 63%
5. 4%
6. 45%
7. 2%
8. 110%
9. 7%
10. 55%
11. 80%
12. 17%
13. 66%
14. 13%
15. 73%
16. 25%
17. 410%
18. 1%
19. 50%
20. 99%
21. 107%

To change a decimal to a percent, move the decimal two places to the right, and add a percent sign. You may need to add a "0". (See 0.8 below.)

Example 6: $0.62 = 62\%$ $0.07 = 7\%$ $0.8 = 80\%$ $0.166 = 16.6\%$ $1.54 = 154\%$

Change the following decimal numbers to percents.

22. 0.15
23. 0.87
24. 1.53
25. 0.22
26. 0.35
27. 0.375
28. 0.648
29. 0.044
30. 0.58
31. 0.86
32. 0.29
33. 0.06
34. 0.48
35. 3.089
36. 0.042
37. 0.375
38. [illegible]
39. 0.75
40. 0.3
41. 2.9
42. 0.06

2.11 Changing Percents to Fractions and Fractions to Percents

Example 7: Change 15% to a fraction.

Step 1: Copy the number without the percent sign. 15 is the top number of the fraction.

Step 2: The bottom number of the fraction is always 100.

$$15\% = \frac{15}{100}$$

Step 3: Reduce the fraction. $\frac{15}{100} = \frac{3}{20}$

Change the following percents to fractions and reduce.

1. 50%	5. 52%	9. 18%	13. 16%	17. 99%
2. 13%	6. 63%	10. 3%	14. 1%	18. 30%
3. 22%	7. 75%	11. 25%	15. 79%	19. 42%
4. 95%	8. 91%	12. 5%	16. 40%	20. 84%

Example 8: Change $\frac{7}{8}$ to a percent.

Step 1: Divide 7 by 8. Add as many 0's as necessary.

$$\begin{array}{r} 0.875 \\ 8\overline{)\,7.000} \\ -\underline{64} \\ 60 \\ -\underline{56} \\ 40 \\ -\underline{40} \\ 0 \end{array}$$

Step 2: Change the decimal answer, 0.875, to a percent by moving the decimal point 2 places to the right.

$$\frac{7}{8} = 0.875 = 87.5\%$$

Change the following fractions to percents.

1. $\frac{1}{5}$	4. $\frac{3}{8}$	7. $\frac{1}{10}$	10. $\frac{3}{4}$	13. $\frac{1}{16}$	16. $\frac{7}{20}$
2. $\frac{5}{8}$	5. $\frac{3}{16}$	8. $\frac{4}{5}$	11. $\frac{1}{8}$	14. $\frac{1}{4}$	17. $\frac{2}{5}$
3. $\frac{7}{16}$	6. $\frac{19}{100}$	9. $\frac{15}{16}$	12. 5/[illegible]	15. $\frac{4}{100}$	18. $\frac{16}{25}$

2.12 Changing Percents to Mixed Numbers and Mixed Numbers to Percents

Example 9: Change 218% to a fraction.

Step 1: Copy the number without the percent sign. 218 is the top number of the fraction.

Step 2: The bottom number of the fraction is always 100.

$$218\% = \frac{218}{100}$$

Step 3: Reduce the fraction, and convert to a mixed number. $\frac{218}{100} = \frac{109}{50} = 2\frac{9}{50}$

Change the following percents to mixed numbers.

1. 150%	6. 163%	11. 205%	16. 340%
2. 113%	7. 275%	12. 405%	17. 199%
3. 222%	8. 191%	13. 516%	18. 300%
4. 395%	9. 108%	14. 161%	19. 125%
5. 252%	10. 453%	15. 179%	20. 384%

Example 10: Change $5\frac{3}{8}$ to a percent.

Step 1: Divide 3 by 8. Add as many 0's as necessary.

$$\begin{array}{r} 0.375 \\ 8\overline{)\,3.000} \\ -\,2\,4 \\ \hline 60 \\ -56 \\ \hline 40 \\ -40 \\ \hline 0 \end{array}$$

Step 2: So, $5\frac{3}{8} = 5.375$. Change the decimal answer to a percent by moving the decimal point 2 places to the right.

$5\frac{3}{8} = 5.375 = 537.5\%$

Change the following mixed and whole numbers to percents.

1. $5\frac{1}{2}$	4. $3\frac{1}{4}$	7. $1\frac{3}{10}$	10. $2\frac{13}{25}$	13. $1\frac{3}{16}$	16. $4\frac{4}{5}$
2. $8\frac{3}{4}$	5. $4\frac{7}{8}$	8. $6\frac{1}{5}$	11. $1\frac{1}{8}$	14. $1\frac{1}{16}$	17. $3\frac{2}{5}$
3. 1	6. 3	9. 4	12. 2	15. 5	18. 6

2.13 Changing to Percent Word Problems

Example 11: Three out of four students prefer pizza for lunch over hot dogs. What percent of the students prefer pizza?

Step 1: Change three out of four to a fraction. $\frac{3}{4}$

Step 2: Change $\frac{3}{4}$ to a percent. $\frac{3}{4} = 75\%$ of the students prefer pizza.

For each of the following, find the percent.

1. 18 out of 24 students take a music class in middle school.
2. 6 out of the 30 students get a grade of an A in the class.
3. 3 out of 10 students ride the bus to school.
4. 16 out of 2[illegible] students in my first period ate breakfast.
5. Out of 1[illegible] cats, 3 of them have stripes.
6. In a neighborhood store, 8 out of 50 customers have an infant in their cart.
7. [illegible]40 out of 500 people at the fair are over 21.
8. In Alaska, 9 months out of 12 have temperatures below 0° C.
9. He hits the ball 360 times out of 500 times at bat.
10. She scores 5[illegible]2 out of a possible 600 points on her math test.
11. 16 out of 40 pieces in the box of candy are covered in dark chocolate.
12. 680 out of 800 students have not had a cold this year.
13. 7 out of 8 of the flowers are pink.
14. 312 out of 600 men at the baseball game have on earmuffs.
15. 824 out of 1000 tulips are red.
16. Kristen sells 475 out of her 500 boxes of Girl Scout cookies.
17. 459 out of 675 students in the 6th grade went to the spring dance this year.
18. In a bag of 60 candy-coated chocolate pieces, 9 are yellow.

2.14 Finding the Percent of the Total

Example 12: There are 75 customers at Billy's gas station this morning. Thirty-two percent use a credit card to make their purchases. How many customers used credit cards this morning at Billy's?

Step 1: Change 32% to a decimal. 0.32

Step 2: Multiply by the total number mentioned.

$$\begin{array}{r} 0.3\,2 \\ \times 7\,5 \\ \hline 16\,0 \\ 224 \\ \hline 24.00 \end{array}$$

24 customers used credit cards.

Answer the following questions.

1. Eighty-five percent of Mrs. Coomer's math class pass her final exam. There are 40 students in her class. How many pass?

2. Fifteen percent of a bag of chocolate candies have a red coating on them. How many red pieces are in a bag of 60 candies?

3. Sixty-eight percent of Valley Creek School students attend this year's homecoming dance. There are 675 students. How many attend the dance?

4. Out of the 4,500 people who attend the rock concert, forty-six percent purchase a T-shirt. How many people buy T-shirts?

5. Nina sells ninety-five percent of her 500 cookies at the bake sale. How many cookies does she sell?

6. Twelve percent of yesterday's customers purchased premium-grade gasoline from GasCo. If GasCo had 200 customers, how many purchased premium-grade gasoline?

7. The Candy Shack sells 138 pounds of candy on Tuesday. Fifty-two percent of the candy is jelly beans. How many pounds of jelly beans are sold Tuesday?

8. A fund-raiser at the school raises $617.50. Ninety-four percent goes to local charities. How much money goes to charities?

9. Out of the company's $6.5 million profit, eight percent will be paid to shareholders. How much will be paid to the shareholders?

10. Ted's Toys sells seventy-five percent of its stock of stuffed bean animals on Saturday. If Ted's Toys had 620 originally in stock, how many are sold on Saturday?

2.15 Finding the Percent Increase or Decrease

Example 13: Office Supply Co. purchases paper wholesale for $18.00 per case. They sell the paper for $20.00 per case. By what percent does the store increase the price of the paper (or what is the percent markup)?

$$\textbf{Percent Change} = \frac{\textbf{Amount of Change}}{\textbf{Original Amount}}$$

Step 1: Find the amount of change. In this problem, the price was marked up $2.00. The amount of change is 2.

Step 2: Divide the amount of change, 2, by the wholesale cost, 18. $\frac{2}{18} = 0.111$

Step 3: Change the decimal, 0.111, to a percent. $0.111 = 11.1\%$

Example 14: The price of gas goes from $2.40 per gallon to $1.30. What is the percent of decrease in the price of gas?

$$\textbf{Percent Change} = \frac{\textbf{Amount of Change}}{\textbf{Original Amount}}$$

Step 1: Find the amount of change. In this problem, the price is decreased $1.10. The amount of change is $1.10.

Step 2: Divide the amount of change, 1.10, by the original cost, 2.40. $\frac{1.10}{2.40} = 0.46$

Step 3: Change the decimal, 0.46, to a percent. $0.46 = 46\%$, the price of gas has decreased 46%.

Find the percent increase or decrease to the nearest percent for each of the problems below.

1. Mary was making $25,000 per year. Her boss gives her a $3,000 raise. What percent increase is that?

2. Last week Matt's total sales were $12,000. This week his total sales were only $2,000. By what percent did his sales for this week decrease?

3. Eric cuts lawns for $16.00. Next year, he will charge $2.00 more per lawn. What percent increase will he charge?

4. At an office supply store, pens are marked down from $1.50 to $1.20. What percent discount is that?

5. Cowboys buys boots wholesale for $103.35 a pair. They sell the boots in their store for $159. What percent is the markup on the boots?

6. Blakeville has a population of 1,600 by recent count. According to a previous census, Blakeville had a population of 1,850. What has been the percent decrease in population?

7. Last year Roswell Elementary School had 680 graduates. This year they graduated 812. What has been the percent increase in graduates?

8. Michi's father gets a new job that pays $52,000 per year. That is $16,000 more than his last job. What percent pay increase is that?

2.16 Finding the Amount of a Discount

Sale prices are sometimes marked 30% off, or better yet 50% off. A 30% **discount** means you will pay 30% less than the original price. How much money you will save is also known as the amount of the discount. Read the example below to learn to figure the amount of a discount.

Example 15: A $179.00 chair is on sale for 30% off. How much can I save if I buy it now?

Step 1: Change 30% to a decimal. 30% = 0.30

Step 2: Multiply the original price by the discount.

	ORIGINAL PRICE		**$179.00**
×	**% DISCOUNT**	×	**0.30**
	SAVINGS		**$53.70**

Practice finding the amount of discount. Round off answers to the nearest penny.

1. Tubby Telephones is offering a 25% discount on phones purchased on Tuesday. How much can you save if you buy a phone on Tuesday regularly priced at $2[illegible]5.00 any other day of the week?

2. The regular price for a garden rake is $10.97 at Sly's Super Store. This week, Sly is offering a 30% discount. How much is the discount on the rake?

3. Christine buys a sweater regularly priced at $26.80 with a coupon for 20% off any sweater. How much does she save?

4. The software that Myochi needs for her computer is priced at $69.85. If she waits until a store offers it for 20% off, how much will she save?

5. Ty purchases jeans that are priced at $23.97. He receives a 15% employee discount. How much does he save?

6. The Bakery Company offers a 60% discount on all bread made the day before. How much can you save on a $2.40 loaf made today if you wait until tomorrow to buy it?

7. A furniture store advertises a 40% off sale on all items. How much would the discount be on a $2,[illegible]30 bedroom set?

8. Sharta buys a $4.00 bottle of nail polish on sale for 30% off. What is the dollar amount of the discount?

9. How much is the discount on a $350 racing bike marked 15% off?

10. Raymond receives a 2% discount from his credit card company on all purchases made with the credit card. What is his discount on $1,575.50 worth of purchases?

2.17 Finding the Discounted Sale Price

To find the discounted sale price, you must go one step further than shown on the previous page. Read the example below to learn how to figure **discount** prices.

Example 16: A $74.00 chair is on sale for 25% off. How much will it cost if I buy it now?

Step 1: Change 25% to a decimal.

$25\% = 0.25$

Step 2: Multiply the original price by the discount.

	ORIGINAL PRICE		**$74.00**
×	**% DISCOUNT**	×	**0.25**
	SAVINGS		**$18.50**

Step 3: Subtract the savings amount from the original price to find the sale price.

	ORIGINAL PRICE		$74.00
−	**SAVINGS**	−	18.50
	SALE PRICE		$55.50

Figure the sale price of the items below. The first one is done for you.

ITEM	PRICE	%OFF	MULTIPLY	SUBTRACT	SALE PRICE
1. pen	$1.50	20%	$1.50 \times 0.2 = \$0.30$	$1.50 - 0.30 = 1.20$	$1.20
2. recliner	$325	25%			
3. juicer	$55	15%			
4. blanket	$14	10%			
5. earrings	$2.40	20%			
6. figurine	$8	15%			
7. boots	$159	35%			
8. calculator	$80	30%			
9. candle	$6.20	50%			
10. camera	$445	20%			
11. DVD player	$235	25%			
12. video game	$25	10%			

2.18 Sales Tax

Example 17: The total price of a sofa is $560.00 × 6% **sales tax**. How much is the sales tax? What is the total cost?

Step 1: You will need to change 6% to a decimal.

$6\% = 0.06$

Step 2: Simply multiply the cost, $560, by the tax rate, 6%. $560 \times 0.06 = 33.6$
The answer will be $33.60. (You need to add a 0 to the answer. When dealing with money, there must be two places after the decimal point.)

	COST		**$560**
×	**6% TAX**	×	**0.06**
	SALES TAX		**$33.60**

Step 3: Add the sales tax amount, $33.60, to the cost of the item sold, $560. This is the total cost.

	COST		**$560.00**
+	**SALES TAX**	+	**33.60**
	TOTAL COST		**$593.60**

Note: When the answer to the question involves money, you always need to round off the answer to the nearest hundredth (2 places after the decimal point). Sometimes you will need to add a zero.

Figure the total costs in the problems below. The first one is done for you.

ITEM		PRICE	% TAX	MULTIPLY	PRICE PLUS TAX	TOTAL
1.	jeans	$42	7%	$42 × 0.07 = $2.94	42 + 2.94 = 44.94	$44.94
2.	truck	$17,495	6%			
3.	film	$3.89	8%			
4.	T-shirt	$12	5%			
5.	football	$36.40	4%			
6.	soda	$1.78	5%			
7.	4 tires	$105.80	10%			
8.	clock	$18	6%			
9.	burger	$2.34	5%			
10.	software	$89.95	8%			

Chapter 2 Review

Find the greatest common factor for the following sets of numbers.

1. 9 and 15
2. 12 and 16
3. 10 and 25
4. 8 and 24

Find the least common multiple for the following sets of numbers.

5. 8 and 12
6. 5 a[illegible]
7. 4 and 10
8. 6 and 8

Perform the following operations and simplify.

9. $\frac{5}{9} + \frac{7}{9}$
10. $7\frac{1}{2} + 3\frac{3}{8}$
11. $4\frac{4}{15} + \frac{1}{5}$
12. $\frac{1}{7} + \frac{3}{[illegible]}$
13. $10 - 5\frac{1}{8}$
14. $3\frac{1}{3} - \frac{3}{4}$
15. $9\frac{3}{4} - 2\frac{3}{8}$
16. $6\frac{1}{5} - 1\frac{3}{10}$
17. $1\frac{1}{3} \times 3\frac{1}{2}$
18. $5\frac{3}{7} \times \frac{7}{8}$
19. $4\frac{4}{6} \times 1\frac{5}{7}$
20. $\frac{2}{3} \times \frac{5}{6}$
21. $\frac{1}{2} \div \frac{4}{5}$
22. $6\frac{6}{7} \div 2\frac{2}{3}$
23. $3\frac{5}{6} \div 11\frac{1}{2}$
24. $1\frac{1}{3} \div 3\frac{1}{5}$

Perform the following operations.

25. $12.589 + [illegible].62 + 0.9$
26. $7.8 + 10.24 + 1.903$
27. $152.6[illegible] + 1[illegible].3 + 0.[illegible]24$
28. $1[illegible].54[illegible] - [illegible].62$
29. $1.[illegible]5 - 0.093$
30. $45.2 - 37.9$
31. 4.58×0.025
32. 0.879×1.7
33. $30.[illegible] \times 0.0041$
34. $1[illegible].28 \div [illegible].054$
35. $174.66 \div 1.23$
36. $2.115 \div 9$

Change to a fraction.

37. 0.55
38. 0.[illegible]4
39. 0.32

Change to a mixed number.

40. [illegible].375
41. 9.6
42. 13.25

Change to a decimal.

43. $5\frac{3}{25}$
44. $\frac{7}{100}$
45. $10\frac{2}{3}$

Change the following percents to decimals.

46. 45% 47. 219% 48. 22% 49. 1.25%

Change the following decimals to percents.

50. 0.52 51. 0.64 52. 1.09 53. 0.625

54. What is 1.65 written as a percent?

55. Change 5.65 to a percent.

Change the following percents to fractions.

56. 25% 57. 3% 58. [illegible]8% 59. 102%

Change the following fractions to percents.

60. $\frac{9}{10}$ 61. $\frac{5}{16}$ 62. $\frac{1}{8}$ 63. $\frac{1}{4}$

64. The Vargas family is hiking a $23\frac{1}{3}$-mile trail. The first day, they hiked $1[illegible]\frac{1}{5}$ miles. How much farther do they have to go to complete the trail?

65. Jena walks $\frac{1}{[illegible]}$ of a mile to a friend's house, $1\frac{1}{3}$ miles to the store, and $\frac{3}{4}$ of a mile back home. How far does Jena walk?

66. Cory uses $2\frac{[illegible]}{5}$ gallons of paint to mark one mile of this year's spring road race. How many gallons will he use to mark the entire $6\frac{1}{4}$ mile course?

67. Gene works for his father sanding wooden rocking chairs. He earns $6.35 per chair. How many chairs does he need to sand in order to buy a portable radio CD player for $146.05?

68. Margo's Mint Shop has a machine that produces 4.35 pounds of mints per hour. How many pounds of mints are produced in each 8-hour shift?

69. Carter's Junior High track team runs the first leg of a 400-meter relay race in 10.23 seconds, the second leg in 11.4 seconds, the third leg in 10.77 seconds, and the last leg in 9.9 seconds. How long does it take for them to complete the race?

70. Spaulding High School decided to sell boxes of oranges to earn money for new football uniforms. They ordered a truckload of 500 boxes of oranges from a California grower for $1[illegible].[illegible]0 per box. They sold 450 boxes for $19.00 per box. On the last day of the sale, they sold the oranges they had left for $17.00 per box. How much profit did they make?

71. A school receives 56% of the total sale at the end of a fund-raiser. If a fund-raiser makes $564,000, how much money does the school receive?

72. Peeler's Jewelry is offering a 30% off sale on all bracelets. How much will you save if you buy a $45.00 bracelet during the sale?

73. How much would an employee pay for a $724 stereo if the employee gets a 15% discount?

74. Misha buys a CD for $14.95. If sales tax is 7%, how much does she pay total?

75. The Pep band made $640 during a fund-raiser. The band spent $400 of the money on new uniforms. What percent of the total did the band members spend on uniforms?

76. Hank gets 10 hours per week to play his video games. Since he made straight A's last semester, his parents increased his playing time to 16 hours per week. What percent increase in time does he get?

77. Patton, Patton, and Clark, a law firm, won a malpractice lawsuit for $4,500,000. Sixty-eight percent went to the law firm. How much did the law firm make?

78. A department store is selling all swimsuits for 40% off in August. How much would you pay for a swimsuit that is normally priced at $35.80?

79. High school students voted on where they would go on a field trip. For every 3 students who wanted to see Calaveras Big Trees State Park, 8 students wanted to see Columbia State Historic Park. What percent of the students wanted to go to Columbia State Historic Park?

80. An increase from 20 to 36 is what percent of increase?

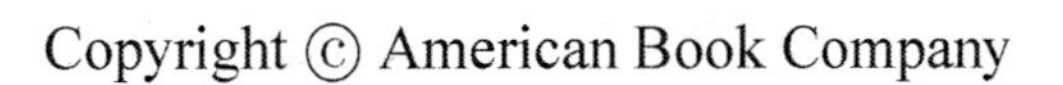

Chapter 3
Exponents and Roots

3.1 Understanding Exponents

Sometimes it is necessary to multiply a number by itself one or more times. For example, a math problem may require multiplying 3×3 or $5 \times 5 \times 5 \times 5$. Instead of writing 3×3, you can write 3^2, or instead of writing $5 \times 5 \times 5 \times 5$, 5^4 means the same thing. The first number is the **base**. The small, raised number is called the **exponent** or **power**. The exponent tells how many times the base should be multiplied by itself.

Example 1: 6^3 ← **exponent (or power)**; 6 ← **base**. This means multiply by 6 three times: $6 \times 6 \times 6$

Example 2: **Negative numbers can be raised to exponents also.**
An **even** exponent will give a **positive** answer: $(-2)^2 = (-2) \times (-2) = 4$
An **odd** exponent will give a **negative** answer: $(-2)^3 = -2 \times -2 \times -2 = -8$

You also need to know two special properties of exponents:

1. Any base number raised to the exponent of 1 equals the base number.
2. Any base number raised to the exponent of 0 equals 1.

Example 3: $4^1 = 4$ $10^1 = 10$ $25^1 = 25$ $4^0 = 1$ $10^0 = 1$ $25^0 = 1$

Rewrite the following problems using exponents.
Example 4: $2 \times 2 \times 2 = 2^3$

1. $7 \times 7 \times 7 \times 7$
2. 10×10
3. $12 \times 12 \times 12$
4. $4 \times 4 \times 4 \times 4$
5. $9 \times 9 \times 9$
6. 25×25
7. $15 \times 15 \times 15$
8. $5 \times 5 \times 5 \times 5 \times 5$
9. $2 \times 2 \times 2 \times 2$
10. 14×1[illegible]
11. $3 \times 3 \times 3 \times 3 \times 3$
12. $11 \times 11 \times 11$

Use a calculator to figure what product each number with an exponent represents.
Example 5: $2^3 = 2 \times 2 \times 2 = 8$

13. $(-8)^3$
14. 12^2
15. 20^1
16. 5^4
17. 15^0
18. 10^2
19. $(-10)^2$
20. 3^5
21. [illegible]
22. [illegible]
23. 4^3
24. 54^1

Express the numbers as a base with an exponent. (Some may have multiple answers.)
Example 6: $4 = 2 \times 2 = 2^2$

25. 9
26. 16
27. 27
28. 36
29. 8
30. 32
31. 1000
32. 12[illegible]
33. 81
34. 64
35. 49
36. 121

3.2 Multiplication with Exponents

Rule 1: To multiply two expressions with the same base, add the exponents together and keep the base the same.

Example 7: $2^3 \times 2^5 = 2^{3+5} = 2^8$

Rule 2: If a power is raised to another power, multiply the exponents together and keep the base the same.

Example 8: $(2^3)^2 = 2^{3\times2} = 2^6$

Rule 3: If a product in parenthesis is raised to a power, then each factor is raised to the power when parenthesis are eliminated.

Example 9: $(2 \times 4)^2 = 2^2 \times 4^2 = 4 \times 16 = 64$
Example 10: $(3a)^3 = 3^3 \times a^3 = 27a^3$
Example 11: $(7b^5)^2 = 7^2b^{10} = 49b^{10}$

Simplify each of the expressions below.

1. $(5^3)^2$	9. $(3^3)^2$	17. $(6 \times 4)^2$	25. $(4^4)^2$
2. $6^3 \times 6^5$	10. $2^5 \times 2^5$	18. $(9a^5)^3$	26. $(2b^3)^4$
3. $4^3 \times 4^3$	11. $(3 \times 3)^2$	19. $4^3 \times 4^4$	27. $(5a^2)^{[illegible]}$
4. $(7^5)^2$	12. $(2a)^4$	20. $(6b^5)^2$	28. $(8^3)^2$
5. $(c^2)^5$	13. $(3^2)^4$	21. $(5^2)^3$	29. $(9^2)^2$
6. $2^5 \times 2^3$	14. $4^5 \times 4^3$	22. $3^7 \times 3^3$	30. $10^5 \times 10^4$
7. $(4 \times 5)^2$	15. $(3 \times 2)^4$	23. $([illegible]a)^2$	31. $(3 \times 5)^2$
8. $(3^4)^{[illegible]}$	16. $(5^2)^2$	24. $(3^4)^2$	32. $(7^3)^2$

3.3 Division with Exponents

Rule 1: Expressions can also have negative exponents. Negative exponents do not indicate negative numbers. They indicate reciprocals, which is 1 over the original number.

Example 12: $2^{-3} = \frac{1}{2^3} = \frac{1}{8}$

Example 13: $3a^{-5} = 3 \times \frac{1}{a^5} = \frac{3}{a^5}$

Rule 2: When dividing expressions with exponents that have the same base, subtract the exponents. Expressions in simplified form have only positive exponents.

Example 14: $\frac{3^5}{3^3} = 3^5 - 3^3 = 3^{5-3} = 3^2 = 9$

Example 15: $\frac{3^5}{3^8} = 3^5 - 3^8 = 3^{5-8} = 3^{-3} = \frac{1}{3^3} = \frac{1}{27}$

Rule 3: If a fraction is raised to a power, then both the numerator and the denominator are raised to the same power.

Example 16: $\left(\frac{3}{4}\right)^3 = \frac{3^3}{4^3} = \frac{27}{64}$

Example 17: $(2x)^{-2} = \frac{1}{(2x)^2} = \frac{1}{4x^2}$

Reduce the following expressions to their simplest form. All exponents should be positive.

1. $5x^{-4}$
2. $\frac{2^2}{2^4}$
3. $\left(\frac{2}{3}\right)^2$
4. $6a^{-2}$
5. $\frac{3^6}{3^3}$
6. $(5a)^{-2}$
7. $\frac{3^4}{3^3}$
8. $\left(\frac{7}{8}\right)^3$
9. $(6a)^{-2}$
10. $\frac{(x^2)^3}{x^4}$
11. $\frac{(3y)^3}{3^2y}$
12. $\frac{(3a^2)^3}{a^{[illegible]}}$
13. $(2x^2)^{-5}$
14. $2x^{-2}$
15. $(a^3)^{-2}$
16. $(2^{-[illegible]})$
17. $\left(\frac{[illegible]}{2}\right)^2$
18. $\frac{1}{3^{-2}}$
19. $(4y)^{-5}$
20. $4y^{-5}$

3.4 Square Root

Just as working with exponents is related to multiplication, so finding square roots is related to division. In fact, the sign for finding the square root of a number looks similar to a division sign. The best way to learn about square roots is to look at examples.

Example 18: This is a square root problem: $\sqrt{64}$
It is asking, "What is the square root of 64?"
It means, "What number multiplied by itself equals 64?"
The answer is 8. $8 \times 8 = 64$.

Example 19: Find the square root of 36 and 144.

$\sqrt{36}$ — $6 \times 6 = 36$ so $\sqrt{36} = 6$

$\sqrt{144}$ — $12 \times 12 = 144$ so $\sqrt{144} = 12$

Find the square roots of the following numbers.

1. $\sqrt{49}$	4. $\sqrt{16}$	7. $\sqrt{100}$	10. $\sqrt{36}$	13. $\sqrt{64}$
2. $\sqrt{81}$	5. $\sqrt{121}$	8. $\sqrt{289}$	11. $\sqrt{4}$	14. $\sqrt{9}$
3. $\sqrt{25}$	6. $\sqrt{625}$	9. $\sqrt{196}$	12. $\sqrt{900}$	15. $\sqrt{144}$

3.5 Simplifying Square Roots

Square roots can sometimes be simplified even if the number under the square root is not a perfect square. One of the rules of roots is that if a and b are two positive real numbers, then it is always true that $\sqrt{a \times b} = \sqrt{a} \times \sqrt{b}$. You can use this rule to simplify square roots.

Example 20: $\sqrt{100} = \sqrt{4 \times 25} = \sqrt{4} \times \sqrt{25} = 2 \times 5 = 10$

Example 21: $\sqrt{200} = \sqrt{100 \times 2} = 10\sqrt{2}$ This is 10 multiplied by the square root of 2.

Example 22: $\sqrt{160} = \sqrt{10 \times 16} = 4\sqrt{10}$

Simplify.

1. $\sqrt{98}$	5. $\sqrt{8}$	9. $\sqrt{54}$	13. $\sqrt{90}$
2. $\sqrt{600}$	6. $\sqrt{63}$	10. $\sqrt{40}$	14. $\sqrt{175}$
3. $\sqrt{50}$	7. $\sqrt{48}$	11. $\sqrt{72}$	15. $\sqrt{18}$
4. $\sqrt{27}$	8. $\sqrt{75}$	12. [illegible]	16. $\sqrt{20}$

3.6 Adding and Subtracting Roots

You can add and subtract terms with square roots only if the number under the square root sign is the same.

Example 23: $2\sqrt{2} + 3\sqrt{2} = 5\sqrt{2}$

Example 24: $12\sqrt{7} - 3\sqrt{7} = 9\sqrt{7}$

Or, look at the following examples where you can simplify the square roots and then add or subtract.

Example 25: $2\sqrt{25} + \sqrt{36}$

Step 1: Simplify. You know that $\sqrt{25} = 5$, and $\sqrt{36} = 6$ so the problem simplifies to $2(5) + 6$

Step 2: Solve: $2(5) + 6 = 10 + 6 = 16$

Example 26: $2\sqrt{72} - 3\sqrt{2}$

Step 1: Simplify what you know. $\sqrt{72} = \sqrt{36 \cdot 2} = 6\sqrt{2}$

Step 2: Substitute $6\sqrt{2}$ for $\sqrt{72}$ and simplify.
$2(6)\sqrt{2} - 3\sqrt{2} = 12\sqrt{2} - 3\sqrt{2} = 9\sqrt{2}$

Simplify the following addition and subtraction problems.

1. $3\sqrt{5} + 9\sqrt{5}$
2. $3\sqrt{25} + 4\sqrt{16}$
3. $4\sqrt{8} + 2\sqrt{2}$
4. $3\sqrt{32} - 2\sqrt{2}$
5. $\sqrt{25} - \sqrt{49}$
6. $2\sqrt{5} + 4\sqrt{20}$
7. $5\sqrt{8} - 3\sqrt{72}$
8. $\sqrt{27} + 3\sqrt{27}$
9. $3\sqrt{20} - 4\sqrt{45}$
10. $4\sqrt{45} - \sqrt{75}$
11. $2\sqrt{28} + 2\sqrt{7}$
12. $\sqrt{64} + \sqrt{81}$
13. $5\sqrt{54} - 2\sqrt{24}$
14. $\sqrt{32} + 2\sqrt{50}$
15. $2\sqrt{7} + 4\sqrt{63}$
16. $8\sqrt{2} + \sqrt{8}$
17. $2\sqrt{8} - 4\sqrt{32}$
18. $\sqrt{36} + \sqrt{100}$
19. $\sqrt{9} + \sqrt{25}$
20. $\sqrt{64} - \sqrt{36}$
21. $\sqrt{75} + \sqrt{108}$
22. $\sqrt{81} + \sqrt{100}$
23. $\sqrt{192} - \sqrt{75}$
24. $3\sqrt{5} + \sqrt{245}$

3.7 Multiplying Roots

You can also multiply square roots. To multiply square roots, you just multiply the numbers under the square root sign and then simplify. Look at the examples below.

Example 27: $\sqrt{2} \times \sqrt{6}$

Step 1: $\sqrt{2} \times \sqrt{6} = \sqrt{2 \times 6} = \sqrt{12}$ Multiply the numbers under the square root sign.

Step 2: $\sqrt{12} = \sqrt{4 \times 3} = 2\sqrt{3}$ Simplify.

Example 28: $3\sqrt{3} \times 5\sqrt{6}$

Step 1: $(3 \times 5)\sqrt{3 \times 6} = 15\sqrt{18}$ Multiply the numbers in front of the square root, and multiply the numbers under the square root sign.

Step 2: $15\sqrt{18} = 15\sqrt{2 \times 9}$
$15 \times 3\sqrt{2} = 45\sqrt{2}$ Simplify.

Example 29: $\sqrt{14} \times \sqrt{42}$ For this more complicated multiplication problem, use the rule of roots that you learned above, $\sqrt{a \cdot b} = \sqrt{a} \cdot \sqrt{b}$.

Step 1: $\sqrt{14} = \sqrt{7} \times \sqrt{2}$ and $\sqrt{42} = \sqrt{2} \times \sqrt{3} \times \sqrt{7}$ Instead of multiplying 14 by 42, divide these numbers into their roots.

$\sqrt{14} \times \sqrt{42} = \sqrt{7} \times \sqrt{2} \times \sqrt{2} \times \sqrt{3} \times \sqrt{7}$

Step 2: Since you know that $\sqrt{7} \times \sqrt{7} = 7$ and $\sqrt{2} \times \sqrt{2} = 2$, the problem simplifies to $(7 \times 2)\sqrt{3} = 14\sqrt{3}$

Simplify the following multiplication problems.

1. $\sqrt{5} \times \sqrt{7}$
2. $\sqrt{32} \times \sqrt{2}$
3. $\sqrt{10} \times \sqrt{14}$
4. $2\sqrt{3} \times 3\sqrt{6}$
5. $4\sqrt{2} \times 2\sqrt{10}$
6. $\sqrt{5} \times 3\sqrt{15}$
7. $\sqrt{45} \times \sqrt{27}$
8. $5\sqrt{21} \times \sqrt{7}$
9. $\sqrt{42} \times \sqrt{21}$
10. $4\sqrt{3} \times 2\sqrt{12}$
11. $\sqrt{56} \times \sqrt{24}$
12. $\sqrt{11} \times 2\sqrt{33}$
13. $\sqrt{13} \times \sqrt{26}$
14. $2\sqrt{2} \times 5\sqrt{5}$
15. $\sqrt{6} \times \sqrt{12}$

3.8 Dividing Roots

When dividing a number or a square root by another square root, you cannot leave the square root sign in the denominator (the bottom number of a fraction). You must simplify the problem so that the square root is not in the denominator. Look at the examples below.

Example 30: $\dfrac{\sqrt{2}}{\sqrt{5}}$

Step 1: $\dfrac{\sqrt{2}}{\sqrt{5}} \times \dfrac{\sqrt{5}}{\sqrt{5}}$ The fraction $\frac{\sqrt{5}}{\sqrt{5}}$ is equal to 1, and multiplying by 1 does not change the value of a number.

Step 2: $\dfrac{\sqrt{2 \times 5}}{5} = \dfrac{\sqrt{10}}{5}$ Multiply and simplify. Since $\sqrt{5} \times \sqrt{5}$ equals 5, you no longer have a square root in the denominator.

Example 31: $\dfrac{6\sqrt{2}}{2\sqrt{10}}$ In this problem, the numbers outside of the square root will also simplify.

Step 1: $\dfrac{6}{2} = 3$ so you have $\dfrac{3\sqrt{2}}{\sqrt{10}}$

Step 2: $\dfrac{3\sqrt{2}}{\sqrt{10}} \times \dfrac{\sqrt{10}}{\sqrt{10}} = \dfrac{3\sqrt{2 \times 10}}{10} = \dfrac{3\sqrt{20}}{10}$

Step 3: $\dfrac{3\sqrt{20}}{10}$ will further simplify because $\sqrt{20} = 2\sqrt{5}$, so you then have $\dfrac{3 \times 2\sqrt{5}}{10}$ which reduces to $\dfrac{3\sqrt{5}}{5}$.

Simplify the following division problems.

1. $\dfrac{[illegible]\sqrt{3}}{\sqrt{5}}$
2. $\dfrac{16}{\sqrt{8}}$
3. $\dfrac{24\sqrt{10}}{12\sqrt{3}}$
4. $\dfrac{\sqrt{121}}{\sqrt{6}}$
5. $\dfrac{\sqrt{40}}{\sqrt{90}}$
6. $\dfrac{33\sqrt{15}}{11\sqrt{2}}$
7. $\dfrac{\sqrt{32}}{\sqrt{12}}$
8. $\dfrac{\sqrt{11}}{\sqrt{5}}$
9. $\dfrac{\sqrt{2}}{\sqrt{6}}$
10. $\dfrac{2\sqrt{7}}{\sqrt{14}}$
11. $\dfrac{5\sqrt{2}}{4\sqrt{8}}$
12. $\dfrac{4\sqrt{21}}{[illegible]}$
13. $\dfrac{[illegible]\sqrt{22}}{2\sqrt{2}}$
14. $\dfrac{\sqrt{35}}{2\sqrt{14}}$
15. $\dfrac{\sqrt{40}}{\sqrt{15}}$
16. $\dfrac{\sqrt{3}}{\sqrt{12}}$

Chapter 3 Review

Simplify the following problems.

1. 15^0

2. $\sqrt{100}$

3. $\sqrt{49}$

4. $(-3)^3$

Write using exponents.

5. $3 \times 3 \times 3 \times 3$

6. $6 \times 6 \times 6 \times 6 \times 6 \times 6$

7. $11 \times 11 \times 11$

8. $2 \times 2 \times 2 \times 2 \times 2 \times 2 \times 2 \times 2$

Simplify the following square root problems.

9. $5\sqrt{27} + 7\sqrt{3}$

10. $\sqrt{40} - \sqrt{10}$

11. $\sqrt{64} + \sqrt{81}$

12. $\dfrac{\sqrt{56}}{\sqrt{35}}$

13. $14\sqrt{5} + 8\sqrt{80}$

14. $\sqrt{63} \times \sqrt{28}$

15. $8\sqrt{50} - 3\sqrt{32}$

16. $\sqrt{8} \times \sqrt{50}$

Chapter 4
Introduction to Algebra

4.1 Algebra Vocabulary

Vocabulary Word	Example	Definition
variable	$4x$ (x is the variable)	a letter that can be replaced by a number
coefficient	$4x$ (4 is the coefficient)	a number multiplied by a variable or variables
term	$5x^2 + x - 2$ ($5x^2$, x, and -2 are terms)	numbers or variables separated by $+$ or $-$ signs
constant	$5x + 2y + 4$ (4 is a constant)	a term that does not have a variable
degree	$4x^2 + 3x - 2$ (the degree is 2)	the largest power of a variable in an expression
leading coefficient	$4x^2 + 3x - 2$ (4 is the leading coefficient)	the number multiplied by the term with the highest power
sentence	$2x = 7$ or $5 \leq x$	two algebraic expressions connected by $=$, $\neq$, $<$, $>$, $\leq$, $\geq$, or $\approx$
equation	$4x = 8$	a sentence with an equal sign
inequality	$7x < 30$ or $x \neq 6$	a sentence with one of the following signs: $\neq$, $<$, $>$, $\leq$, $\geq$, or $\approx$
base	6^3 (6 is the base)	the number used as a factor
exponent	6^3 (3 is the exponent)	the number of times the base is multiplied by itself

4.2 Substituting Numbers for Variables

These problems may look difficult at first glance, but they are very easy. Simply replace the variable with the number the variable is equal to, and solve the problems.

Example 1: In the following problems, substitute 10 for a.

Problem	Calculation	Solution
1. $a+1$	Simply replace the a with 10. $10+1$	11
2. $17-a$	$17-10$	7
3. $9a$	This means multiply. 9×10	90
4. $\frac{30}{a}$	This means divide. $30\div 10$	3
5. a^3	$10\times 10\times 10$	1000
6. $5a+6$	$(5\times 10)+6$	56

Note: Be sure to do all multiplying and dividing before adding and subtracting.

Example 2: In the following problems, let $x=2$, $y=4$, and $z=5$.

Problem	Calculation	Solution
1. $5xy+z$	$5\times 2\times 4+5$	45
2. xz^2+5	$2\times 5^2+5=2\times 25+5$	55
3. $\frac{yz}{x}$	$(4\times 5)\div 2=20\div 2$	10

In the following problems, $t=7$. Solve the problems.

1. $t+3=$
2. $18-t=$
3. $\frac{21}{t}=$
4. $3t-5=$
5. $t^2+1=$
6. $2t-4=$
7. $9t\div 3=$
8. $\frac{t^2}{7}=$
9. $5t+6=$
10. $\frac{(t^2-7)}{6}=$
11. $4t+5t=$
12. $\frac{6t}{3}=$

In the following problems $a=4$, $b=-2$, $c=5$, and $d=10$. Solve the problems.

13. $4a+2c=$
14. $3bc-d=$
15. $\frac{ac}{d}=$
16. $d-2a=$
17. $a^2-b=$
18. $abd=$
19. $5c-ad=$
20. $cd+bc=$
21. $\frac{6b}{a}=$
22. $9a+b=$
23. $5+3bc=$
24. $d^2+d+1=$

4.3 Understanding Algebra Word Problems

The biggest challenge to solving word problems is figuring out whether to add, subtract, multiply, or divide. Below is a list of key words and their meanings. This list does not include every situation you might see, but it includes the most common examples.

Words Indicating Addition	Example	Add
and	6 **and** 8	$6 + 8$
increased	The original price of $15 **increased** by $5.	$15 + 5$
more	3 coins and 8 **more**	$3 + 8$
more than	Josh has 10 points. Will has 5 **more than** Josh.	$10 + 5$
plus	8 baseballs **plus** 4 baseballs	$8 + 4$
sum	the **sum** of 3 and 5	$3 + 5$
total	the **total** of 10, 14, and 15	$10 + 14 + 15$

Words Indicating Subtraction	Example	Subtract
decreased	$16 **decreased** by $5	$16 - 5$
difference	the **difference** between 18 and 6	$18 - 6$
less	14 days **less** 5	$14 - 5$
less than	Jose completed 2 laps **less than** Mike's 9.	*$9 - 2$
left	Ray sold 15 out of 35 tickets. How many did he have **left**?	*$35 - 15$
lower than	This month's rainfall is 2 inches **lower than** last month's rainfall of 8 inches.	*$8 - 2$
minus	15 **minus** 6	$15 - 6$

* In subtraction word problems, you cannot always subtract the numbers in the order that they appear in the problem. Sometimes the first number should be subtracted from the last. You must read each problem carefully.

Words Indicating Multiplication	Example	Multiply
double	Her $1,000 profit **doubled** in in a month.	$1,000 \times 2$
half	**Half** of the $600 collected went to charity.	$\frac{1}{2} \times 600$
product	the **product** of 4 and 8	4×8
times	Li scored 3 **times** as many points as Ted who only scored 4.	3×4
triple	The bacteria **tripled** its original colony of 10,000 in just one day.	$3 \times 10,000$
twice	Ron has 6 CDs. Tom has **twice** as many.	2×6

Words Indicating Division	Example	Divide
divide into, by, or among	The group of 70 **divided into** 10 teams	$70 \div 10$ or $\frac{70}{10}$
quotient	the **quotient** of 30 and 6	$30 \div 6$ or $\frac{30}{6}$

Match the phrase with the correct algebraic expression below. The answers will be used more than once.

A. $y - 2$

B. $2y$

C. $y + 2$

D. $\frac{y}{2}$

E. $2 - y$

1. 2 more than y
2. y divided into 2
3. 2 less than y
4. twice y
5. the quotient of y and 2
6. y increased by 2
7. 2 less y
8. the product of 2 and y
9. y decreased by 2
10. y doubled
11. 2 minus y
12. the total of 2 and y

Now practice writing parts of algebraic expressions from the following word problems.

Example 3: the product of 3 and a number, t Answer: $3t$

13. 3 less than x
14. y divided among 10
15. the sum of t and 5
16. n minus 14
17. 5 times k
18. the total of z and 12
19. double the number b
20. x increased by 1
21. the quotient of t and 4
22. half of a number, y
23. bacteria culture, b, doubled
24. triple John's age, y
25. a number, n, plus 4
26. quantity, t, less 6
27. 18 divided by a number, x
28. n feet lower than 10
29. 3 more than p
30. the product of 4 and m
31. a number, y, decreased by 20
32. [illegible] times as much as x

4.4 Setting Up Algebra Word Problems

So far, you have seen only the first part of algebra word problems. To complete an algebra problem, an equal sign must be added. The words "**is**" or "**are**" as well as "**equal(s)**" signal that you should add an equal sign.

Example 4: Double Jake's age, a, minus 4 is 22.

$$2a \quad - \quad 4 \quad = \quad 22$$

Translate the following word problems into algebra problems. DO NOT find the solutions to the problems yet.

1. Triple the original number, n, is $2,700$.
2. The product of a number, y, and 5 is equal to 15.
3. Four times the difference of a number, x, and 2 is 20.
4. The total, t, divided into 5 groups is 45.
5. The number of parts in inventory, p, minus 54 parts sold today is 329.
6. One-half an amount, x, added to \$5[illegible] is \$262
7. One hundred seeds divided by 5 rows equals n number of seeds per row.
8. A number, y, less than 50 is 82.
9. His base pay of \$200 increased by his commission, x, is \$500.
10. Seventeen more than half a number, h, is 35.
11. This month's sales of \$[illegible], 300 are double January's sales, x.
12. The quotient of a number, w, and 4 is 32.
13. Six less a number, d, is 12.
14. Four times the sum of a number, y, and 10 is 48.
15. We started with a number of students. When 5 moved away, we had 42 left.
16. A number, b, divided into 36 is 12.

4.5 Changing Algebra Word Problems to Algebraic Equations

Example 5: There are 3 people who have a total weight of 595 pounds. Sally weighs 20 pounds less than Jessie. Rafael weighs 15 pounds more than Jessie. How much does Jessie weigh?

Step 1: Notice everyone's weight is given in terms of Jessie. Sally weighs 20 pounds less than Jessie. Rafael weighs 15 pounds more than Jessie. First, we write everyone's weight in terms of Jessie, j.

$$\begin{aligned} \text{Jessie} &= j \\ \text{Sally} &= j - 20 \\ \text{Rafael} &= j + 15 \end{aligned}$$

Step 2: We know that all three together weigh 595 pounds. We write the sum of everyone's weight equal to 595.

$$j + j - 20 + j + 15 = 595$$

We will learn to solve these problems in Chapter 7.

Change the following word problems to algebraic equations.

1. Fluffy, Spot, and Shampy have a combined age in dog years of 91. Spot is 14 years younger than Fluffy. Shampy is 6 years older than Fluffy. What is Fluffy's age, f, in dog years?

2. Jerry Marcosi puts 5% of the amount he makes per week into a retirement account, r. He is paid \$11.00 per hour and works 40 hours per week for a certain number of weeks, w. Write an equation to help him find out how much he puts into his retirement account.

3. A furniture store advertises a 40% off liquidation sale on all items. What would the sale price (p) be on a \$2,530 dining room set?

4. Kyle Thornton buys an item which normally sells for a certain price, x. Today the item is selling for 25% off the regular price. A sales tax of 6% is added to the equation to find the final price, f.

5. Tamika Francois runs a floral shop. On Tuesday, Tamika sold a total of \$600 worth of flowers. The flowers cost her \$100, and she paid an employee to work 8 hours for a given wage, w. Write an equation to help Tamika find her profit, p, on Tuesday.

6. Sharice is a waitress at a local restaurant. She makes an hourly wage of \$3.50, plus she receives tips. On Monday, she works 6 hours and receives tip money, t. Write an equation showing what Sharice makes on Monday, y.

7. Jenelle buys x shares of stock in a company at \$3[illegible] per share. She later sells the shares at \$40.50 per share. Write an equation to show how much money, m, Jenelle has made.

4.6 Substituting Numbers in Formulas

Example 6: Area of a parallelogram: $A = b \times h$
Find the area of the parallelogram if $b = 20$ cm and $h = 10$ cm.

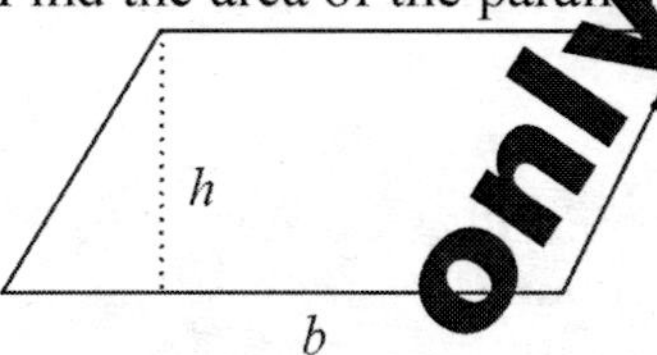

Step 1: Copy the formula with the numbers given in place of the letter in the formula.
$A = 20 \times 10$

Step 2: Solve the problem. $A = 20 \times 10 = 200$. Therefore, $A = 200 \text{ cm}^2$.

Solve the following problems using the formulas given.

1. The volume of a rectangular pyramid is determined by using the following formula:
$V = \dfrac{lwh}{3}$
Find the volume of the pyramid if $l = 6$ in, $w = 6$ in, and $h = 11$ in.

2. Find the volume of a cone with a radius of 30 inches and a height of 60 inches using the formula:
$V = \frac{1}{3}\pi r^2 h$ $\pi = 3.14$

3. Lumber is measured by the following formula:
Number of board feet $= \dfrac{LWT}{12}$
Find the number of board feet if $L = 14$ feet, $W = 8$ feet, and $T = 6$ feet.

4. The perimeter of a square is figured by the formula $P = 4s$.
Find the perimeter if $s = 6$.

5. What is the circumference of a circle with a diameter of 8 cm?
$C = \pi d$ $\pi = 3.14$

6. Find the area of the trapezoid

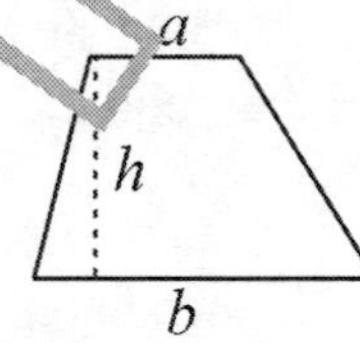

$A = \frac{1}{2}h(a + b)$
$a = 11$ in
$b = 23$ in
$h = 18$ in

7. Find the volume of a sphere with a radius of 6 cm. $\pi = 3.14$

$V = \frac{4}{3}\pi r^3$

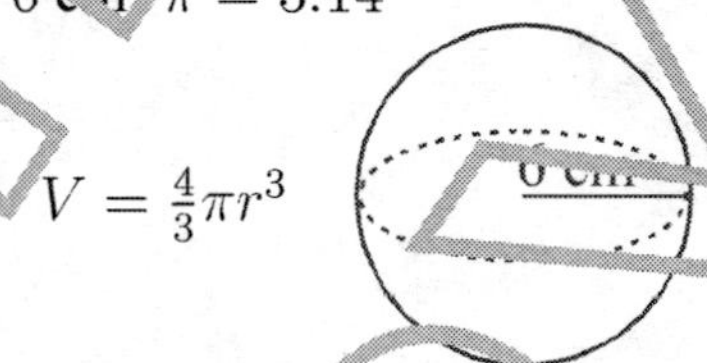

8. Find the area of the following ellipse given by the equation: $A = \pi ab$

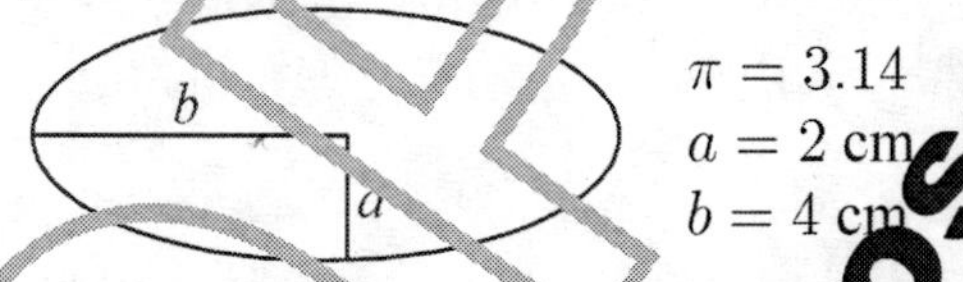

$\pi = 3.14$
$a = 2$ cm
$b = 4$ cm

9. The formula for changing from degrees Fahrenheit to degrees Celsius is:
$$C = \frac{5(F - 32)}{9}$$
If it is 68°F outside, how many degrees Celsius is it?

10. Find the volume. $V = \frac{4}{3}\pi r^3$ $\pi = 3.14$

11. Louise has a cone-shaped mold to make candles. The diameter of the base is 10 cm, and it is 13 cm tall. How many cubic centimeters of liquid wax will it hold?
$\pi = 3.14$
$V = \frac{1}{3}\pi r^2 h$

4.7 Properties of Addition and Multiplication

The Associative, Commutative, and Distributive properties and the Identity of Addition and Multiplication are listed below by example as a quick refresher.

Property	Example
1. Associative Property of Addition	$(a+b)+c=a+(b+c)$
2. Associative Property of Multiplication	$(a \times b) \times c = a \times (b \times c)$
3. Commutative Property of Addition	$a+b=b+a$
4. Commutative Property of Multiplication	$a \times b = b \times a$
5. Distributive Property	$a \times (b+c) = (a \times b) + (a \times c)$
6. Identity Property of Addition	$0+a=a$
7. Identity Property of Multiplication	$1 \times a = a$
8. Inverse Property of Addition	$a+(-a)=0$
9. Inverse Property of Multiplication	$a \times \frac{1}{a} = \frac{a}{a} = 1 \quad a \neq 0$

Write the property listed above that describes each of the following statements.

1. $4+5=5+4$
2. $4+(2+8)=(4+2)+8$
3. $10(4+7)=(10)(4)+(10)(7)$
4. $(2 \times 3) \times 4 = 2 \times (3 \times 4)$
5. $1 \times 12 = 12$
6. $8\left(\frac{1}{8}\right)=1$
7. $1c=c$
8. $18+0=18$
9. $9+(-9)=0$
10. $p \times q = q \times p$
11. $t+0=t$
12. $x(y+z)=xy+xz$
13. $(m)(n \cdot p)=(m \cdot n)(p)$
14. $-y+y=0$

Chapter 4 Review

Solve the following problems using $x = 2$.

1. $3x + 4 =$
2. $\frac{6x}{4} =$
3. $x^2 - 5 =$
4. $\frac{x^3 + 8}{[illegible]} =$
5. $1[illegible] - 3x =$
6. $x - 5 =$
7. $[illegible]5x + 4 =$
8. $9 - x =$
9. $2x + 2 =$

Solve the following problems. Let $w = -1$, $y = 3$, $z = 5$.

10. $5w - y =$
11. $wyz + 2 =$
12. $z - 2w =$
13. $\frac{3z + 5}{[illegible]wz} =$
14. $\frac{6w}{y} + \frac{z}{w} =$
15. $25 - 2yz =$
16. $-2y + 3[illegible]$
17. $4w - (yw) =$
18. $[illegible]y - 5z =$

For questions 19–21, write an equation to match the problem.

19. Calista earns \$450 per week for a 40-hour work week plus \$16.[illegible]3 per hour for each hour of overtime after 40 hours. Write an equation that would be used to determine her weekly wages where w is her wages and v is the number of overtime hours worked.

20. Daniel purchased a 1-year CD, c, from a bank. He bought it at an annual interest rate of 6%. After 1 year, Daniel cashes in the CD. What is the total amount it is worth?

21. Omar is a salesman. He earns an hourly wage of \$8.[illegible] per hour plus he receives a commission of 7% on the sales he makes. Write an equation which would be used to determine his weekly salary, w, where x is the number of hours worked, and y is the amount of sales for the week.

22. Tom earns \$500 per week before taxes are taken out. His employer takes out a total of 33% for state, federal, and Social Security taxes. Which expression below will help Tom figure his net pay?

(A) $500 - 0.33$
(B) $500 \div 0.33$
(C) $500 + 0.33\,(500)$
(D) $500 - 0.33\,(500)$

23. Rosa has to pay the first $100 of her medical expenses each year before she qualifies for her insurance company to begin paying. After paying the $100 "deductible," her insurance company will pay 80% of her medical expenses. This year, her total medical expenses came to $960.00. Which expression below shows how much her insurance company will pay?

(A) $0.80\,(960-100)$
(B) $100+(960\div 0.80)$
(C) $960\,(100-0.80)$
(D) $0.80\,(960+100)$

24. A plumber charges $45 per hour plus a $25.00 service charge. If a represents his total charges in dollars and b represents the number of hours worked, which formula below could the plumber use to calculate his total charges?

(A) $a=45+25b$
(B) $a=45+25+b$
(C) $a=45b+25$
(D) $a=(45)(25)+b$

25. In 2005, B[illegible] Computers informed its sales force to expect a 2.6% price increase on all computer equipment in the year 2006. A certain sales representative wanted to see how much the increase would be on a computer, c, that sold for $2200 in 2005. Which expression below will help him find the cost of the computer in the year 2006?

(A) [illegible] (2200)
(B) $2200-0.026\,(2200)$
(C) $2200+0.026\,(2200)$
(D) $0.026\,(2200)-2200$

26. Juan sells a boat that he bought 5 years ago. He sells it for 60% less than he originally paid for it. If the original cost is b, write an expression that shows how much he sells the boat for.

27. Toshi is going to get a 7% raise after he works at his job for 1 year. If s represents his starting salary, write an expression that shows how much he will make after his raise, x.

28. Lumber is measured with the following formula:

Number of board feet $=\dfrac{LWT}{12}$
L = Length of the board in feet
W = Width of the board in feet
T = Thickness of the board in feet
Find the number of board in feet if $L=12$ feet, $W=$ [illegible] feet, and $T=6$ feet.

29. To convert from degrees Celsius to degrees Fahrenheit, use the following formula:

$$\mathrm{F}=\frac{9\mathrm{C}}{5}+32$$

If it is 15°C outside, what is the temperature in degrees Fahrenheit?

Chapter 5
Introduction to Graphing

5.1 Graphing on a Number Line

Number lines allow you to graph values of positive and negative numbers as well as zero. Any real number, whether it is a fraction, decimal, or integer can be plotted on a number line. Number lines can be horizontal or vertical. The examples below illustrate how to plot different types of numbers on a number line.

5.2 Graphing Fractional Values

Example 1: What number does point A represent on the number line below?

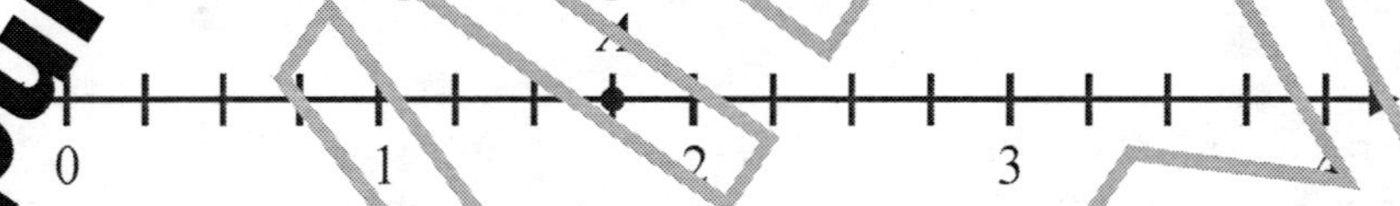

Step 1: Point A is between the numbers 1 and 2, so it is greater than 1 but less than 2. We can express the value of A as a fractional value that falls between 1 and 2. To do so, copy the integer that point A falls between which is closer to zero on the number line. In this case, copy the 1 because 1 is closer to zero on the number line than the 2.

Step 2: Count the number of spaces between each integer. In this case, there are 4 spaces between the 1 and the 2. Put this number as the bottom number in your fraction.

Step 3: Count the number of spaces between the 1 and the point A. Point A is 3 spaces away from number 1. Put this number as the top number in your fraction.

The integer that point A falls between that is closest to 0

Point A is at $1\frac{3}{4}$ ← The number of spaces between 1 and A; ← The number of spaces between 1 and 2

Example 2: What number does point B represent on the number line below?

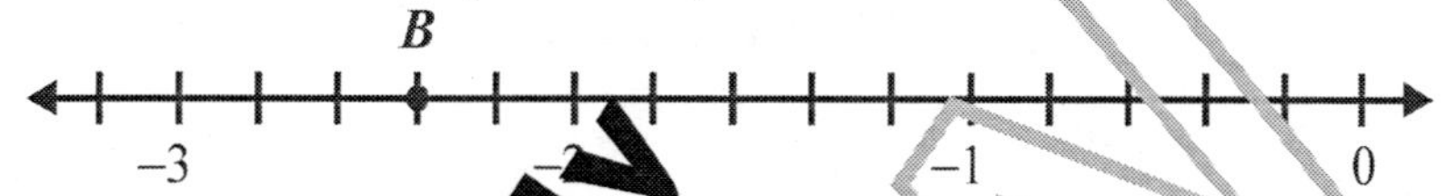

Step 1: Point B is between -2 and -3. Again, we can express the value of B as a fraction that falls between -2 and -3. Copy the integer that point B falls between which is closer to zero. The -2 is closer to zero than -3, so copy -2.

Step 2: In this example there are 5 spaces between each integer. Five will be the bottom number in the fraction.

Step 3: There are 2 spaces between -2 and point B. Two will be the top number in the fraction.
Point B is at $-2\frac{2}{5}$.

Determine and record the value of each point on the number lines below.

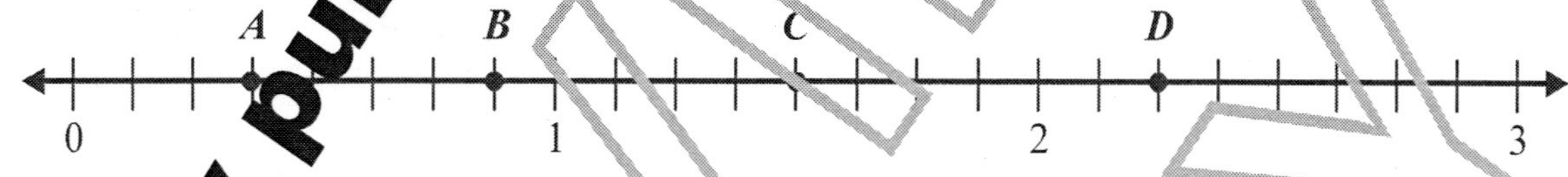

1. $A =$ _____ $B =$ _____ $C =$ _____ $D =$ _____

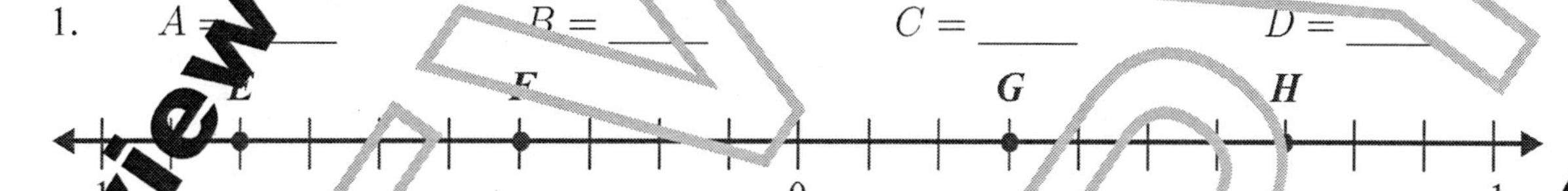

2. $E =$ _____ $F =$ _____ $G =$ _____ $H =$ _____

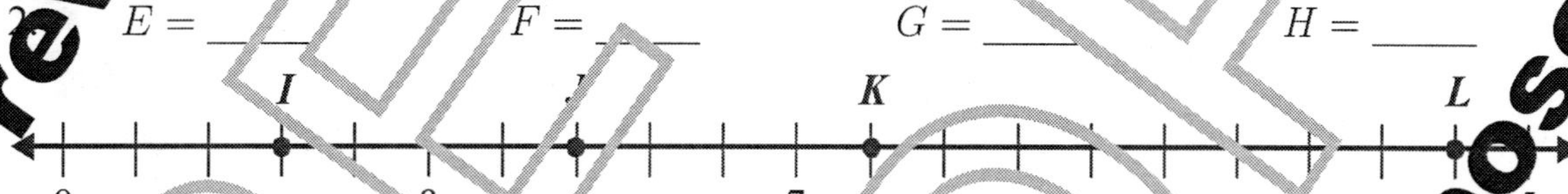

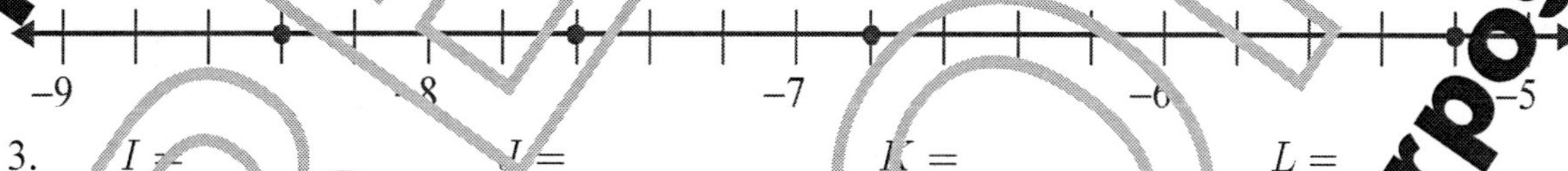

3. $I =$ _____ $J =$ _____ $K =$ _____ $L =$ _____

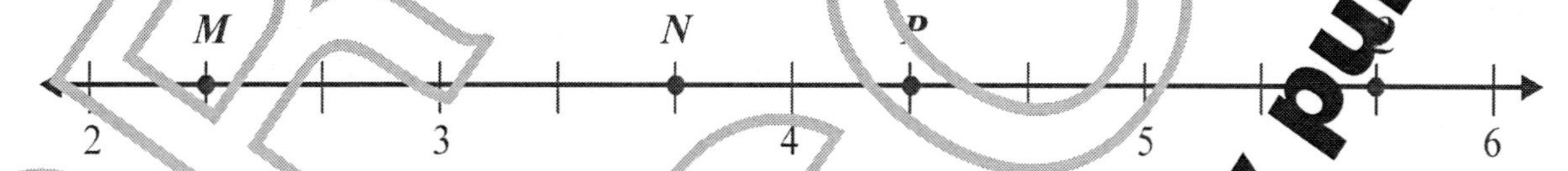

4. $M =$ _____ $N =$ _____ $P =$ _____ $Q =$ _____

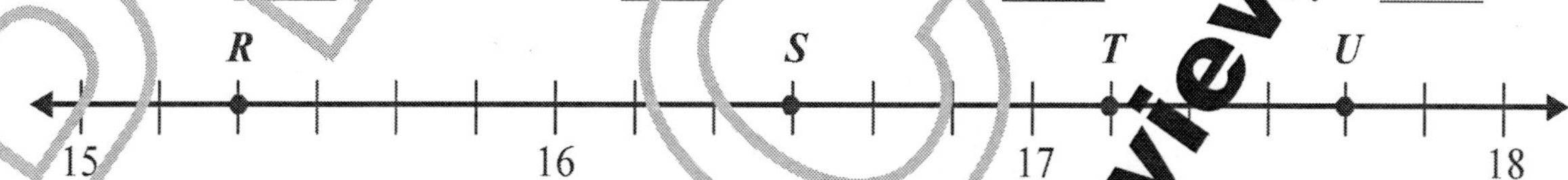

5. $R =$ _____ $S =$ _____ $T =$ _____ $U =$ _____

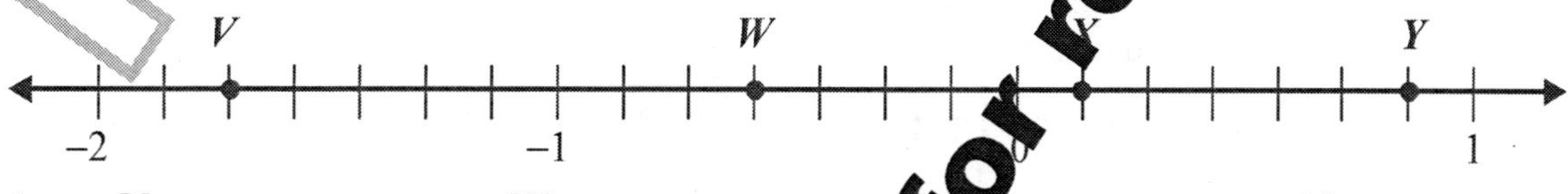

6. $V =$ _____ $W =$ _____ $X =$ _____ $Y =$ _____

5.3 Recognizing Improper Fractions, Decimals, and Square Root Values on a Number Line

Improper fractions, decimal values, and square root values can also be plotted on a number line. Study the examples below.

Example 3: Where would $\frac{4}{3}$ fall on the number line below?

Step 1: Convert the improper fraction to a mixed number. $\frac{4}{3} = 1\frac{1}{3}$

Step 2: $1\frac{1}{3}$ is $\frac{1}{3}$ of the distance between the numbers 1 and 2. Estimate this distance by dividing the distance between points 1 and 2 into thirds. Plot the point at the first division.

Example 4: Plot the value of -1.75 on the number line below.

Step 1: Convert the value -1.75 to a mixed fraction. $-1.75 = -1\frac{3}{4}$

Step 2: $-1\frac{3}{4}$ is $\frac{3}{4}$ of the distance between the numbers -1 and -2. Estimate this distance by dividing the distance between points -1 and -2 into fourths. Plot the point at the third division.

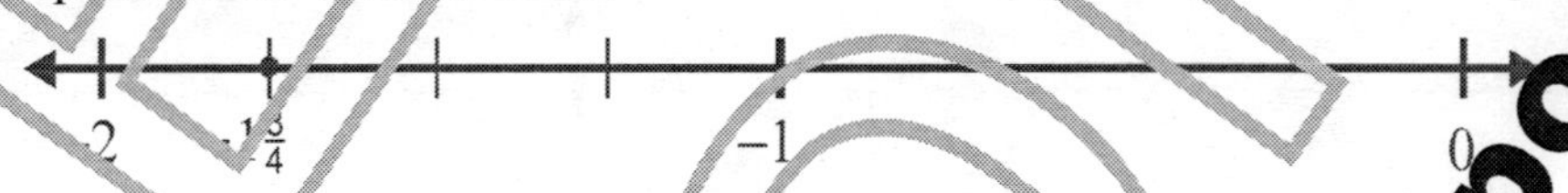

Example 5: Plot the value of $\sqrt{3}$ on the number line below.

Step 1: Estimate the value of $\sqrt{3}$ by using the square root of values that you know. $\sqrt{1} = 1$ and $\sqrt{4} = 2$, so the value of $\sqrt{3}$ is going to be between 1 and 2.

Step 2: To estimate a little closer, try squaring 1.5. $1.5 \times 1.5 = 2.25$, so $\sqrt{3}$ has to be greater than 1.5. If you do further trial and error calculations, you will find that $\sqrt{3}$ is greater than 1.7 ($1.7 \times 1.7 = 2.89$) but less than 1.8 ($1.8 \times 1.8 = 3.24$).

Step 3: Plot $\sqrt{3}$ around 1.75.

0 1 2 3 4

Plot and label the following values on the number lines given below.

1. $A = \frac{5}{4}$ $\quad$ $B = \frac{1?}{3}$ $\quad$ $C = \frac{2}{3}$ $\quad$ $D = -\frac{3}{2}$

2. $E = 1.4$ $\quad$ $F = -2.25$ $\quad$ $G = -0.6$ $\quad$ $H = 0.625$

3. $I = \sqrt{2}$ $\quad$ $J = \sqrt{5}$ $\quad$ $K = \sqrt{6}$ $\quad$ $L = \sqrt{8}$

Match the correct value for each point on the number line below.

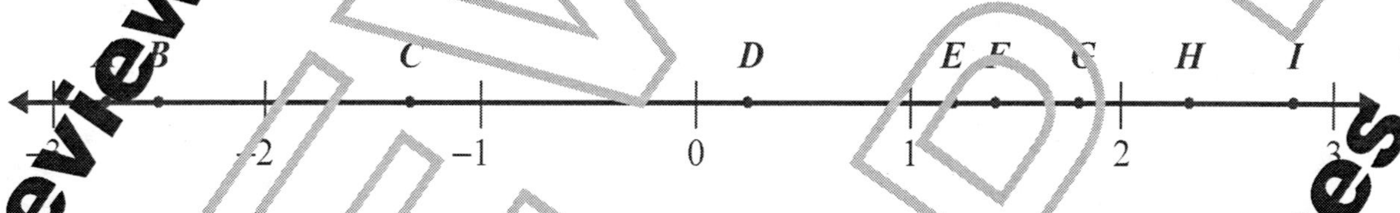

4. $1.8 =$ ____
5. $\frac{7}{3} =$ ____
6. $\sqrt{2} =$ ____
7. $-\frac{5}{2} =$ ____
8. $-2.75 =$ ____
9. $-\frac{4}{3} =$ ____
10. $\sqrt{8} =$ ____
11. $\frac{6}{5} =$ ____
12. $0.25 =$ ____

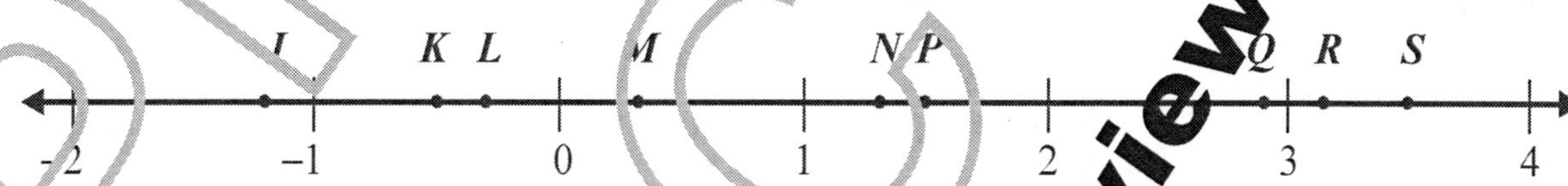

13. $\sqrt{12} =$ ____
14. $-0.5 =$ ____
15. $\frac{5}{4} =$ ____
16. $\frac{1}{3} =$ ____
17. $1.5 =$ ____
18. $-0.3 =$ ____
19. $-\frac{6}{5} =$ ____
20. $\sqrt{10} =$ ____
21. $2.9 =$ ____

5.4 Plotting Points on a Vertical Number Line

Number lines can also be drawn up and down **(vertical)** instead of across the page **(horizontal)**. You plot points on a vertical number line the same way as you do on a horizontal number line.

Record the value represented by each point on the number lines below.

1. $A =$ ____
2. $B =$ ____
3. $C =$ ____
4. $D =$ ____
5. $E =$ ____
6. $F =$ ____
7. $G =$ ____
8. $H =$ ____

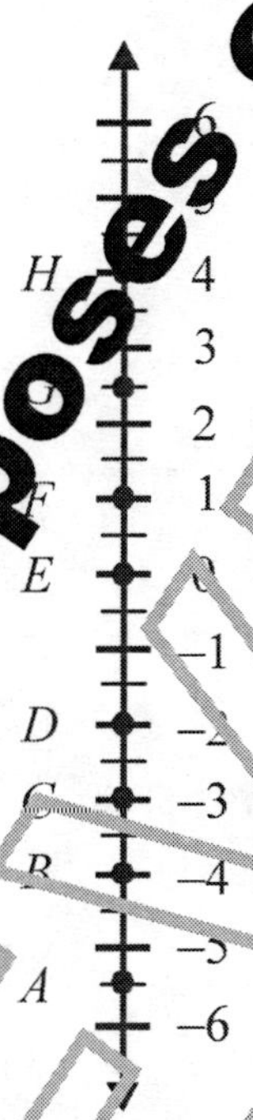

17. $Q =$ ____
18. $R =$ ____
19. $S =$ ____
20. $T =$ ____
21. $V =$ ____
22. $W =$ ____
23. $X =$ ____
24. $Y =$ ____

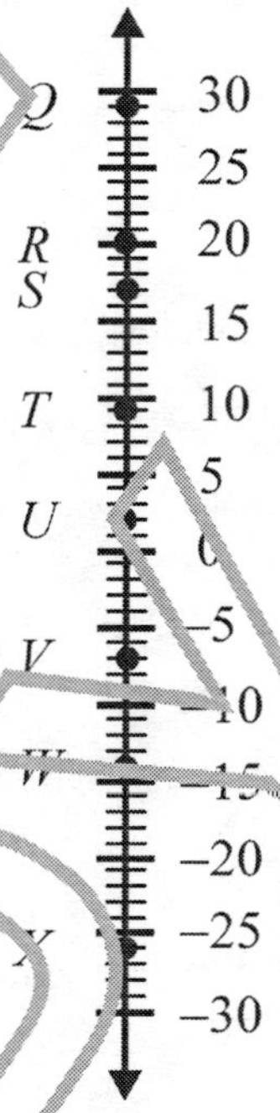

9. $I =$ ____
10. $J =$ ____
11. $K =$ ____
12. $L =$ ____
13. $M =$ ____
14. $N =$ ____
15. $P =$ ____
16. $Q =$ ____

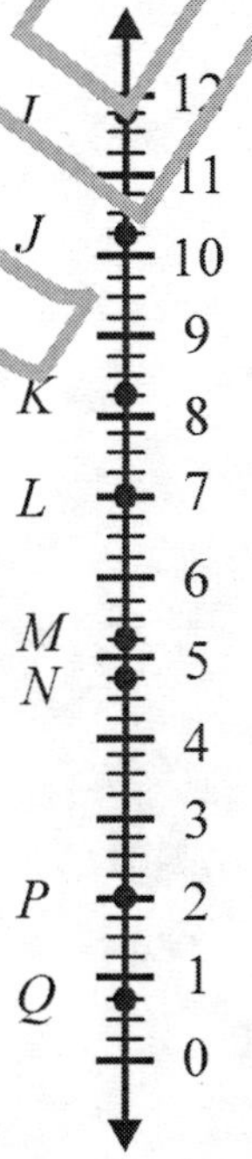

25. $A =$ ____
26. $B =$ ____
27. $C =$ ____
28. $D =$ ____
29. $E =$ ____
30. $G =$ ____
31. $H =$ ____
32. $I =$ ____

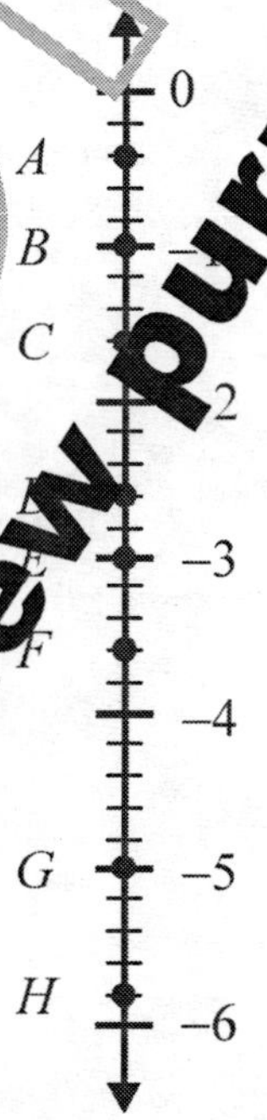

5.5 Cartesian Coordinates

A **Cartesian coordinate plane** allows you to graph points with two values. A Cartesian coordinate plane is made up of two number lines. The horizontal number line is called the x-**axis**, and the vertical number line is called the y-**axis**. The point where the x and y axes intersect is called the **origin**. The x and y axes separate the Cartesian coordinate plane into four quadrants that are labeled I, II, III, and IV. The quadrants are labeled and explained on the graph below. Each point graphed on the plane is designated by an **ordered pair** of coordinates. For example, $(2, -1)$ is an ordered pair of coordinates designated by point B on the plane below. The first number, 2, tells you to go over positive two on the x-axis. The -1 tells you to then go down negative one on the y-axis.

Remember: The first number always tells you how far to go right or left of 0, and the second number always tells you how far to go up or down from 0.

Quadrant II:
The x-coordinate is negative, and the y-coordinate is positive $(-, +)$.

Quadrant III:
Both coordinates in the ordered pair are negative $(-, -)$.

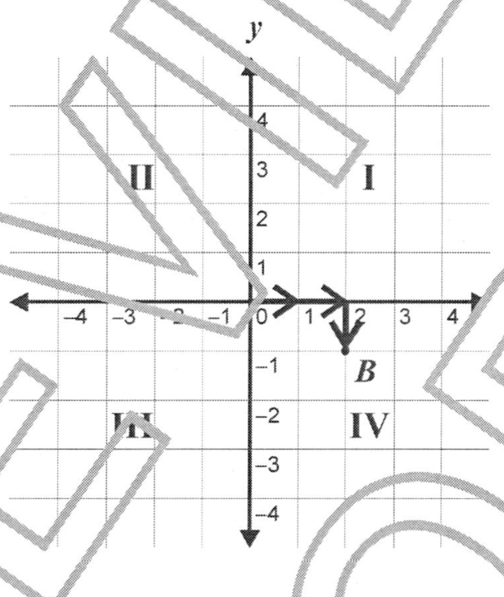

Quadrant I:
Both coordinates in the ordered pair are positive $(+, +)$.

Quadrant IV:
The x-coordinate is positive and the y-coordinate is negative $(+, -)$.

Plot and label the following points on the Cartesian coordinate plane provided.

A. $(2, 4)$	F. $(-3, -5)$	K. $(-1, -1)$	P. $(0, 4)$
B. $(-1, 5)$	G. $(-2, 5)$	L. $(3, -3)$	Q. $(2, 0)$
C. $(3, -4)$	H. $(5, -1)$	M. $(5, 5)$	R. $(-4, 0)$
D. $(-5, -2)$	I. $(4, -4)$	N. $(-2, -2)$	S. $(0, -2)$
E. $(5, 3)$	J. $(5, 2)$	O. $(0, 0)$	T. $(5, 1)$

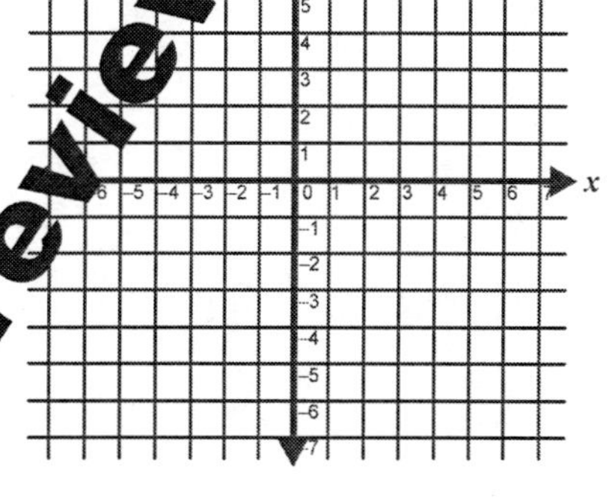

5.6 Identifying Ordered Pairs

When identifying ordered pairs, count how far left or right of 0 to find the x-coordinate and then how far up or down from 0 to find the y-coordinate.

Point A: Left (negative) two and up (positive) three $= (-2, 3)$ in quadrant II

Point B: Right (positive) one and up (positive) one $= (1, 1)$ in quadrant I

Point C: Left (negative) three and down (negative) one $= (-3, -1)$ in quadrant III

Point D: Right (positive) one and down (negative) three $= (1, -3)$ in quadrant IV

Fill in the ordered pair for each point, and tell which quadrant it is in.

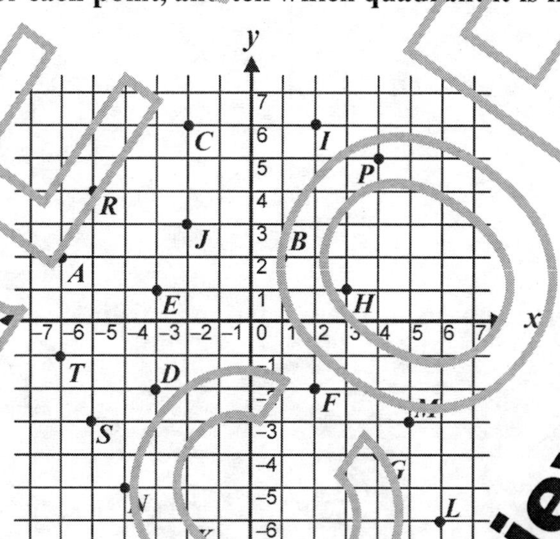

1. point A	4. point D	7. point G	10. point J	13. point M	16. point R
2. point B	5. point E	8. point H	11. point K	14. point N	17. point S
3. point C	6. point F	9. point I	12. point L	15. point P	18. point T

Sometimes, points on a coordinate plane fall on the x or y axis. If a point falls on the x-axis, then the second number of the ordered pair is 0. If a point falls on the y-axis, the first number of the ordered pair is 0.

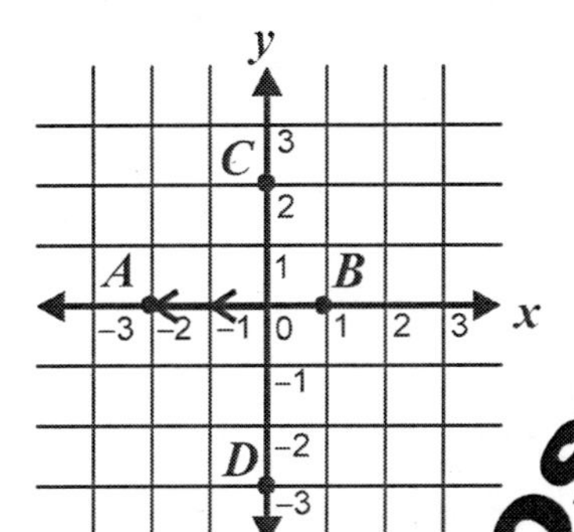

Point A: Left (negative) two and up zero = $(-2, 0)$
Point B: Right (positive) one and up zero = $(1, 0)$
Point C: Left/right zero and up (positive) two = $(0, 2)$
Point D: Left/right zero and down (negative) three = $(0, -3)$

Fill in the ordered pair for each point.

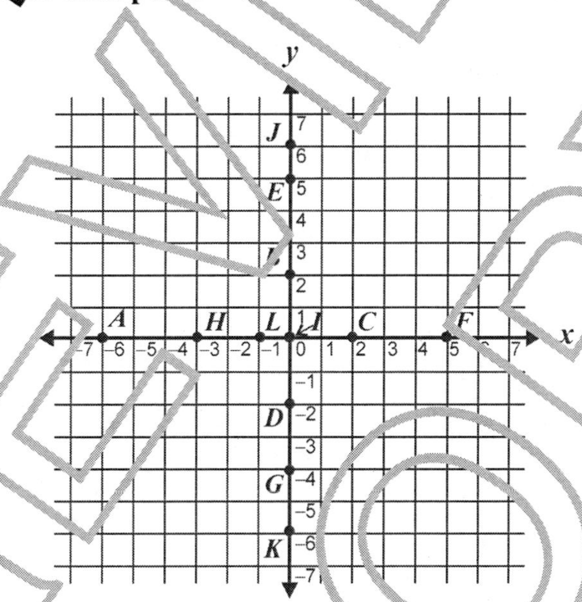

1. point A = (,)
2. point B = (,)
3. point C = (,)
4. point D = (,)
5. point E = (,)
6. point F = (,)
7. point G = (,)
8. point H = (,)
9. point I = (,)
10. point J = (,)
11. point K = (,)
12. point L = (,)

Chapter 5 Review

1. (number line: 4, 5, 6)

   Plot and label $5\frac{3}{5}$ on the number line above.

2. (number line: -4, -3, -2, -1, 0)

   Plot and label $-3\frac{1}{2}$ on the number line above.

3. (number line: 6, 7, 8, 9, 10)

   Plot and label 7.8 on the number line above.

4. (number line: -3, -2, -1)

   Plot and label -2.3 on the number line above.

Record the value represented by the point on the number line for questions 5 -10.

5. $A =$ ______
6. $B =$ ______
7. $C =$ ______
8. $D =$ ______
9. $E =$ ______
10. $F =$ ______

Answer the following questions.

11. In which quadrant does the point $(2, 3)$ lie?
12. In which quadrant does the point $(-5, -2)$ lie?

Record the coordinates and quadrants of the following points.

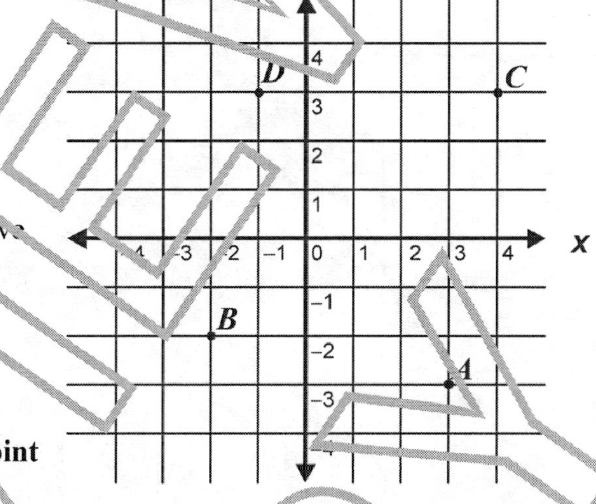

	Coordinates	Quadrants
13. $A =$	______	______
14. $B =$	______	______
15. $C =$	______	______
16. $D =$	______	______

On the same plane above, label these additional coordinates.

17. $E = (0, -3)$
18. $F = (-3, 1)$
19. $G = (4, 0)$
20. $H = (2, 2)$

Chapter 6
Solving One-Step Equations and Inequalities

6.1 One-Step Algebra Problems with Addition and Subtraction

You have been solving algebra problems since second grade by filling in blanks. For example, $5 + \_\_ = 8$. The answer is 3. You can solve the same kind of problems using algebra. The problems only look a little different because the blank has been replaced with a letter. The letter is called a **variable**.

Example 1: **Arithmetic** $5 + \_\_ = 14$
Algebra $5 + x = 14$

The goal in any algebra problem is to move all the numbers to one side of the equal sign and have the letter (called the **variable**) on the other side. In this problem the 5 and the "x" are on the same side. The 5 is added to x. To move it, do the **opposite** of **add**. The **opposite** of **add** is **subtract**, so subtract 5 from both sides of the equation. Now the problem looks like this:

$$\begin{array}{r} 5 + x = 14 \\ -5 \quad\quad -5 \\ \hline x = 9 \end{array}$$

To check your answer, put 9 in place of x in the original problem. Does $5 + 9 = 14$? Yes, it does.

Example 2:

$$\begin{array}{r} y - 16 = 27 \\ +16 \quad +16 \\ \hline y = 43 \end{array}$$

Again, the 16 has to move. To move it to the other side of the equation, we do the **opposite** of **subtract**. We **add** 16 to both sides. Check by putting 43 in place of the y in the original problem. Does $43 - 16 = 27$? Yes.

Solve the problems below.

1. $n + 9 = 27$	6. $15 + x = 24$	11. $k - 5 = 29$	16. $t - 16 = 2$[illegible]	21. $r - 12 = 37$
2. $12 + y = 55$	7. $w - 14 = 89$	12. $a + 17 = 45$	17. $m + 14 = 37$	22. $h - 17 = 22$
3. [illegible]$1 + v = 67$	8. $t - 26 = 20$	13. $d + 26 = 56$	18. y [illegible] $= 29$	23. $x - 37 = 46$
4. $f + 16 = 31$	9. $m - 12 = 17$	14. $15 + x = 56$	19. $f + 7 = 31$	24. $r - 11 = 28$
5. $5 + x = 23$	10. $c - 7 = 21$	15. $y + 19 = 32$	20. $h - 12 = 18$	25. $t - 5 = 52$

6.2 One-Step Algebra Problems with Multiplication and Division

Solving one-step algebra problems with multiplication and division are just as easy as adding and subtracting. Again, you perform the **opposite** operation. If the problem is a **multiplication** problem, you **divide** to find the answer. If it is a **division** problem, you **multiply** to find the answer. Carefully read the examples below, and you will see how easy they are.

Example 3: $4x = 20$ ($4x$ means 4 times x. 4 is the coefficient of x.)

The goal is to get the numbers on one side of the equal sign and the variable x on the other side. In this problem, the 4 and the x are on the same side of the equal sign. The 4 has to be moved over. $4x$ means 4 times x. The opposite of **multiply is divide**. If we divide both sides of the equation by 4, we will find the answer.

$4x = 20$ **We need to divide both sides by 4.**

This means divide by 4. $\longrightarrow \frac{\overset{1}{\cancel{4}x}}{\underset{1}{\cancel{4}}} = \frac{\overset{5}{\cancel{20}}}{\underset{1}{\cancel{4}}}$ **We see that** $1x = 5$, **so** $x = 5$

When you put 5 in place of x **in the original problem, it is correct.** $4 \times 5 = 20$

Example 4: $\frac{y}{4} = 2$

This problem means y divided by 4 is equal to 2. In this case, the opposite of divide is multiply. We need to multiply both sides of the equation by 4.

$\cancel{4} \times \frac{y}{\cancel{4}} = 2 \times 4$ so $y = 8$

When you put 8 in place of y **in the original problem, it is correct.** $\frac{8}{4} = 2$

Solve the problems below.

1. $2x = 14$
2. $\frac{w}{5} = 11$
3. $3h = 45$
4. $\frac{x}{4} = 36$
5. $\frac{x}{3} = 9$
6. $6d = 66$
7. $\frac{w}{9} = 3$
8. $7r = 98$
9. $\frac{y}{[illegible]} = 2$
10. $10y = 30$
11. $\frac{r}{4} = 7$
12. $8t = 96$
13. $\frac{z}{2} = 15$
14. $\frac{n}{9} = 5$
15. $4[illegible] = 24$
16. $6d = 84$
17. $\frac{t}{3} = 3$
18. $\frac{m}{6} = 9$
19. $9p = 72$
20. $5a = 60$

Sometimes the answer to the algebra problem is a **fraction**. Read the example below, and you will see how easy it is.

Example 5: $4x = 5$

Problems like this are solved just like the problems above and those on the previous page. The only difference is that the answer is a fraction.
In this problem, the 4 is **multiplied** by x. To solve, we need to divide both sides of the equation by 4.

$4x = 5$ Now **divide** by 4. $\frac{4x}{4} = \frac{5}{4}$ Now cancel. $\frac{\not{4}x}{\not{4}} = \frac{5}{4}$ So $x = \frac{5}{4}$

When you put $\frac{5}{4}$ in place of x in the original problem, it is correct.

$4 \times \frac{5}{4} = 5$ Now cancel. ⟶ $\not{4} \times \frac{5}{\not{4}} = 5$ So $5 = 5$

Solve the problems below. Some of the answers will be fractions. Some answers will be integers.

1. $2x = 3$
2. $4y = 5$
3. $5[illegible] = 2$
4. $12b = 144$
5. $9a = 72$
6. $8y = 16$
7. $7x = 21$
8. $4z = 64$
9. $7x = 126$
10. $6p = 10$
11. $2n = 9$
12. $5x = 11$
13. $15m = 180$
14. $5h = 21$
15. $3y = 8$
16. $2[illegible] = 10$
17. $3b = 2$
18. $5c = 14$
19. $4d = 3$
20. $5z = 75$
21. $9y = [illegible]$
22. $7d = 12$
23. $2[illegible] = 13$
24. $9g = 81$
25. $6a = 18$
26. $2p = 16$
27. $15w = 3$
28. $5x = 13$

6.3 Multiplying and Dividing with Negative Numbers

Example 6: $-3x = 15$

Step 1: In the problem, -3 is **multiplied** by x. To find the solution, we must do the opposite. The opposite of **multiply** is **divide**. We must divide both sides of the equation by -3.

$$\frac{-3x}{-3} = \frac{15}{-3}$$

Step 2: Then cancel.

$$\frac{\cancel{-3}x}{\cancel{-3}} = \frac{\overset{5}{\cancel{15}}}{\cancel{-3}} \qquad x = -5$$

Example 7: $\frac{y}{-4} = -20$

Step 1: In this problem, y is **divided** by -4. To find the answer, do the opposite. **Multiply** both sides by -4.

$$\cancel{-4} \times \frac{y}{\cancel{-4}} = (-20) \times (-4) \qquad \text{so } y = 80$$

Example 8: $-6a = 2$

Step 1: The answer to an algebra problem can also be a negative fraction.

$$\frac{\cancel{6}a}{\cancel{6}} = \frac{2}{-6} \longleftarrow \text{reduce to get } a = \frac{1}{-3} \text{ or } -\frac{1}{3}$$

Note: A negative fraction can be written several different ways.

$$\frac{1}{-3} = \frac{-1}{3} = -\frac{1}{3} = -\left(\frac{1}{3}\right)$$

All mean the same thing

Solve the problems below. Reduce any fractions to lowest terms.

1. $2z = -6$
2. $\frac{y}{-5} = 20$
3. $-6k = 54$
4. $4x = -24$
5. $\frac{t}{7} = -4$
6. $\frac{r}{-2} = -10$
7. $9x = 72$
8. $\frac{x}{-6} = 3$
9. $\frac{w}{-11} =$ [illegible]
10. $5y = -35$
11. $\frac{x}{-4} = -9$
12. $7t = -49$
13. $-14x =$ [illegible]
14. $\frac{m}{3} = -12$
15. [illegible] $= -6$
16. $\frac{d}{8} = -7$
17. $\frac{y}{-9} = -4$
18. $-15u = -60$
19. [illegible]$12v = 36$
20. $-8z =$ [illegible]32
21. $-4x = -3$
22. $-12y = 7$
23. $\frac{a}{-2} = 22$
24. $-18b = 6$
25. $13a = -36$
26. $\frac{o}{-2} = -14$
27. $-24x = -6$
28. $\frac{y}{-9} = -6$
29. $\frac{x}{-23} = -1$
30. $7x = -7$
31. $-9y = -1$
32. $\frac{d}{[illegible]} = -10$
33. $\frac{z}{-13} = -2$
34. $-5c = 45$
35. $2d = -3$
36. $-8d = -12$
37. $-24w = 9$
38. $-6p = 42$
39. $-9a = -18$
40. $\frac{p}{-2} = 15$

6.4 Variables with a Coefficient of Negative One

The answer to an algebra problem should not have a negative sign in front of the variable. For example, the problem $-x = 5$ is not completely solved. Study the examples below to learn how to finish solving this problem.

Example 9: $-x = 5$

$-x$ means the same thing as $-1x$ or -1 times x. To solve this problem, **multiply** both sides by -1.

$(-1)(-1x) = (-1)(5)$ so $x = -5$

Example 10: $-y = -3$ Solve the same way.

$(-1)(-y) = (-1)(-3)$ so $y = 3$

Solve the following problems.

1. $-w = 14$
2. $-a = 20$
3. $-x = -15$
4. $-x = -25$
5. $-y = -16$
6. $-t = 62$
7. $-p = -34$
8. $-m = 81$
9. [illegible] $= 17$
10. $-v = -9$
11. $-k = 13$
12. $-q = 7$

6.5 Graphing Inequalities

An inequality is a sentence that contains a $\neq$, $<$, $>$, $\leq$, or $\geq$ sign. Look at the following graphs of inequalities on a number line.

NUMBER LINE

$x < 3$ is read "x is less than 3."

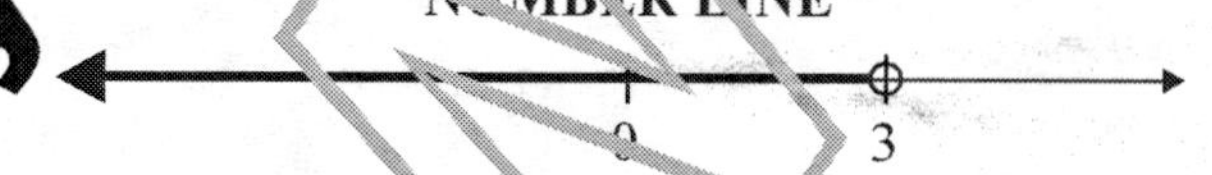

There is no line under the $<$ sign, so the graph uses an **open** endpoint to show x is less than 3 but does not include 3.

$x \leq 5$ is read "x is less than or equal to 5."

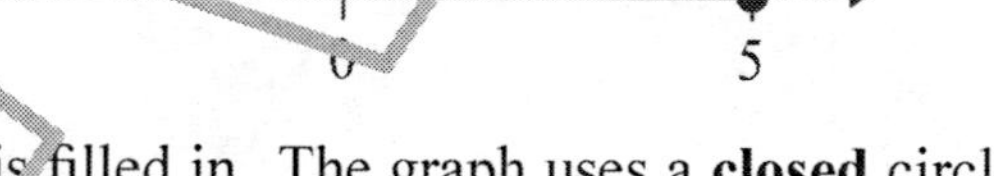

If you see a line under $<$ or $>$ ($\leq$ or $\geq$), the endpoint is filled in. The graph uses a **closed** circle because the number 5 is included in the graph.

$x > -2$ is read "x is greater than -2."

$x \geq 1$ is read "x is greater than or equal to 1."

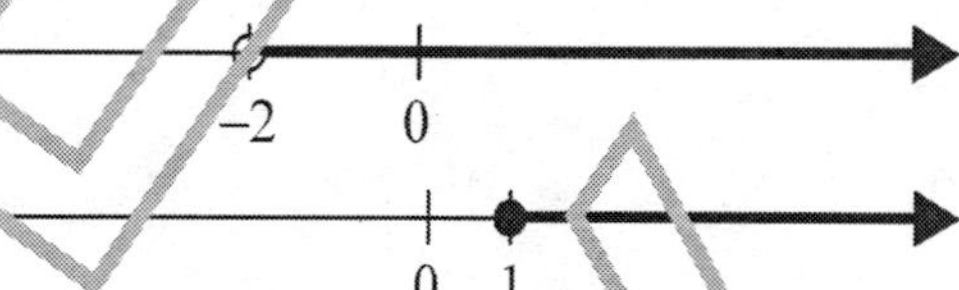

There can be more than one inequality sign. For example:
$-2 \leq x \leq 4$ is read "-2 is less than or equal to x and x is less than 4."

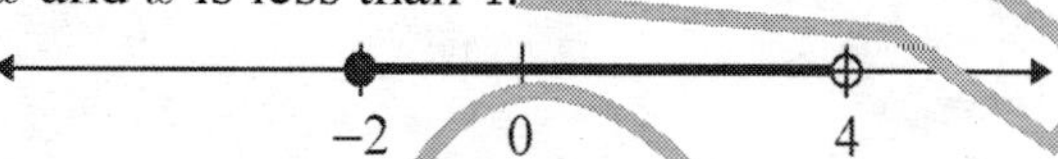

$x < 1$ or $x \geq 4$ is read "x is less than 1 or x is greater than or equal to 4."

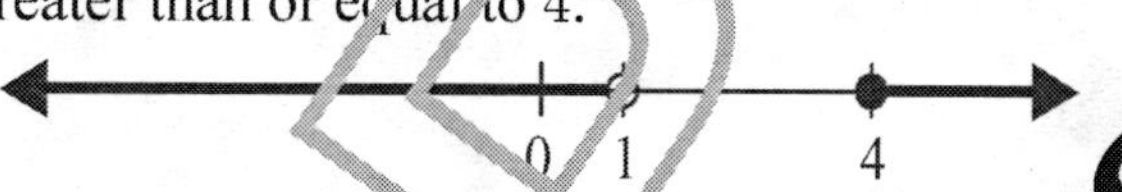

Graph the solution sets of the following inequalities.

1. $x > 8$
2. $x \leq 5$
3. $-5 < x <$ [illegible]
4. x [illegible] 7
5. $1 \leq x < 4$
6. $x < -2$ or $x > 1$
7. $x \geq 10$
8. $x < 4$
9. $x \leq 3$ or $x \geq$ [illegible]
10. $x < -1$ or $x > 1$

Give the inequality represented by each of the following number lines.

11. 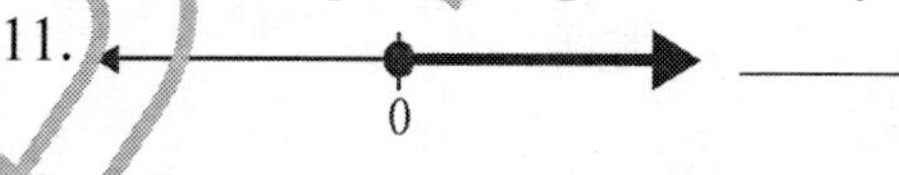 ________

12. 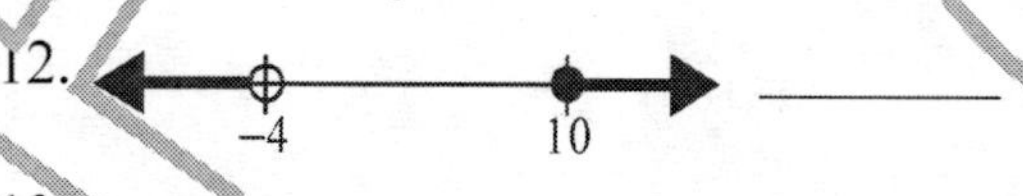 ________

13. 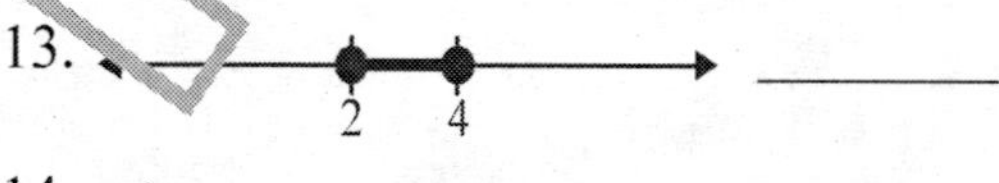 ________

14. 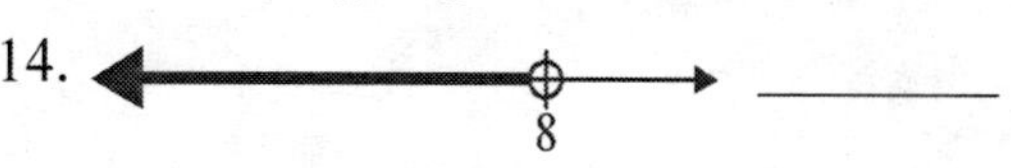 ________

15.  ________

16.  ________

17. 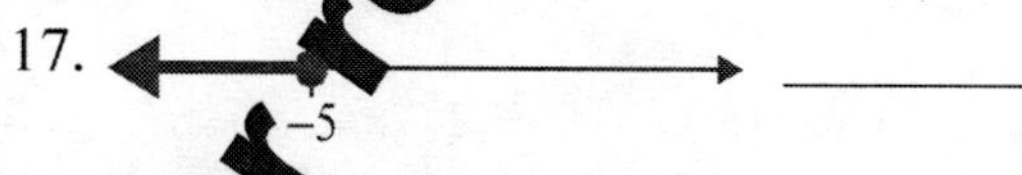 ________

18. 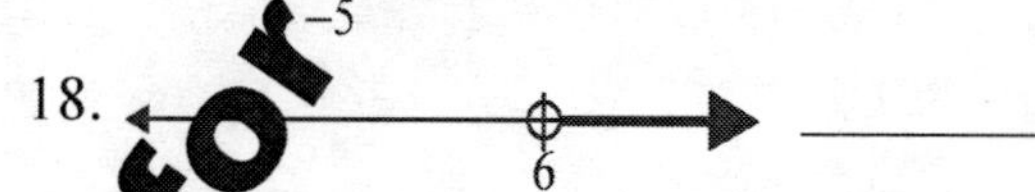 ________

6.6 Solving Inequalities by Addition and Subtraction

If you add or subtract the same number to both sides of an inequality, the inequality remains the same. It works just like an equation.

Example 11: Solve and graph the solution set for $x - 2 \leq 5$.

Step 1: Add 2 to both sides of the inequality.

$$\begin{array}{r} x - 2 \leq 5 \\ +2 \;\; +2 \\ \hline x \leq 7 \end{array}$$

Step 2: Graph the solution set for the inequality.

0 7

Solve and graph the solution set for the following inequalities.

1. $x + 5 > 3$
2. $x + 10 < 5$
3. $x - 2 \leq 1$
4. $9 + x \geq 7$
5. $x - 4 > -2$
6. $x + 11 \leq 20$
7. $x - 3 < -12$
8. $x + 6 \geq -3$
9. $x + 12 \leq 8$
10. $15 + x > 5$
11. $x - 6 < -2$
12. $x + 7 \geq 4$
13. $14 + x \leq 8$
14. $x - 8 > 24$
15. $x + 1 \leq 12$
16. $11 + x \geq 11$
17. $x - 3 < 17$
18. $x + 9 > -4$
19. $x + 6 \geq 14$
20. $x - 3 \geq 19$

6.7 Solving Inequalities by Multiplication and Division

If you multiply or divide both sides of an inequality by a **positive** number, the inequality symbol stays the same. However, if you multiply or divide both sides of an inequality by a **negative** number, **you must reverse the direction of the inequality symbol.**

Example 12: Solve and graph the solution set for $4x \le 20$.

Step 1: Divide both sides of the inequality by 4. $\frac{\overset{1}{\cancel{4}x}}{\underset{1}{\cancel{4}}} \le \frac{\overset{5}{\cancel{20}}}{\underset{1}{\cancel{4}}}$

Step 2: Graph the solution. $x \le 5$

0 5

Example 13: Solve and graph the solution set for $6 > -\frac{x}{3}$.

Step 1: Multiply both sides by -3 and **reverse the direction of the symbol.**

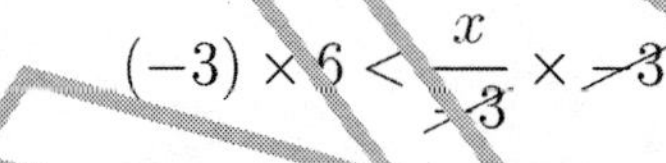

$(-3) \times 6 < \frac{x}{\cancel{-3}} \times \cancel{-3}$

Step 2: Graph the solution. $-18 < x$

Solve and graph the following inequalities.

1. $\frac{x}{5} > 4$
2. $2x \le 24$
3. $-6x \ge 36$
4. $\frac{x}{10} > -2$
5. $-\frac{x}{4} > 8$
6. $-7x \le -49$
7. $-3x > 18$
8. $-\frac{x}{7} \ge 9$
9. $9x \le 54$
10. $\frac{x}{8} > 1$
11. $-\frac{x}{9} \le 3$
12. $-4x < -12$
13. $-\frac{x}{2} \ge -20$
14. $10x \le 30$
15. $\frac{x}{12} > -4$
16. $6x < 24$

Chapter 6 Review

Solve the following one-step algebra problems.

1. $5y = -25$
2. $x + 4 = 24$
3. $d - 11 = 14$
4. $\frac{a}{6} = -8$
5. $-t = 2$
6. $-14b = 12$
7. $\frac{c}{-10} = -3$
8. $z - 15 = -19$
9. [illegible] $= 4$
10. $\frac{x}{-14} = 2$
11. $-4k = -1$[illegible]
12. $y + 13 =$ [illegible]
13. 15 [illegible] $= 4$
14. [illegible]$p = 2$
15. $\frac{b}{4} = 11$
16. $p - 26 = 12$
17. $x + (-2) = 5$
18. $m + 17 = 2$[illegible]
19. $\frac{k}{-4} = 13$
20. -18[illegible] $=$ [illegible]7
21. $21t = -7$
22. z [illegible] $(-9) = 1$[illegible]
23. $23 + w = 2$[illegible]
24. n [illegible] $35 = -16$
25. $-a = 26$
26. [illegible]19 [illegible] $f = -9$
27. $\frac{w}{11} = 3$
28. $-7y = 28$
29. $x + 23 = 20$
30. $z - 12 =$ [illegible]7
31. $-16 + g = 40$
32. $\frac{m}{-3} = -9$
33. $d + (-6) = 17$
34. $-p = 47$
35. $k - 16 = 5$
36. [illegible]$y = -3$
37. [illegible] $= -36$
38. $10h = 12$
39. $w - 16 = 4$
40. $y + 10 = -8$

Graph the solution sets of the following inequalities.

41. $x \leq -3$

42. $x > 6$

43. $x < -2$

44. $x \geq 4$

Give the inequality represented by each of the following number lines.

45.  ________

46. 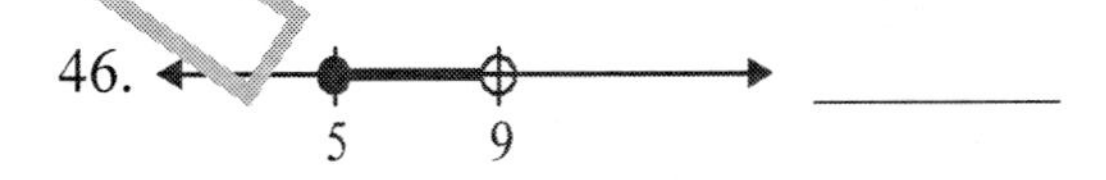 ________

47.  ________

48. 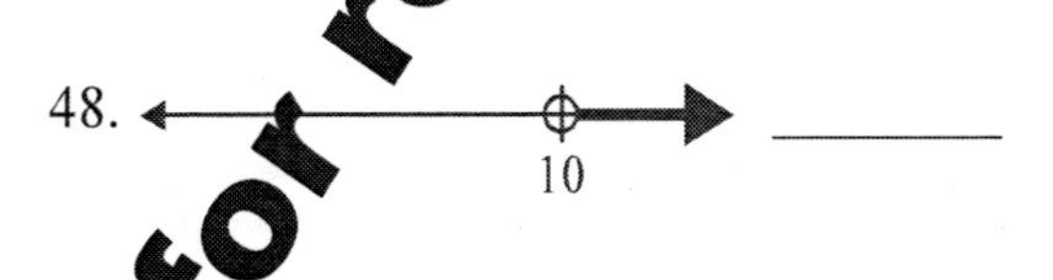 ________

Solve and graph the solution set for the following inequalities.

49. $x - 2 > 8$

50. $4 + x < -1$

51. $6x \geq 54$

52. $-2x \leq 8$

53. $\frac{x}{2} > -1$

54. $-x < -9$

55. $\frac{x}{3} \leq 5$

56. $x + 10 \leq 4$

57. $x - 6 \geq -2$

58. $7x < -14$

59. $-3x > -12$

60. $-\frac{x}{6} \leq -3$

Chapter 7
Solving Multi-Step Equations and Inequalities

7.1 Two-Step Algebra Problems

In the following two-step algebra problems, **additions** and **subtractions** are performed first and then **multiplication** and **division**.

Example 1: $-4x + 7 = 31$

Step 1: Subtract 7 from both sides.

$$\begin{array}{rcr} -4x + 7 & = & 31 \\ -7 & & -7 \\ \hline -4x & = & 24 \end{array}$$

Step 2: Divide both sides by -4.

$$\frac{-4x}{-4} = \frac{24}{-4} \qquad \text{so } x = -6$$

Example 2: $-8 - y = 12$

Step 1: Add 8 to both sides.

$$\begin{array}{rcr} -8 - y & = & 12 \\ +8 & & +8 \\ \hline -y & = & 20 \end{array}$$

Step 2: To finish solving a problem with a negative sign in front of the variable, multiply both sides by -1. The variable needs to be positive in the answer.

$$(-1)(-y) = (-1)(20) \text{ so } y = -20$$

Solve the two-step algebra problems below.

1. $6x - 4 = -34$
2. $5y - 3 = 32$
3. $8 - t = 1$
4. $10p - 6 = -36$
5. $11 - 9m = -70$
6. $4x - 12 = 24$
7. $3x - 17 = -41$
8. $9d - 5 = 49$
9. $10h + 8 = 78$
10. $-6b - 8 = 10$
11. $-g - 24 = -17$
12. $-7k - 12 =$ [illegible]0
13. $9 - 5r = -64$
14. $6y - 4 = 34$
15. 12[illegible] $+ 15 = 51$
16. $21t + 17 = 80$
17. $20y + 9 = 149$
18. $15p - 27 = 33$
19. $22h + 9 = 97$
20. $-5 + 36w = 175$

7.2 Two-Step Algebra Problems with Fractions

An algebra problem may contain a fraction. Study the following example to understand how to solve algebra problems that contain a fraction.

Example 3: $\frac{x}{2} + 4 = 3$

Step 1:

$$\begin{array}{rl} \frac{x}{2} + 4 & = 3 \\ -4 & -4 \\ \hline \frac{x}{2} & = -1 \end{array}$$

Subtract 4 from both sides.

Step 2: $\frac{x}{2} = -1$ Multiply both sides by 2 to eliminate the fraction.

$\frac{x}{2} \times 2 = -1 \times 2,\ x = -2$

Simplify the following algebra problems.

1. $4 + \frac{y}{[illegible]} = 7$
2. $[illegible] + 5 = 12$
3. $\frac{w}{5} - 3 = 6$
4. $\frac{x}{[illegible]} - 9 = [illegible]5$
5. $\frac{b}{6} + 2 = -4$
6. $7 + \frac{z}{2} = -1[illegible]$
7. $\frac{[illegible]}{[illegible]} - 7 = 3$
8. $\frac{c}{5} + 6 = -2$
9. $[illegible] + \frac{x}{11} = 7$
10. $16 + \frac{m}{6} = 14$
11. $\frac{[illegible]}{3} + 5 = -2$
12. $\frac{t}{8} + 9 = 3$
13. $\frac{v}{7} - 8 = -1$
14. $[illegible] + \frac{h}{10} = 8$
15. $\frac{k}{7} - [illegible] = 1$
16. $\frac{y}{4} + 13 = 8$
17. $15 + \frac{z}{14} = 13$
18. $\frac{b}{6} - 9 = -14$
19. $\frac{[illegible]}{3} + 7 = 12$
20. $1[illegible] + \frac{b}{6} = 4[illegible]$
21. $2 + \frac{p}{4} = [illegible]$
22. $\frac{t}{[illegible]} [illegible] 9 = -5$
23. $\frac{a}{10} - 1 = 3$
24. $\frac{a}{8} + 16 = 9$

7.3 More Two-Step Algebra Problems with Fractions

Study the following example to understand how to solve algebra problems that contain a different type of fraction.

Example 4: $\frac{x+2}{4} = 3$ — In this example, "$x + 2$" is divided by 4, and not just the x or the 2.

Step 1: $\frac{x+2}{4} \times 4 = 3 \times 4$ — First multiply both sides by 4 to eliminate the fraction.

Step 2: $\begin{array}{r} x + 2 = 12 \\ -2 \quad -2 \\ \hline x = 10 \end{array}$ — Next, subtract 2 from both sides.

Solve the following problems.

1. $\frac{x+1}{5} =$ [illegible]
2. $\frac{\text{[illegible]}}{2} = 7$
3. $\frac{b-4}{4} = -$[illegible]
4. $\frac{y-9}{3} =$ [illegible]
5. $\frac{d-10}{-2} = 12$
6. $\frac{w-10}{-8} = -$[illegible]
7. $\frac{\text{[illegible]}-1}{-2} = -5$
8. $\frac{c+40}{-5} = -7$
9. $\frac{13+\text{[illegible]}}{2} = 12$
10. $\frac{k-10}{3} = 9$
11. $\frac{a+11}{-4} = 4$
12. $\frac{x-20}{7} = 6$
13. $\frac{t+2}{6} = -5$
14. $\frac{\text{[illegible]}}{-7} = 2$
15. $\frac{f-9}{3} = 8$
16. $\frac{4+w}{6} = -6$
17. $\frac{3+t}{3} = 10$
18. $\frac{x+5}{\text{[illegible]}} = -3$
19. $\frac{g+\text{[illegible]}}{2} = 11$
20. $\frac{\text{[illegible]}+1}{-6} = 5$
21. $\frac{y-14}{2} =$ [illegible]8
22. $\frac{z}{\text{[illegible]}} = 13$
23. $\frac{w+2}{15} = -1$
24. $\frac{3+h}{3} = 6$

7.4 Combining Like Terms

In algebra problems, separate **terms** by $+$ and $-$ signs. The expression $5x - 4 - 3x + 7$ has 4 terms: $5x$, 4, $3x$, and 7. Terms having the same variable can be combined (added or subtracted) to simplify the expression. $5x - 4 - 3x + 7$ simplifies to $2x + 3$.

$$5x - 3x \quad -4 + 7 \quad = 2x + 3$$

Simplify the following expressions.

1. $7x + 12x$
2. $8y - 5y + 8$
3. $4 - 2x + 9$
4. $11a - 16 - a$
5. $9w + 3w +$ [illegible]
6. $-5x +$ [illegible] $+ 2x$
7. $w - 15 + 9w$
8. $21 - 10t + 9 - 2t$
9. $-3 + x - 4x + 9$
10. $7b + 12 + 4b$
11. $4h - h + 2 - 5$
12. $-6k + 10 - 4k$
13. $2a + 12a - 5 + a$
14. $5 + 9c - 10$
15. $-d + 1 + 2d - 4$
16. $-8 + 4h + 1 - h$
17. $12x - 4x + 7$
18. $10 + 3z + z - 5$
19. $14 + 3y - y - 2$
20. $11p - 4 + p$
21. $11m + 2 - m + 1$

7.5 Solving Equations with Like Terms

When an equation has two or more like terms on the same side of the equation, combine like terms as the **first** step in solving the equation.

Example 5: $7x + 2x - 7 = 21 + 8$

Step 1: Combine like terms on both sides of the equation.

Step 2: Solve the two-step algebra problem as explained previously.

$$
\begin{aligned}
7x + 2x - 7 &= 21 + 8 \\
9x - 7 &= 29 \\
+7 \quad & \quad +7 \\
9x \div 9 &= 36 \div 9 \\
x &= 4
\end{aligned}
$$

Solve the equations below combining like terms first.

1. $3w - 2w + 4 = 6$
2. $7x + 3 + x = 16 + 3$
3. $5 - 6y + 9y = -15 + 5$
4. $-14 + 7a + 2a = -5$
5. $-2t + 4t - 7 = 9$
6. $9d + d - 3d = 14$
7. $-6c - 4 - 5c = 10 +$ [illegible]
8. $15m - 9 - 6m =$ [illegible]
9. $-4 - 3x - x = -16$
10. $9 - 12p + 5p = 14 + 2$
11. $10y + 4 - 7y = -17$
12. $-8a - 15 - 4a = 9$

If the equation has like terms on both sides of the equation, you must get all of the terms with a **variable** on one side of the equation and all of the **integers** on the other side of the equation.

Example 6: $3x + 2 = 6x - 1$

Step 1: Subtract $6x$ from both sides to move all the **variables** to the left side.

Step 2: Subtract 2 from both sides to move all the **integers** to the right side.

Step 3: Divide by -3 to solve for x.

$$\begin{aligned} 3x + 2 &= 6x - 1 \\ -6x \quad & \quad -6x \\ -3x + 2 &= -1 \\ -2 \quad & \quad -2 \\ \hline \frac{-3x}{-3} &= \frac{-3}{-3} \\ x &= 1 \end{aligned}$$

Solve the following problems.

1. $3a + 1 = a + 9$
2. $2d - 12 =$ [illegible] 3
3. $5x + 6 = 14 - 3x$
4. $15 - 4y = 2y - 3$
5. $9w - 7 = 12w - 13$
6. $10b + 19 = 4b - 5$
7. $-7m + 3 = 29 - 2m$
8. $5x - 26 = 13x - 2$
9. $19 - p = 3p - 9$
10. $-7p - 14 = -2p + 11$
11. $16y + 12 = 9y + 33$
12. $13 - 11w = 3 -$ [illegible]w
13. $-17b + 23 = -4 - 8b$
14. $k + 5 = 20 - 2k$
15. $12 + m = 4m + 21$
16. $7p - 30 = p + 6$
17. $19 - 13z = 9 - 12z$
18. $8y - 2 = 4y + 22$
19. $5 + 16w = 6w - 45$
20. $-27 - 7x = 2x + 18$
21. $-12x + 14 = 8x - 46$
22. $27 - 11h = 5 - 9h$
23. $5t + 36 = -6 - 2t$
24. $17y + 42 = 10y + 7$
25. $22x - 24 = 14x - 8$
26. $p - 1 =$ [illegible]$p + 17$
27. $4d + 14 = 3d - 1$
28. $7w - 5 = 8w + 12$
29. $-3y - 2 = 9y + 22$
30. [illegible] $+ 9m = m - 23$

7.6 Removing Parentheses

The distributive principle is used to remove parentheses.

Example 7: $2(a+6)$

You multiply 2 by each term inside the parentheses. $2 \times a = 2a$ and $2 \times 6 = 12$. The 12 is a positive number so use a plus sign between the terms in the answer.

$2(a+6) = 2a + 12$

Example 8: $4(-5c+2)$

The first term inside the parentheses could be negative. Multiply in exactly the same way as the examples above. $4 \times (-5c) = -20c$ and $4 \times 2 = 8$

$4(-5c+2) = -20c + 8$

Remove the parentheses in the problems below.

1. $7(n+6)$
2. $8(2g-5)$
3. $11(5z$ [illegible]$)$
4. $6($[illegible]$-4)$
5. [illegible]$(-3k+5)$
6. $4(d-8)$
7. $2(-4x+6)$
8. $7(4+6p)$
9. $5(-4w-8)$
10. $6(11x+2)$
11. $10(9-y)$
12. $9(c-9)$
13. $12(-3t+1)$
14. $3(4y+9)$
15. $8(b+3)$

The number in front of the parentheses can also be negative. Remove these parentheses the same way.

Example 9: $-2(b-4)$

First, multiply $-2 \times b = -2b$

Second, multiply $-2 \times -4 = 8$

Copy the two products. The second product is a positive number so put a plus sign between the terms in the answer.

$-2(b-4) = -2b + 8$

Remove the parentheses in the following problems.

16. $-7(x+2)$
17. $-5(4-y)$
18. $-4(2b-2)$
19. $-2(8c+6)$
20. $-5(-w-8)$
21. $-3(4x-2)$
22. $-2(-z+2)$
23. $-4(7p+7)$
24. $-9(t-6)$
25. $-10(2w+4)$
26. $-3(9-7p)$
27. $-9(-k-3)$
28. $-1(7b-9)$
29. $-6(-5t-2)$
30. $-7(-v+4)$

7.7 Multi-Step Algebra Problems

You can now use what you know about removing parentheses, combining like terms, and solving simple algebra problems to solve problems that involve three or more steps. Study the examples below to see how easy it is to solve multi-step problems.

Example 10: $3(x+6) = 5x - 2$

Step 1: Use the distributive property to remove parentheses. $3x + 18 = 5x - 2$

Step 2: Subtract $5x$ from each side to move the terms with variables to the left side of the equation.

$$\begin{array}{rr} 3x + 18 & = 5x - 2 \\ -5x & -5x \\ \hline -2x + 18 & = -2 \end{array}$$

Step 3: Subtract 18 from each side to move the integers to the right side of the equation.

$$\begin{array}{rr} -18 & -18 \\ \hline \frac{-2x}{-2} & = \frac{-20}{-2} \end{array}$$

Step 4: Divide both sides by -2 to solve for x.

$$x = 10$$

Example 11: $\frac{3(x-3)}{2} = 9$

Step 1: Use the distributive property to remove parentheses. $\frac{3x-9}{2} = 9$

Step 2: Multiply both sides by 2 to eliminate the fraction. $\frac{2(3x-9)}{2} = 2(9)$

Step 3: Add 9 to both sides, and combine like terms.

$$\begin{array}{rr} 3x - 9 & = 18 \\ +9 & +9 \\ \hline \end{array}$$

Step 4: Divide both sides by 3 to solve for x.

$$\frac{3x}{3} = \frac{27}{3}$$

$$x = \text{[illegible]}$$

Solve the following multi-step algebra problems.

1. $2(y-3) = 4y+6$
2. $\frac{2(a+4)}{2} = 12$
3. $\frac{10(x-2)}{5} = 14$
4. $\frac{12y-18}{6} = 4y+3$
5. $2x + 3x = 30 - x$
6. $\frac{2a+1}{3} = a+5$
7. $5(b-4) = 10b+5$
8. $-8(y+4) = 10y+4$

9. $\frac{x+4}{-3} = 6 - x$

10. $\frac{4(n+3)}{5} = n - 3$

11. $3(2x - 5) = 8x - 9$

12. $7 - 10a = 9 - 9a$

13. $7 - 5x = 10 - (6[illegible]7)$

14. $4(x - 3) - [illegible] = x - 6$

15. $4a + 4 = 3a - 4$

16. $[illegible]3(x - 4) + [illegible] = -2x [illegible] 2$

17. $5b - 11 = 1[illegible] - b$

18. $\frac{-[illegible]x + [illegible]}{[illegible]x} = \frac{[illegible]}{2x}$

19. $-(x + 1) = -2(5 - x)$

20. $[illegible](2x + 3) - 7 = 13$

21. $[illegible] - 3a = 9 - 2(2a + 5)$

22. $-5x + 9 = -3x + 11$

23. $3y + 2 - [illegible]y - 5 = 4y + 3$

24. $3y [illegible] 10 = [illegible] - 4[illegible]$

25. $-(a + 3) = -[illegible](2a + 1) - 7$

26. $5[illegible] - 2(m + 1) = m - 10$

27. $\frac{1}{2}(b - 2) = 5$

28. $-3(b - 4) = -2b$

29. $4x + 12 = -2(x + 3)$

30. $\frac{7x+4}{3} = 2x - 1$

31. $9x - 5 = 8x - 7$

32. $7x - 5 = 4x + 10$

33. $\frac{4x+8}{2} = 6$

34. $2(c + 4) + 8 = 1[illegible]$

35. $y - (y + 3) = y + 6$

36. $4 + [illegible] - 2(x - 6) = 8$

7.8 Solving Radical Equations

Some multi-step equations contain radicals. An example of a radical is a square root, $\sqrt{}$.

Example 12: Solve the following equation for x. $\sqrt{4x-3}+2=5$

Step 1: The first step is to get the constants that are not under the radical on one side. Subtract 2 from both sides of the equation.
$\sqrt{4x-3}+2-2=5-2$
$\sqrt{4x-3}+0=3$
$\sqrt{4x-3}=3$

Step 2: Next, you must get rid of the radical sign by squaring both sides of the equation.
$\left(\sqrt{4x-3}\right)^2=(3)^2$
$4x-3=9$

Step 3: Add 3 to both sides of the equation to get the constants on just one side of the equation.
$4x-3+3=9+3$
$4x+0=12$
$4x=12$

Step 4: Last, get x on one side of the equation by itself by dividing both sides by 4.
$\frac{4x}{4}=\frac{12}{4}$
$x=3$

Solve the following equations.

1. $\sqrt{x+3}-13=-8$
2. $3+\sqrt{7t-3}=5$
3. $\sqrt{3q+12}-4=5$
4. $\sqrt{11f+3}+2=8$
5. $5=\sqrt{6g-5}+(-2)$
6. $2=\sqrt{x-3}$
7. $\sqrt{-8t}-3=1$
8. $\sqrt{-d+1}-9=-6$
9. $10-\sqrt{8x+2}=9$
10. $\sqrt{15y+4}+4=12$
11. $\sqrt{r+14}-$ [illegible] $=9$
12. $3-\sqrt{\text{[illegible]}q-1}=0$
13. $\sqrt{5t+16}+4=13$
14. $17=\sqrt{23-f}+15$
15. $19-\sqrt{7x-5}=16$

7.9 Multi-Step Inequalities

Remember that adding and subtracting with inequalities follow the same rules as equations. When you multiply or divide both sides of an inequality by the same positive number, the rules are also the same as for equations. However, when you multiply or divide both sides of an inequality by a **negative** number, you must **reverse** the inequality symbol.

Example 13: $-x > 4$
$(-1)(-x) < (-1)(4)$
$x < -4$

Example 14: $-4x < 2$

$\frac{-4x}{-4} > \frac{2}{-4}$

$x > -\frac{1}{2}$

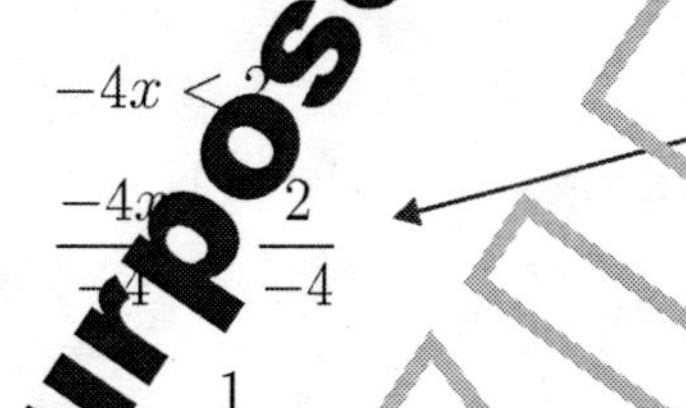

Reverse the symbol when you multiply or divide by a negative number.

When solving multi-step inequalities, first add and subtract to isolate the term with the variable. Then multiply and divide.

Example 15: $2x - 8 > 4x + 1$

Step 1: Add 8 to both sides.

$2x - 8 + 8 > 4x + 1 + 8$
$2x > 4x + 9$

Step 2: Subtract $4x$ from both sides.

$2x - 4x > 4x + 9 - 4x$
$-2x > 9$

Step 3: Divide by -2. Remember to change the direction of the inequality sign.

$\frac{-2x}{-2} < \frac{9}{-2}$

$x < -\frac{9}{2}$

Solve each of the following inequalities.

1. $8 - 3x \leq 7x - 2$

2. $3(2x - 5) \geq 8x - 5$

3. $\frac{1}{3}b - 2 > 5$

4. $7 + 3y > 2y - 5$

5. $3a + 5 < 2a -$ [illegible]

6. $3(a - 2) > 5a - 2(3 - a)$

7. $2x - 7 \geq 4(x - 3) + 3x$

8. $6x - 2 \leq 5x + 5$

9. $-\frac{x}{4} > 12$

10. $-\frac{2x}{3} \leq 6$

11. $3b + 5 < 2b - 8$

12. $4x - 5 \leq 7x + 13$

13. $4x + 5 \leq -2$

14. $2y - 5 > 7$

15. $4 + 2(3 - 2y) \leq 6y - 20$

16. $-4c + 6 \leq$ [illegible]

17. $-\frac{1}{2}x + 2 > 9$

18. $\frac{1}{4}y - 3 \leq 1$

19. $-3x + 4 > 5$

20. $\frac{y}{2} - 2 \geq 10$

21. $7 + 4c < -2$

22. $2 - \frac{a}{2} > 1$

23. $10 + 4b \leq -2$

24. $-\frac{1}{2}x + 3$ [illegible]

7.10 Solving Equations and Inequalities with Absolute Values

When solving equations and inequalities which involve variables placed in absolute values, remember that there will be two or more numbers that will work as correct answers. This is because the absolute value variable will signify both positive and negative numbers as answers.

Example 16: $5 + 3|k| = 8$ Solve as you would any equation.

Step 1: $3|k| = 3$ Subtract 5 from each side.

Step 2: $|k| = 1$ Divide by 3 on each side.

Step 3: $k = 1$ or $k = -1$ Because k is an absolute value, the answer can be 1 or -1.

Example 17: $2|x| - 3 < 7$ Solve as you normally would an inequality.

Step 1: $2|x| < 10$ Add 3 to both sides.

Step 2: $|x| < 5$ Divide by 2 on each side.

Step 3: $x < 5$ or $x > -5$ or $-5 < x < 5$ Because x is an absolute value, the answer is a set of both positive and negative numbers.

Read each problem, and write the number or set of numbers which solves each equation or inequality.

1. $7 + 2|y| = 15$
2. $4|x| - 9 < 3$
3. $6|k| + 2 = 14$
4. $10 - 4|n| > -14$
5. $-3 = 5|z| + 12$
6. $-4 + 7|m| < 10$
7. $5|x| - 12 > 13$
8. $21|g| + 7 = 49$
9. $-9 + 6|x| = 15$
10. $12 - 6|w| > -12$
11. $31 > 13 + 9|r|$
12. $-30 = 21 - 3|$[illegible]$|$
13. $9|x| - 19 < 35$
14. $-13|c| + 21 \geq$ [illegible]
15. $5 - 11|k| < -17$
16. $-42 + 14|p| =$ [illegible]
17. $15 < 3|$[illegible]$| + 6$
18. $9 + 5|q| = 29$
19. $-14|y| - 38 < -45$
20. $36 = 4|s| + 20$
21. $20 \leq -60 + 8|e|$

7.11 More Solving Equations and Inequalities with Absolute Values

Now, look at the following examples in which numbers and variables are added or subtracted within the absolute value symbols ($||$).

Example 18: $|3x - 5| = 10$

Remember an equation with absolute value symbols has two solutions.

Step 1: To find the first solution, remove the absolute value symbol and solve the equation.

$3x - 5 = 10$
$3x - 5 + 5 = 10 + 5$
$\frac{\not{3}x}{\not{3}} = \frac{15}{3}$
$x = 5$

Step 2: To find the second solution, solve the equation for the negative of the expression in absolute value symbols.

$-(3x - 5) = 10$
$-3x + 5 = 10$
$-3x + 5 - 5 = 10 - 5$
$-3x = 5$
$x = -\frac{5}{3}$

Solutions: $x = \{5, -\frac{5}{3}\}$

Example 19: $|5z - 10| < 20$

Remove the absolute value symbols and solve the inequality.

Step 1:

$5z - 10 < 20$
$5z - 10 + 10 < 20 + 10$
$\frac{\not{5}z}{\not{5}} < \frac{30}{5}$
$z < 6$

Step 2: Next, solve the equation for the negative of the expression in the absolute value symbols.

$-(5z - 10) < 20$
$-5z + 10 < 20$
$-5z + 10 - 10 < 20 - 10$
$\frac{-\not{5}z}{\not{5}} < \frac{10}{5}$
$-z < 2$
$z > -2$

Solution: $-2 < z < 6$

Example 20: $|4y + 7| - 5 > 18$

Step 1:
$4y + 7 - 5 + 5 > 18 + 5$
$4y + 7 > 23$
$4y + 7 - 7 > 23 - 7$
$4y > 16$
$y > 4$

Remove the absolute value symbols and solve the inequality.

Step 2:
$-(4y + 7) - 5 > 18$
$-4y - 7 - 5 + 5 > 18 + 5$
$-4y - 7 + 7 > 23 + 7$
$-4y > 30$
$y < -7\frac{1}{2}$

Solve the equation for the negative of the expression in the absolute value symbols.

Solutions: $y > 4$ or $y < -7\frac{1}{2}$

Solve the following equations and inequalities below.

1. $-4 + |2[illegible] + 4| = 14$
2. $|4[illegible] - 7| + 3 > 12$
3. $[illegible] + |12e + 3| < 39$
4. $-15 + |8f - 14| > 35$
5. $|-9b + 13| - 12 = 10$
6. $-25 + |7b + 11| < 35$
7. $|7w + 2| - 60 > 30$
8. $63 + |3d - 12| = 21$
9. $|-23 + 8x| - 12 > +37$
10. $|61 + 20x| + 32 > 51$
11. $|4[illegible] + 13| + 31 = 50$
12. $4 + |4k - 32| < 51$
13. $8 + |4x + 3| = 21$
14. $|28 + 7v| - 28 < 77$
15. $|62p + 31| - 43 = 136$
16. $18 - |6v + 22| < 22$
17. $12 = 4 + |42 + 10m|$
18. $53 < 18 + |12e + 31|$
19. $38 > -39 + |7j + 14|$
20. $9 = |14 + 15u| + 7$
21. $11 - |2j + 50| > 45$
22. $|35 + 6[illegible]| - 3 = 14$
23. $|26 - 8r| - 9 > 41$
24. $|25 + 6z| - 21 = 28$
25. $12 < |2t + 6| - [illegible]$
26. $50 > |9q - [illegible]| + 6$
27. $12 + |8v - 18| > 26$
28. $-38 + |16i - 33| = 41$
29. $|-14 + 6p| - 9 < 7$
30. $28 > |25 - 5f| - 12$

7.12 Inequality Word Problems

Inequality word problems involve staying under a limit or having a minimum goal one must meet.

Example 21: A contestant on a popular game show must earn a minimum of 800 points by answering a series of questions worth 40 points each per category in order to win the game. The contestant will answer questions from each of four categories. Her results for the first three categories are as follows: 160 points, 200 points, and 240 points. Write an inequality which describes how many points, (p), the contestant will need on the last category in order to win.

Step 1: Add to find out how many points she already has. $160 + 200 + 240 = 600$

Step 2: Subtract the points she already has from the minimum points she needs. $800 - 600 = 200$. She must get at least 200 points in the last category to win. If she gets more than 200 points, that is okay, too. To express the number of points she needs, use the following inequality statement:

$p \geq 200$ The points she needs must be greater than or equal to 200.

Solve each of the following problems using inequalities.

1. Stella wants to place her money in a high interest money market account. However, she needs at least $\$1,000$ to open an account. Each month, she set aside some of her earnings in a savings account. In January through June, she added the following amounts to her savings: \$121, \$206, \$158, \$212, \$109, and \$134. Write an inequality which describes the amount of money she can set aside in July to qualify for the money market account.

2. A high school band program will receive $\$2,000.00$ for selling $\$10,000.00$ worth of coupon books. Six band classes participate in the sales drive. Classes 1–5 collect the following amounts of money: $\$1,400$, $\$2,600$, $\$1,800$, $\$2,450$, and $\$1,550$. Write an inequality which describes the amount of money the sixth class must collect so that the band will receive $\$2,000$.

3. A small elevator has a maximum capacity of $1,000$ pounds before the cable holding it in place snaps. Six people get on the elevator. Five of their weights follow: 146, 180, 130, 262, and 135. Write an inequality which describes the amount the sixth person can weigh without snapping the cable.

4. A small high school class of 9 students were told they would receive a pizza party if their class average was 92% or higher on the next exam. Students 1–8 scored the following on the exam: 86, 91, 98, 83, 97, 89, 99, and 96. Write an inequality which describes the score the ninth student must make for the class to qualify for the pizza party.

5. Raymond wants to spend his entire credit limit on his credit card. His credit limit is $\$2,000$. He purchases items costing \$600, \$800, \$50, \$168, and \$3. Write an inequality which describes the amounts Raymond can put on his credit card for his next purchases.

Chapter 7 Review

Solve each of the following equations.

1. $4a - 8 = 28$
2. $5 + \frac{x}{8} = -4$
3. -7 [illegible] $= 108$
4. $\frac{y \text{ [illegible]}}{6} = 7$
5. [illegible] $- 13 = 5$
6. $\frac{\text{[illegible]} + 9}{12} = -3$

Solve.

7. $19 - 8d = d - 17$
8. $-\frac{x}{3} + 11 = -1$
9. $7w - 8w = -4w$ [illegible]
10. $6 - 2x = 4$
11. $\frac{12}{f} - 7 = -5$
12. $6 + 16x = -2x - 12$
13. $w + 11 = 15$
14. $6 - \frac{q}{2} = 4$
15. $7k - 3 = 11$

Remove parentheses.

16. $3(-4x + 7$[illegible]
17. $11(2y + 5)$
18. $6(8 - 9b)$
19. $-8(-2 + 3a)$
20. $-2(5c - 3)$
21. [illegible]$(7y - 1)$

Solve for the variable.

22. If $3x - y = 15$, then $y =$

23. If $7a + 2b = 1$, then $b =$

Solve each of the following equations and inequalities

24. $\frac{11c - 35}{4} = 4c - 2$
25. $5 + x - 3(x + 4) = -17$
26. $4(2x + 3) \geq 2x$
27. $7 - 3x \leq 6x -$ [illegible]
28. $\frac{5(n + 4)}{3} = n - 8$
29. $-y > 14$
30. $2(3x - 1) \geq 3x - 7$
31. $3(x + 2) < 7x - 10$

32. Jim takes great pride in decorating his float for the homecoming parade for his high school. With the $5,000 he has to spend, Jim bought 5,000 carnations at $0.25 each, 4,000 tulips at $0.50 each, and 300 irises at $0.90 each. Write an inequality which describes how many roses, r, Jim can buy if roses cost $0.80 each.

33. Mr. Chan wants to sell some or all of his shares of stock in a company. He purchased the 80 shares for $0.50 last month, and the shares are now worth $4.50 each. Write an inequality which describes how much profit, p, Mr. Chan can make by selling his shares.

Chapter 8
Rates, Ratios, and Proportions

8.1 Time of Travel

Example 1: Katrina drove 384 miles at an average of 64 miles per hour. How many hours did she travel?

Divide the number of miles by the miles per hour. $\dfrac{384 \text{ miles}}{64 \text{ miles/hour}} = 6 \text{ hours}$

Katrina traveled 6 hours.

Find the hours of travel in each problem below.

1. Bobbi drove 342 miles at an average speed of 57 miles per hour. How many hours did she drive?

2. Jan set her speed control at 55 miles per hour and drove for 165 miles. How many hours did she drive?

3. John traveled 2, 092 miles in a jet that flew an average of 523 miles per hour. How long was he in the air?

4. How long will it take a bus averaging 54 miles per hour to travel 378 miles?

5. Kyle drove his motorcycle in a 225 mile race, and he averaged 75 miles per hour. How long did it take for him to complete the race?

6. Stacy drove 576 miles at an average speed of 48 miles per hour. How many hours did she drive?

7. Kendra flew 250 miles in a glider and averaged 125 miles per hour in speed. How many hours did she fly?

8. Travis traveled 496 miles at an average speed of 62 miles per hour. How long did he travel?

9. Wanda rode her bicycle an average of 15 miles an hour for 60 miles. How many hours did she ride?

10. Kami drove 184 miles at an average speed of 46 miles per hour. How many hours did he drive?

11. A train traveled at a constant 85 miles per hour for 425 miles. How many hours did the train travel?

12. How long was Amy on the road if she drove 195 miles at an average of 65 miles per hour?

8.2 Rate

Example 2: Laurie traveled 312 miles in 6 hours. What was her average rate of speed?

Divide the number of miles by the number of hours. $\frac{312 \text{ miles}}{6 \text{ hours}} = 52$ miles/hour

Laurie's average rate of speed was 52 miles per hour (or 52 mph).

Find the average rate of speed in each problem below.

1. A race car went 5[illegible]0 miles in 4 hours. What was its average rate of speed?

2. Carrie drove [illegible]24 miles in 2 hours. What was her average speed?

3. After [illegible] hours of driving, Chad had gone 364 miles. What was his average speed?

4. Anna drove 360 miles in 8 hours. What was her average speed?

5. After 3 hours of driving, Paul had gone 183 miles. What was his average speed?

6. Nicole ran 25 miles in 5 hours. What was her average speed?

7. A train traveled 492 miles in 6 hours. What was its average rate of speed?

8. A commercial jet traveled 1,572 miles in 3 hours. What was its average speed?

9. Jillian drove 195 miles in 3 hours. What was her average speed?

10. Greg drove 8 hours from his home to a city 336 miles away. At what average speed did he travel?

11. Caleb drove 128 miles in two hours. What was his average speed in miles per hour?

12. After 9 hours of driving, Kate had traveled 405 miles. What speed did she average?

8.3 More Rates

Rates are often discussed in terms of miles per hour, but a rate can be any measured quantity divided by another measurement such as feet per second, kilometers per minute, mass per unit volume, etc. A rate can be how fast something is done. For example, a bricklayer may lay 80 bricks per hour. Rates can also be used to find measurements such as density. For example, 35 grams of salt in 1 liter of water gives the mixture a density of 35 grams/liter.

Example 3: Nathan entered his snail in a race. His snail went 18 feet in 6 minutes. How fast did his snail move?

In this problem, the units given are feet and minutes, so the rate will be feet per minute (or feet/minute).

You need to find out how far the snail went in one minute.

Rate equals $\frac{\text{distance}}{\text{time}}$ so $\frac{18 \text{ feet}}{6 \text{ minutes}} = \frac{3 \text{ feet}}{1 \text{ minute}}$

Nathan's snail went an average of 3 feet per minute or 3 $\frac{\text{ft}}{\text{min}}$.

Find the average rate for each of the following problems.

1. Tewanda read a 2,000-word news article in 8 minutes. How fast did she read the news article?

2. Chandler rides his bike to school every day. He travels 2,560 feet in 640 seconds. How many feet did he travel per second?

3. Mr. Molier is figuring out the semester averages for his history students. He can figure the average for 20 students in an hour. How long does it take him to figure the average for each student?

4. In 1908, John Hurlinger of Austria walked 1,400 kilometers from Vienna to Paris on his hands. The journey took 55 days. What was his average speed per day?

5. Spectators at the Super Circus were amazed to watch a cannon shoot a clown 212 feet into a net in 4 seconds. How many feet per second did the clown travel?

6. Marcus Page, star receiver for the Big Bulls, was awarded a 5-year contract for 105 million dollars. How much will his annual rate of pay be if he is paid the same amount each year?

7. Duke Delaney scored 28 points during the 4 quarters of the basketball playoffs. What was his average score per quarter?

8. The new McDonald's in Moscow serves 11,208 customers during a 24-hour period. What is the average number of customers served per hour?

8.4 Distance

Example 4: Jessie traveled for 7 hours at an average rate of 58 miles per hour. How far did she travel?

Multiply the number of hours by the average rate of speed.

$7 \text{ hours} \times 58 \frac{\text{miles}}{\text{hour}} = 406 \text{ miles}$

Find the distance in each of the following problems.

1. Myra traveled for 9 hours at an average rate of 45 miles per hour. How far did she travel?

2. A tour bus drove 4 hours, averaging 58 miles per hour. How many miles did it travel?

3. Tina drove for 7 hours at an average speed of 53 miles per hour. How far did she travel?

4. Dustin raced for 3 hours, averaging 176 miles per hour. How many miles did he race?

5. Kris drove 5 hours and averaged 49 miles per hour. How far did she travel?

6. Oliver drove at an average of 93 miles per hour for 3 hours. How far did he travel?

7. A commercial airplane traveled 514 miles per hour for 2 hours. How far did it fly?

8. A train traveled at 125 miles per hour for 4 hours. How many miles did it travel?

9. Carmen drove a constant 65 miles an hour for 3 hours. How many miles did he drive?

10. Jasmine drove for 5 hours, averaging 40 miles per hour. How many miles did she drive?

11. Roger flew his glider for 2 hours at 87 miles per hour. How many miles did his glider fly?

12. Beth traveled at a constant 65 miles per hour for 4 hours. How far did she travel?

8.5 Ratio Problems

In some word problems, you may be asked to express answers as a **ratio**. Ratios can look like fractions. Numbers must be written in the order they are requested. In the following problem, 8 cups of sugar is mentioned before 6 cups of strawberries. But in the question part of the problem, you are asked for the ratio of STRAWBERRIES to SUGAR. The amount of strawberries IS THE FIRST WORD MENTIONED, so it must be the **top** number of the fraction. The amount of sugar, THE SECOND WORD MENTIONED, must be the **bottom** number of the fraction.

Example 5: The recipe for jam requires 8 cups of sugar for every 6 cups of strawberries. What is the ratio of strawberries to sugar in this recipe?

First number requested $\frac{6}{8}$ $\frac{\text{cups strawberries}}{\text{cups sugar}}$
Second number requested

Answers may be reduced to lowest terms. $\frac{6}{8} = \frac{3}{4}$

Practice writing ratios for the following word problems and reduce to lowest terms. DO NOT CHANGE ANSWERS TO MIXED NUMBERS. Ratios should be left in fraction form.

1. Out of the 248 seniors, 112 are boys. What is the ratio of boys to the total number of seniors?

2. It takes 7 cups of flour to make 2 loaves of bread. What is the ratio of cups of flour to loaves of bread?

3. A skyscraper that stands 620 feet tall casts a shadow that is 125 feet long. What is the ratio of the shadow to the height of the skyscraper?

4. Twenty boxes of paper weigh 520 pounds. What is the ratio of boxes to pounds?

5. The newborn weighs 8 pounds and is 22 inches long. What is the ratio of weight to length?

6. Jack paid $6.00 for 10 pounds of apples. What is the ratio of the price of apples to the pounds of apples?

7. Jordan spends $45 on groceries. Of that total, $23 is for steaks. What is the ratio of steak cost to the total grocery cost?

8. Madison's flower garden measures 8 feet long by 6 feet wide. What is the ratio of length to width?

8.6 Writing Ratios Using Variables

Ratios can be written using variables instead of just numbers.

Example 6: Timothy has a bag of marbles. He only has red, r, marbles and blue, b, marbles. If he gives three of the blue marbles to his little brother, what fractional part of the marbles remaining in the bag are blue?

Step 1: The total number of marbles in the bag before Timothy gives three to his brother is $r + b$. So the fractional part of the marbles that are blue before Timothy gives some away is $\frac{b}{r + b}$.

Step 2: Since Timothy takes three blue marbles out of the bag, the total number of marbles that are left are $r + b - 3$. The total number of blue marbles left is $b - 3$. So, the fractional part of the marbles that are blue after Timothy gives three to his brother is $\frac{b - 3}{r + b - 3}$.

Use the following for questions 1 through 4.

Sancho has a box of bouncy balls. He has green, g, bouncy balls, yellow, y, bouncy balls, and blue-and-white striped, b, bouncy balls. He gives four yellow bouncy balls and 1 green bouncy ball to a friend.

1. What is the ratio of green bouncy balls to the total number of bouncy balls after he gave the 4 yellow balls and 1 green ball to a friend?
2. What is the ratio of yellow and blue and white striped bouncy balls to the total number of bouncy balls after he gave the 4 yellow balls and 1 green ball to a friend?
3. If Sancho originally had 5 green bouncy balls, 7 yellow bouncy balls, and 3 blue and white striped bouncy balls, what is the numerical ratio of yellow bouncy balls to the total number of bouncy balls after he gave the 4 yellow balls and the 1 green ball to a friend?
4. Using variables, what fractional part of the remaining bouncy balls are yellow?

Use the following for questions 5 and 6.

Callie has a bag of lollipops. She has cherry lollipops, designated by the letter c, and blue raspberry lollipops, designated by the letter b. She gave two cherry lollipops to her best friend.

5. What fractional part of the remaining lollipops are cherry?
6. If Callie originally had 8 cherry lollipops and 3 blue raspberry lollipops, what is the numerical ratio of blue raspberry to the total number of lollipops after she gave two cherry lollipops to her best friend?

8.7 Solving Proportions

Two **ratios (fractions)** that are **equal** to each other are called **proportions. For example,** $\frac{1}{4} = \frac{2}{8}$**. Read the following example to see how to find a number missing from a proportion.**

Example 7: $\frac{5}{15} = \frac{8}{x}$

Step 1: To find x, you first multiply the two numbers that are diagonal to each other.

$\frac{5}{\{15\}} = \frac{\{8\}}{x}$

$15 \times 8 = 120$

$5 \times x = 5x$

Therefore, $5x = 120$

Step 2: Then divide the product (120) by the other number in the proportion (5).

$120 \div 5 = 24$

Therefore, $\frac{5}{15} = \frac{8}{24}$ **and** $x = 24$.

Practice finding the number missing from the following proportions. First, multiply the two numbers that are diagonal from each other. Then divide by the other number.

1. $\frac{2}{5} = \frac{6}{x}$
2. $\frac{9}{3} = \frac{x}{5}$
3. $\frac{x}{12} = \frac{3}{4}$
4. $\frac{7}{x} = \frac{3}{9}$
5. $\frac{12}{x} = \frac{2}{5}$
6. $\frac{12}{x} = \frac{4}{3}$
7. $\frac{27}{3} = \frac{x}{2}$
8. $\frac{1}{x} = \frac{3}{12}$
9. $\frac{15}{2} = \frac{x}{4}$
10. $\frac{7}{14} = \frac{x}{6}$
11. $\frac{5}{6} = \frac{10}{x}$
12. $\frac{4}{x} = \frac{3}{6}$
13. $\frac{x}{5} = \frac{9}{15}$
14. $\frac{9}{18} = \frac{x}{2}$
15. $\frac{5}{7} = \frac{35}{x}$
16. $\frac{x}{[illegible]} = \frac{8}{4}$
17. $\frac{15}{20} = \frac{x}{8}$
18. $\frac{x}{40} = \frac{5}{100}$

8.8 Ratio and Proportion Word Problems

Example 8: A stick one meter long is held perpendicular to the ground and casts a shadow 0.4 meters long. At the same time, an electrical tower casts a shadow 112 meters long. Use ratio and proportion to find the height of the tower.

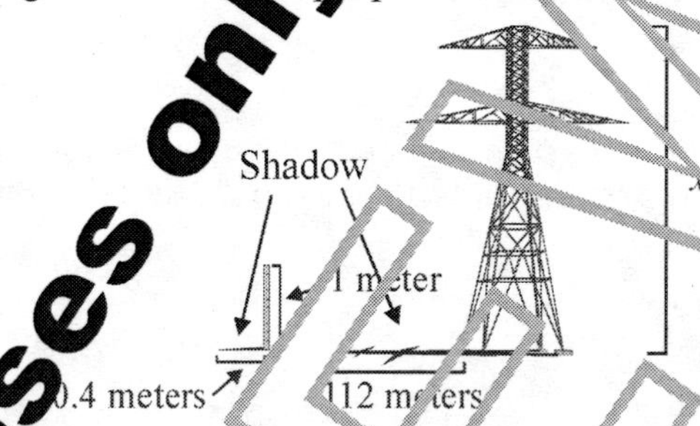

Step 1: Set up a proportion using the numbers in the problem. Put the shadow lengths on one side of the equation and put the heights on the other side. The 1 meter height is paired with the 0.4 meter length, so let them both be top numbers. Let the unknown height be x.

$$\begin{array}{ccc} \text{shadow length} & & \text{object height} \\ \dfrac{0.4}{112} & = & \dfrac{1}{x} \end{array}$$

Step 2: Solve the proportion as you did on page 101.

$112 \times 1 = 112 \qquad 112 \div 0.4 = 280$

Answer: The tower height is 280 meters.

Use ratio and proportion to solve the following problems.

1. Rudolph can mow a lawn that measures 1,000 square feet in 2 hours. At that rate, how long would it take him to mow a lawn 3,500 square feet?

2. Faye wants to know how tall her school building is. On a sunny day, she measures the shadow of the building to be 6 feet. At the same time she measures the shadow cast by a 5-foot statue to be 2 feet. How tall is her school building?

3. Out of every 5 students surveyed, 2 listen to country music. At that rate, how many students in a school of 800 listen to country music?

4. Butterfly, a Labrador retriever, has a litter of 8 puppies. Four are black. At that rate, how many puppies in a litter of 10 would be black?

5. According to the instructions on a bag of fertilizer, 5 pounds of fertilizer are needed for every 100 square feet of lawn. How many square feet will a 25-pound bag cover?

6. A race car can travel 2 laps in 5 minutes. At this rate, how long will it take the race car to complete 100 laps ?

7. If it takes 7 cups of flour to make 4 loaves of bread, how many loaves of bread can you make from 35 cups of flour?

8. If 3 pounds of jelly beans cost $6.30, how much would 2 pounds cost?

9. For the first 4 home football games, the concession stand sold a total of 600 hotdogs. If that ratio stays constant, how many hotdogs will sell for all 10 home games?

8.9 Proportional Reasoning

Proportional reasoning can be used when a selected number of individuals are tagged in a population in order to estimate the total population.

Example 9: A team of scientists capture, tag, and release 50 deer in a particular national forest. One week later, they capture another 50 deer, and 2 of the deer are ones that were tagged previously. What is the approximate deer population in the national forest?

Solution: Use proportional reasoning to determine the total deer population. You know that 50 deer out of the total deer population in the forest were tagged. You also know that 2 out of those 50 were recaptured. These two ratios should be equal because they both represent a fraction of the total deer population.

$$\frac{50 \text{ deer tagged}}{x \text{ deer tagged}} = \frac{2 \text{ deer tagged}}{50 \text{ deer tagged}}$$

$$2x = 2,500$$

$$x = 1,250 \text{ total deer}$$

Use proportional reasoning to solve the following problems.

1. Dr. Wolf, a biologist, captures 20 fish out of a small lake behind his college. He fastens a marker onto each of these and throws them back into the lake. A week later, he again captures 20 fish. Of these, 2 have markers. How many fish could Dr. Wolf estimate are in the pond?

2. Tawanda drew 20 cards from a box. She marked each one, returned them to the box, and shook the box vigorously. She then drew 20 more cards and found that 5 of them were marked. Estimate how many cards were in the box.

3. Maureen pulls 100 pennies out of her money jar, which contains only pennies. She marks each of these, puts them back in the bank, shakes vigorously, and again pulls 100 pennies. She discovers that 2 of them are marked. Estimate how many pennies are in her money jar.

4. Mr. Kizer has a ten-acre wooded lot. He catches 20 squirrels, tags them, and releases them. Several days later, he catches another 20 squirrels. One of these squirrels has a tag. Estimate the number of squirrels living on Mr. Kizer's ten acres. Assume the squirrels just stay on his property.

8.10 Direct and Indirect Variation

The graphs shown below represent functions where x varies with y directly or indirectly. In direct variation, when y increases, x increases, and when y decreases, x decreases. In indirect variation, also called inverse variation, when y increases, x decreases, and when y decreases, x increases.

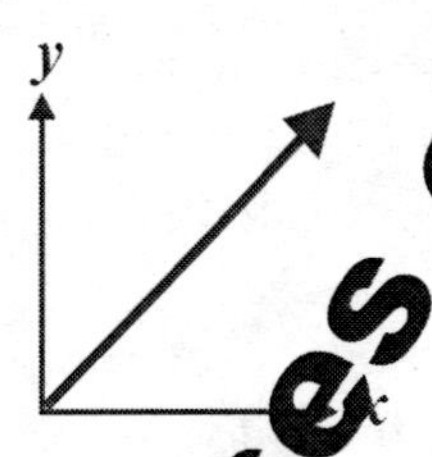

Direct Variation

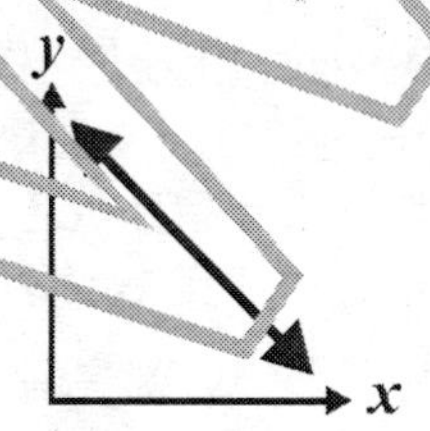

Indirect Variation

Example 10: Direct and indirect variation can be demonstrated with function tables.

Table 1

x	y
0	3
1	4
2	7
3	12
4	19

Table 2

x	y
0	20
1	18
2	16
3	14
4	12

Notice in Table 1, as x increases, y increases also. This means that function Table 1 represents a direct variation between x and y. On the other hand, Table 2 shows a decrease in y when x increases. This means that function Table 2 represents an indirect variation between x and y.

Direct variation occurs in a function when y varies directly, or in the same way, as x varies. The two values vary by a proportional factor, k. The variation is treated just like a proportion.

Example 11: If y varies directly with x, and $y = 18$ when $x = 12$, what is the value of y when $x = 6$?

Step 1: Set up the values in a proportion like you did in the previous two sections. Be sure to put the correct corresponding x and y values on the same line within the fractions.

$$\frac{x \text{ values}}{} \qquad \frac{y \text{ values}}{}$$

$$\frac{12}{6} = \frac{18}{y}$$

Step 2: Solve for y by multiplying the diagonals together and setting them equal to one another.

$12 \times y = 6 \times 18$

$\frac{12y}{12} = \frac{108}{12}$ Divide both sides by 12.

$y = 9$

For an **indirect variation**, y varies inversely with, or opposite of, x. With indirect variation, when x increases, y decreases, and when x decreases, y increases.

Example 12: In a function, y varies inversely as x varies. If $y = 18$ when $x = 12$, what is the value of y when $x = 6$?

Step 1: Set up the problem as a regular proportion problem, like Example 5.

$$\frac{x \text{ values}}{} \quad \frac{y \text{ values}}{}$$
$$\frac{12}{6} = \frac{18}{y}$$

Step 2: Now switch the numerator and denominator of the y values. This allows the x and y values to vary indirectly with each other. (Turn the fraction upside down.)

$$\frac{x \text{ values}}{} \quad \frac{y \text{ values}}{}$$
$$\frac{12}{6} = \frac{y}{18}$$

Step 3: Solve for y by multiplying the diagonals together.

$12 \times 18 = 6 \times y$

$216 = 6y$ Divide both sides by 6.

$36 = y$

Note: In an indirect variation problem, the reciprocal may be used for either side of the equation. In this case the x values or the y values could be switched to get the same value for y.

Example 13: It takes 45 minutes for 2 copiers to finish a printing job. If 5 copiers work together to print a job, how long would it take to finish?

Step 1: It will take less time to finish a job if more copiers work together. As the number of copiers increases, the number of minutes to complete the job decreases. Therefore, this is an indirect variation problem.

Step 2: Let y represent the number of minutes to complete the job. Let x represent the number of copiers. The old values are $y = 45$ minutes and $x = 2$ copiers, and the new value of x is 5 copiers. We are looking for the new y value.

Step 3: Set up the problem as a regular proportion problem.

$$\frac{x \text{ values}}{} \quad \frac{y \text{ values}}{}$$
$$\frac{2}{5} = \frac{45}{y}$$

Step 4: Now switch the numerator and denominator of the y values. This allows the x and y values to vary indirectly with each other. (Turn the fraction upside down.)

$$\frac{x \text{ values}}{} \quad \frac{y \text{ values}}{}$$
$$\frac{2}{5} = \frac{y}{45}$$

x values represent number of copiers
y values represent number of minutes

Step 5: Solve for y by multiplying the diagonals together.

$2 \times 45 = 5 \times y$

$90 = 5y$ Divide both sides by 5.

$18 = y$

It will take 5 copiers only 18 minutes to complete the printing job.

Solve these direct variation problems.

1. If $y = 6$ and $x = 3$, what is the value of y when $x = 5$?
2. If $y = 10$ and $x = 5$, what is the value of y when $x = 4$?
3. If $y = 6$ and $x = 2$, what is the value of y when $x = 7$?
4. If $y = 8$ and $x = 4$, what is the value of y when $x = 6$?
5. If $y = 15$ and $x = 3$, what is the value of y when $x = 5$?

Solve these indirect variation problems.

6. If $y = 6$ and $x = 4$, what is the value of y when $x = 8$?
7. If $y = 12$ and $x = 6$, what is the value of y when $x = 8$?
8. If $y = 9$ and $x = 6$, what is the value of y when $x = 3$?
9. If $y = 6$ and $x = 5$, what is the value of y when $x = 3$?
10. If $y = 3$ and $x = 12$, what is the value of y when $x = 9$?

Solve the following indirect word problems.

11. It takes an average person 60 minutes to type 8 pages on the computer. If three average typists work together to type up an 8-page paper, how long will it take them?

12. Sandra has $40 saved from her allowance this month to rent movies and buy books. If she buys 6 books, she will only be able to rent 2 movies. How many movies will she be able to rent if she only buys 4 books?

Solve the following direct and indirect word problems.

13. At the local grocery store, 2 pineapples cost $2.78. How much do 5 pineapples cost?

14. When Samuel rides his bike at a speed of 22 mph, it takes him 30 minutes to get home from school. Today he needs to be home in 25 minutes. How fast must he ride his bike to get home in time?

15. Jim must help his father carry bags of soil to the backyard. There are 45 bags of soil. Normally this would take Jim 30 minutes to do by himself, but today three of his friends stopped by and offered to help. How long will it take the four boys to carry all of those bags to the backyard?

16. It normally takes Jessica 45 minutes to get to her friend's house 20 miles away. Tomorrow she is meeting her friend at the mall, which is 28 miles away from her house. If she travels at the same rate she normally travels, how long will it take her to get to the mall?

Chapter 8 Review

Solve the following proportions and ratios.

1. $\frac{8}{x} = \frac{1}{2}$ 2. $\frac{2}{5} = \frac{x}{10}$ 3. $\frac{x}{6} = \frac{3}{[illegible]}$ 4. $\frac{4}{9} = \frac{8}{x}$

5. Out of 100 coins, 45 are in mint condition. What is the ratio of mint condition coins to the total number of coins?

6. The ratio of boys to girls in the ninth grade is 6 : 5. If there are 135 girls in the class, how many boys are there?

7. Twenty out of the total 2[illegible]5 seniors graduate with honors. What is the ratio of seniors graduating with honors to the total number of seniors?

8. Aunt Bess uses 3 cups of oatmeal to bake 6-dozen oatmeal cookies. How many cups of oatmeal would she need to bake 15-dozen cookies?

9. On a map, 2 centimeters represents 150 kilometers. If a line between two cities measures 5 centimeters, how many kilometers apart are they?

10. When Rick measures the shadow of a yard stick, it is 5 inches. At the same time, the shadow of the tree he would like to chop down is 45 inches. How tall is the tree in yards?

11. Jamal wonders how many ants are in his ant farm. He puts a stick in the container, and when he pulls it out, there are 15 ants on it. He gently sprays these ants with a mixture of water and green food coloring, then puts them back into the container. The next day his stick draws [illegible] ants, 1 of which is green. Estimate how many ants Jamal has.

12. The animal keeper feeds Mischief, the monkey, 5 pounds of bananas per day. The gorilla eats 4 times as many bananas as the monkey. How many pounds of bananas does the animal keeper need to feed both animals for a week?

13. Jonathan can assemble 47 widgets per hour. How many can he assemble in an [illegible]-hour day?

14. Jacob drove 252 miles, and his average speed was 42 miles per hour. How many hours did he drive?

15. The Jones family traveled 300 miles in 5 hours. What was their average speed?

16. Alisha climbed a mountain that was 4,760 feet high in 14 hours. What was her average speed per hour?

17. Last year Rikki sang 960 songs with his Latin rock band. How many songs did he sing per month?

18. Connie drove for 2 hours at a constant speed of 55 miles per hour. How many total miles did she travel?

Chapter 9
Polynomials

Polynomials are algebraic expressions which include **monomials** containing one term, **binomials** which contain two terms, and **trinomials**, which contain three terms. Expressions with more than three terms are called **polynomials. Terms** are separated by plus and minus signs.

EXAMPLES

Monomials	Binomials	Trinomials	Polynomials
$4f$	$4t + 9$	$x^2 + 2x + 3$	$x^3 - 3x^2 + 3x - 9$
$3x^3$	$9 - 7g$	$5x^2 - 6x - 1$	$p^4 - 2p^3 + p^2 - 5 + p9$
$4g^2$	$5x^2 + 7x$	$y^4 + 15y^2 + 100$	
2	$6x^3 - 8x$		

9.1 Adding and Subtracting Monomials

Two **monomials** are added or subtracted as long as the **variable and it's exponent** are the **same**. This is called combining like terms. Use the same rules you used for adding and subtracting integers

Example 1: $4x + 5x = 9x$ $\quad \begin{array}{r} 3x^4 \\ -8x^4 \\ \hline -5x^4 \end{array}$ $\quad 2x^2 - 9x^2 = -7x^2$ $\quad \begin{array}{r} 5y \\ +2y \\ \hline 7y \end{array}$ $\quad 6y^3 - 5y^3 = y^3$

Remember: When the integer in front of the variable is "1", it is usually not written. $1x^2$ is the same as x^2, and $-1x$ is the same as $-x$.

Add or subtract the following monomials.

1. $2x^2 + 5x^2 =$
2. $5t + 8t =$
3. $9y^3 - 2y^3 =$
4. $6g - 8g =$
5. $7y^2 + 8y^2 =$
6. $s^5 + s^5 =$
7. $-2x - 4x =$
8. $4w^2 - w^2 =$
9. $z^4 + 9z^4 =$
10. $-k + 2k =$
11. $3x^2 - 5x^2 =$
12. $9t + 2t =$
13. $-7v^3 + 10v^3 =$
14. $-2x^3 + x^3 =$
15. $2y^4 - 5y^4 =$

16. $\begin{array}{r} y^4 \\ +2y^4 \\ \hline \end{array}$

17. $\begin{array}{r} 4x^3 \\ -9x^3 \\ \hline \end{array}$

18. $\begin{array}{r} 8t^2 \\ +7t^2 \\ \hline \end{array}$

19. $\begin{array}{r} -2y \\ -4y \\ \hline \end{array}$

20. $\begin{array}{r} 5w^2 \\ +8w^2 \\ \hline \end{array}$

21. $\begin{array}{r} 11t^3 \\ -4t^3 \\ \hline \end{array}$

22. $\begin{array}{r} -5z \\ +9z \\ \hline \end{array}$

23. $\begin{array}{r} 4w^5 \\ +w^5 \\ \hline \end{array}$

24. $\begin{array}{r} 7t^3 \\ -6t^3 \\ \hline \end{array}$

25. $\begin{array}{r} 3x \\ +8x \\ \hline \end{array}$

9.2 Adding Polynomials

When adding **polynomials,** make sure the exponents and variables are the same on the terms you are combining. The easiest way is to put the terms in columns with **like exponents** under each other. Each column is added as a separate problem. Fill in the blank spots with zeros if it helps you keep the columns straight. You never carry to the next column when adding polynomials.

Example 2: Add $3x^2 + 14$ and $5x^2 + 2x$

$$\begin{array}{r} 3x^2 + 0x + 14 \\ (+)\ 5x^2 + 2x + 0 \\ \hline 8x^2 + 2x + 14 \end{array}$$

Example 3: $(4x^3 - 2x) + (-x^3 - 4)$

$$\begin{array}{r} 4x^3 - 2x + 0 \\ (+) - x^3 + 0x - 4 \\ \hline 3x^3 - 2x - 4 \end{array}$$

Add the following polynomials.

1. $y^2 + 3y + 2$ and $2y^2 + 4$
2. $(5y^2 + 4y - [illegible]) + (2y^2 - 5y + 8)$
3. $5x^3 - 2x^2 + 4x - 1$ and $3x^2 - x + 2$
4. $[illegible]p + 4$ and $5p^2 - 2p + 2$
5. $(w - 2) + (w^2 + 2)$
6. $4t^2 - 5t - 7$ and $8t + 2$
7. $t^2 + t + 8$ and $2t^2 + 4t - 4$
8. $(3s^3 + [illegible] - 2) + (-2s^3 + 4)$
9. $(-v^2 + 7v - 8) + (4v^3 - 6v + 4)$
10. $6m^2 - 2m + 10$ and $m^2 - m - 8$
11. $[illegible]x + 4$ and $3x^2 + x - 2$
12. $(8t^2 + 3t) + (-7t^2 - t + 4)$
13. $(3p^4 + 2p^2 - 1) + (-5p^2 - p + 8)$
14. $12s^3 + 9s^2 + 2s$ and $s^3 + s^2 + s$
15. $(-9b^2 + 7b + 2) + (-b^2 + 6b + 9)$
16. $15c^2 - 11c + 5$ and $-7c^2 + 3c - 9$
17. $5c^3 + 2c^2 + 3$ and $2c^3 + 4c^2 + 1$
18. $-14x^3 + 3x^2 + 15$ and $7x^3 - 12$
19. $(-x^2 + 2x - 4) + (3x^2 - 3)$
20. $(y^2 - 11y + 10) + (-13y^2 + [illegible] - 4)$
21. $3d^5 - 4d^3 + 7$ and $2d^4 [illegible] - 2$
22. $(6t^5 - t^3 + 17) + (4[illegible] + 7t^3)$
23. $4p^2 - 8p + 9$ and $[illegible]p^2 - 3p - 5$
24. $20b^3 + 15b$ and $-4b^2 - 5b + 14$
25. $(-2w + 11) + (w^3 + w - 4)$
26. $(25[illegible]^2 + 13z + 8) + (z^2 - 2z - 10)$

9.3 Subtracting Polynomials

When you subtract polynomials, it is important to remember to change all the signs in the subtracted polynomial (the subtrahend) and then add.

Example 4: $(4y^2 + 8y + 9) - (2y^2 + 6y - 4)$

Step 1: Copy the subtraction problem into vertical form.

$$\begin{array}{r} 4y^2 + 8y + 9 \\ (-)\ 2y^2 + 6y - 4 \\ \hline \end{array}$$

Make sure you line up the terms with like exponents under each other.

Step 2: Change the subtraction sign to addition and all the signs of the subtracted polynomial to the opposite sign.

$$\begin{array}{r} 4y^2 + 8y + 9 \\ (+)\ -2y^2 - 6y + 4 \\ \hline \end{array}$$

Subtract the following polynomials.

1. $(2x^2 + 5x$ [illegible]$) - (x^2 + 3x + 1)$
2. $(8y -$ [illegible]$) - (4y + 3)$
3. $(11t^3 - 4t^2 + 3) - (-t^3 + 4t^2 - 5)$
4. $(-3w^2 + 9w - 5) - (-5w^2 - 5)$
5. $(6a^5 - a^3 + a) - (7a$[illegible]$+ a^2 - 3a)$
6. $(14c^4 + 20c^2 + 10) - (7c^4 + 5c^2 + 12)$
7. $(5x^2 - 9x) - (-7x^2 + 4x + 8)$
8. $(12y^3 - 8y^2 - 10) - (3y^3 + y + 9)$
9. $(-3h^2 - 7h + 7) - (5h^2 + 4h + 10)$
10. $(10k^3 - 8) - (-4k^3 + k^2 + 5)$
11. $(x^2 - 5x + 9) - (6x^2 - 5x + 7)$
12. $(12p^2 + 4p) - (9p - 2)$
13. $(-2m - 8) - (6m + 2)$
14. $(13y^3 + 2y^2 - 8y) - (2y^3 + 4y^2 - 7y)$
15. $(7g + 3) - (g^2 + 4g -$ [illegible]$)$
16. $(-8w^3 + 4w) - (-1$[illegible]$w^3 - 4w^2 - w)$
17. $(12x^3 + x^2 - 10) - (3x^3 + 2x^2 +$ [illegible]$)$
18. $(2a^2 + 2a + 2) - (-a^2 + 3a +$ [illegible]$)$
19. $(c + 19) - (3c^2 - 7c + 2)$
20. $(-6v^2 + 12v) - (3v^2 + 2v + 6)$
21. $(4b^3 - 3b^2 + 5) -$ [illegible]$b^3 - 8)$
22. $(15x^3 + 5x^2$ [illegible]$) - (4x^3 - 4x^2)$
23. $(8y^2 -$ [illegible]$) - (11y^2 - 2y - 3)$
24. $(-$[illegible]$5z - 8) - (3z^2 - 5z + 5)$

9.4 Multiplying Monomials

When two monomials have the **same variable**, you can multiply them. Multiply the coefficients together. Then add the **exponents** of the variables together. If the variable has no exponent, it is understood that the exponent is 1.

Example 5: $4x^4 \times 3x^2 = 12x^6$ $2y \times 5y^2 = 10y^3$

Multiply the following monomials.

1. $6a \times 9a^5$
2. $2x^6 \times 5x^3$
3. $4y^3 \times 3y^2$
4. $10t^2 \times 2$[illegible]
5. $2p^5 \times 4p^2$
6. [illegible]$b^2 \times 8b$
7. $3c^3 \times 3c$[illegible]
8. $2d^8 \times 9d^2$
9. $6k^3 \times 5k^2$
10. $7m^5 \times m$
11. $11z \times 2z$[illegible]
12. $3w^4 \times 6w^5$
13. $4x^4 \times 5x^3$
14. $5n^2 \times 3n^3$
15. $8w^7 \times w$
16. $10s^6 \times 5s^3$
17. $4d^5 \times 4d$[illegible]
18. $5y^2 \times 8y^3$
19. $7t^{10} \times 3t^5$
20. $6p^8 \times 2p^3$
21. [illegible]$^3 \times 2x^3$

When problems include negative signs, follow the rules for multiplying integers.

22. $-7s^4 \times 5s^3$
23. $-6a \times -9a^5$
24. $4x \times -x$
25. $-3y^2 \times -y^3$
26. $-5b^2 \times 3b^5$
27. $9c^4 \times -2c$
28. $-4t^3 \times 8t^3$
29. $10d \times -8d^7$
30. $-3g^6 \times -2g^3$
31. $-7s$[illegible] $\times 7s^3$
32. $-d^3 \times -2d$
33. $11p \times -2p$[illegible]
34. $-5x^7 \times -3x^3$
35. $8z^4 \times 7z^4$
36. $-4w \times$ [illegible]
37. $-5y^4 \times$ [illegible]y^2
38. 9[illegible] $\times -7x^5$
39. [illegible]$a^4 \times -a$
40. $-7k^2 \times 3k$
41. $-15t^2 \times -t^4$
42. $3x^8 \times 9x^2$

9.5 Multiplying Monomials by Polynomials

In Chapter 7, you learned to remove parentheses by multiplying the number outside the parentheses by each term inside the parentheses: $2(4x - 7) = 8x - 14$. Multiplying monomials by polynomials works the same way.

Example 6: $-5t\,(2t^2 - 7t + 9)$

Step 1: Multiply $-5t \times 2t^2 = \mathbf{-10t^3}$

Step 2: Multiply $-5t \times -7t = \mathbf{35t^2}$

Step 3: Multiply $-5t \times 9 = \mathbf{-45t}$

Step 4: Arrange the answers horizontally in order: $\mathbf{-10t^3 + 35t^2 - 45t}$

Remove parentheses in the following problems.

1. $3x\,([illegible]^2 + 4x - 1)$
2. [illegible] $(y^3 - 7)$
3. $7a^2\,(2a^2 + [illegible] + 2)$
4. $-5[illegible]\,(d^2 - [illegible]d)$
5. $2[illegible]\,(-4w^2 + 3[illegible] - 8)$
6. $8p\,([illegible]^3 - 6[illegible] + 5)$
7. $-9[illegible]^2\,(-2b + [illegible])$
8. $2[illegible]\,(t^2 - 4t - 10)$
9. $10c\,([illegible]c^2 + 3c - 7)$
10. $6z\,(2z^4 - 5z^2 - 4)$
11. [illegible]$t^2\,(3t^2 + 5t + 6)$
12. $c\,(-3c - 5)$
13. [illegible]$p\,(p^3 - p^2 - 9)$
14. $-k^2\,(2k + 4)$
15. $-3\,(4m^2 - 5m + 8)$
16. $6x\,(-7x^3 + 10)$
17. [illegible]$w\,(w^2 - 4w + [illegible])$
18. $2y\,(5y^2 - [illegible])$
19. $3d\,(d^5 - 7d^3 + 4)$
20. $-5t\,(-4t^2 - 8t + 1)$
21. [illegible] $(2w^2 - 9w + 4)$
22. $3[illegible]^2\,([illegible] - 11)$
23. [illegible]$^2\,(v^2 + 3v + 3)$
24. [illegible]$x\,(2x^3 + 3x + [illegible])$
25. $-5d\,(4d^2 [illegible] - 2)$
26. $-k^2\,(-3k + 6)$
27. $3x\,(-x^2 - 5x + 5)$
28. $4z\,(4z^4 - z - 7)$
29. $-5y\,(9y^3 - 3)$
30. $2b^2\,(7b^2 + 4b + 4)$

9.6 Dividing Polynomials by Monomials

Example 7: $\dfrac{-8wx + 6x^2 - 16wx^2}{2wx}$

Step 1: Rewrite the problem. Divide each term from the top by the denominator, $2wx$.

$$\frac{-8wx}{2wx} + \frac{6x^2}{2wx} + \frac{-16wx^2}{2wx}$$

Step 2: Simplify each term in the problem. Then combine like terms.

$$-4 + \frac{3x}{w} - 8x$$

Simplify each of the following.

1. $\dfrac{bc^2 - 8bc - [illegible]c^2}{2bc}$

2. $\dfrac{3jk^2 + [illegible]jk + 9j^2k}{3jk}$

3. $\dfrac{4x^2y - 8xy^2 + 2y^3}{2xy}$

4. $\dfrac{16st^2 + st - 12s}{4st}$

5. $\dfrac{4wx^2 + 6wx - 12w^3}{2wx}$

6. $\dfrac{cd^2 + 10cd^3 + 16c^2}{2cd}$

7. $\dfrac{y^2z^3 - 2yz - 8z^2}{-2yz^2}$

8. $\dfrac{a^2b + 2ab^2 - 14ab^3}{2a^2}$

9. $\dfrac{pr^2 + 6pr + 8p^2r^2}{2pr^2}$

10. $\dfrac{6xy^2 - 3xy + 18x^2}{-3xy}$

11. $\dfrac{6x^2y + 12xy - 24y^2}{6xy}$

12. $\dfrac{5m^2n - 10mn - 25n^2}{5mn}$

13. $\dfrac{st^2 - 10st - 16s^2t^2}{2st}$

14. $\dfrac{7jk^2 - 14jk - [illegible]}{7jk}$

9.7 Removing Parentheses and Simplifying

In the following problem, you must multiply each set of parentheses by the numbers and variables outside the parentheses, and then add the polynomials to simplify the expressions.

Example 8: $8x\left(2x^2-5x+7\right)-3x\left(4x^2+3x-8\right)$

Step 1: Multiply to remove the first set of parentheses.

$8x\left(2x^2-5x+7\right)=16x^3-40x^2+56x$

Step 2: Multiply to remove the second set of parentheses.

$-3x\left(4x^2+3x-8\right)=-12x^3-9x^2+24x$

Step 3: Copy each polynomial in columns, making sure the terms with the same variable and exponent are under each other. Add to simplify.

$$\begin{array}{r} 16x^3-40x^2+56x \\ (+)\ -12x^3-9x^2+24x \\ \hline 4x^3-49x^2+80x \end{array}$$

Remove the parentheses and simplify the following problems.

1. $4t\left(t+7\right)+5t\left(2t^2-4t+1\right)$
2. $-5y\left(3y^2-5y+3\right)-6y\left(y^2-4y-4\right)$
3. $-3\left(3x^2+4x\right)+5x\left(x^2+3x+2\right)$
4. $2b\left(5b^2-8b-1\right)-3b\left(4b+3\right)$
5. $8d^2\left(3d+4\right)-7d\left(3d^2+4d+5\right)$
6. $5a\left(3a^2+3a+1\right)-\left(-2a^2+5a-\right.$ [illegible]
7. $3m\left(m+7\right)+8\left(4m^2+m+4\right)$
8. $-4c^2\left(-6c^2-3c+2\right)-7c\left(5c^3+2c\right)$
9. $-8w\left(-w+1\right)-4w\left(3w-5\right)$
10. $6p\left(2p^2-4p-6\right)+3p\left(p^2+6p+9\right)$

9.8 Multiplying Two Binomials

When you multiply two binomials such as $(x+6)(x-5)$, you must multiply each term in the first binomial by each term in the second binomial. The easiest way is to use the **FOIL** method. If you can remember the word **FOIL**, it can help you keep order when you multiply. The "**F**" stands for **first**, "**O**" stands for **outside**, "**I**" stands for **inside**, and "**L**" stands for **last**.

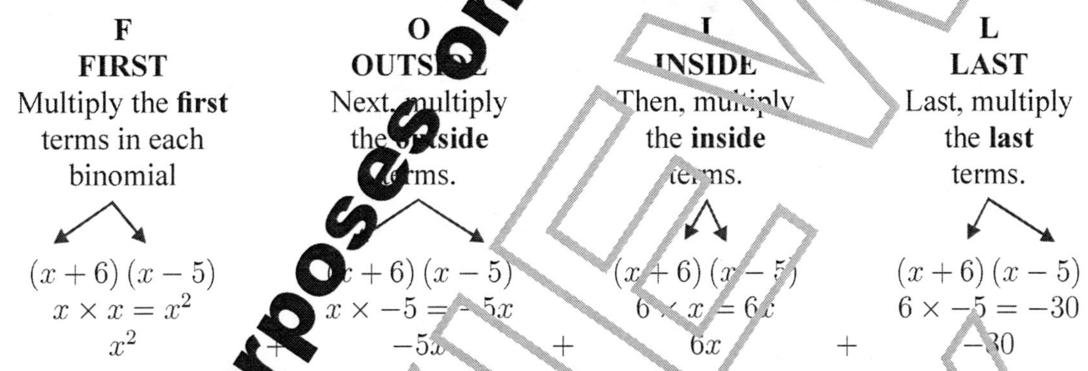

F
FIRST
Multiply the **first** terms in each binomial
O
OUTSIDE
Next, multiply the **outside** terms.
I
INSIDE
Then, multiply the **inside** terms.
L
LAST
Last, multiply the **last** terms.
$(x+6)(x-5)$
$x \times x = x^2$
x^2
$(x+6)(x-5)$
$x \times -5 = -5x$
$-5x$
+
$(x+6)(x-5)$
$6 \times x = 6x$
$6x$
+
$(x+6)(x-5)$
$6 \times -5 = -30$
-30

Now just combine like terms, $6x - 5x = x$ and write your answer.
$(x+6)(x-5) = x^2 + x - 30$.

Note: It is customary for mathematicians to write polynomials in descending order. That means that the term with the highest-number exponent comes first in a polynomial. The next highest exponent is second, and so on. When you use the **FOIL** method, the terms will always be in the customary order. You just need to combine like terms and write your answer.

Multiply the following binomials.

1. $(y-7)(y+3)$
2. $(2x+4)(x+9)$
3. $(4b-3)(3b-4)$
4. $(6g+2)(g-9)$
5. $(7k-5)(-4k-3)$
6. $(8v-2)(3v+4)$
7. $(10p+2)(4p+3)$
8. $(3h-9)(-2h+5)$
9. $(w-4)(w-7)$
10. $(5x+1)(x-2)$
11. $(5t+3)(2t-1)$
12. $(4y-9)(4y+9)$
13. $(a+6)(3a+5)$
14. $(3z-8)(z-4)$
15. $(5c+2)(6c+5)$
16. $(y+3)(y-3)$
17. $(2w-5)(4w+6)$
18. $(7x+1)(x+4)$
19. $(6t-9)(4t-4)$
20. $(5b+6)(6b+2)$
21. $(2z+1)(10z+4)$
22. $(11w-8)(w+3)$
23. $(5d-9)(9d+9)$
24. $(9g+2)(g-2)$
25. $(4p+7)(2p+3)$
26. $(m+5)(m-5)$
27. $(8b-8)(2b-1)$
28. $(z+3)(3z+5)$
29. $(7y-5)(y-3)$
30. $(9x+5)(3x-1)$
31. $(3t+1)(t+10)$
32. $(2w-9)(8w+7)$
33. $(5s-2)(5s+4)$
34. $(4k-1)(8k+9)$
35. $(h+2)(h-2)$
36. $(3x+7)(7x+3)$
37. $(2v-6)(2v+6)$
38. $(2x+8)(2x-3)$
39. $(k-1)(6k+12)$
40. $(3w+11)(2w+2)$
41. $(8y-10)(5y-3)$
42. $(6d+13)(d-1)$

9.9 Simplifying Expressions with Exponents

Example 9: **Simplify** $(2a+5)^2$

When you simplify an expression such as $(2a+5)^2$, write the expression as two binomials and use FOIL to simplify.

$(2a+5)^2 = (2a+5)(2a+5)$

Using FOIL we have $4a^2 + 10a + 10a + 25 = 4a^2 + 20a + 25$

Example 10: **Simplify** $4(3a+2)^2$

Using order of operations, we must simplify the exponent first.

$4(3a+2)^2$

$4(3a+2)(3a+2)$

$4(9a^2+6a+6a+4)$

$4(9a^2+12a+4)$ Now multiply by 4.

$4(9a^2+12a+4) = 36a^2+48a+16$

Multiply the following binomials.

1. $(y+3)^2$
2. $2(2x+4)^2$
3. $5(4b-3)^2$
4. $5(6g+2)^2$
5. $(-4k-3)^2$
6. $3(-2h-5)^2$
7. $-2(8v-2)^2$
8. $(10p+2)^2$
9. $6(-2h-5)^2$
10. $6(w-7)^2$
11. $2(6x+1)^2$
12. $(9x+2)^2$
13. $(5t+3)^2$
14. $3(4y-9)^2$
15. $8(a+6)^2$
16. $4(3z-8)^2$
17. $3(5c+2)^2$
18. $4(3x+9)^2$

Chapter 9 Review

Simplify.

1. $3a^2 + 9a^2$
2. $(7x^2y^4)(9xy^5)$
3. $-6z^2(z+3)$
4. $(4b^2)(5b^3)$
5. $7x^2 - 9x^2$
6. $(5p - 4) - (3p + 2)$
7. $-6t(3t+9)^2$
8. $(3w^3y^2)(4wy^5)$
9. $3(2y+3)^2$
10. $14d^4 - 9d^4$
11. $(7w + 4)(w - 8)$
12. $(9x+2)(x+5)$
13. $4y(4y^2 - 9y + 2)$
14. $(8a^4b)(2ab^3)(ab)$
15. $(5w^6)(9w^9)$
16. $8w^3 + 12w^3$
17. $15p^5 - 11p^5$
18. $(3s^4t^2)(4st^3)$
19. $(4d+9)(2d+7)$
20. $4w(-3w^2 + 7w - 5)$
21. $24z^6 - 10z^6$
22. $-7y^3 - 8y^3$
23. $(a^2v)(2av)(a^3v^6)$
24. $4(6y-5)^2$
25. $(4x^5y^3)(2xy^3)$
26. $24z^6 - 10z^6$
27. $(3p^3 - 1)(p+5)$
28. $2b(b-4) - (b^2 + 2b + 1)$
29. $(6k^2 + 5k) - (k^2 + k + 9)$
30. $(q^2r^3)(3qr^2)(2q^4r)$

Chapter 10
Factoring

In a multiplication problem, the numbers multiplied together are called **factors**. The answer to a multiplication problem is a called the **product**.

In the multiplication problem $5 \times 4 = 20$, 5 and 4 are factors and 20 is the product.

If we reverse the problem, $20 = 5 \times 4$, we say we have **factored** 20 into 5×4.

In this chapter, we will factor **polynomials**.

Example 1: Find the greatest common factor of $2y^3 + 6y^2$.

Step 1: Look at the whole numbers. The greatest common factor of 2 and 6 is 2. Factor the 2 out of each term.

$2\left(y^3 + 3y^2\right)$

Step 2: Look at the remaining terms, $y^3 + 3y^2$. What are the common factors of each term?

$$\begin{array}{ccccccc} y^3 & = & y & \times & \boxed{y \times y} \\ 3y^2 & = & 3 & \times & \boxed{y \times y} \end{array} \longleftarrow \text{common factor} = y^2$$

Step 2: Factor 2 and y^2 out of each term: $2y^2\left(y + 3\right)$

Check: $2y^2\left(y + 3\right) = 2y^3 + 6y^2$

Find the greatest common factor of each of the following.

1. $6x^4 + 18x^{[illegible]}$
2. $14y^3 + 7y$
3. $4b^5 + 12b^3$
4. $10a^{[illegible]} + 5$
5. $2y^3 + 8y^2$
6. $6x^4 - 12x^2$
7. $18y^2 - 12y$
8. $15a^3 - 25a^{[illegible]}$
9. $4x^3 + 16x^2$
10. $6b^2 + 21b^5$
11. $27m^{[illegible]} + 18m^4$
12. $100x^{[illegible]} - 25x^3$
13. $4b^4 + 12b^3$
14. $18c^2 + 2[illegible]c$
15. $20y^3 + 30y^5$
16. $16x^2 - 24x^5$
17. $15a^4 - 25a^2$
18. $24b^3 + 16b^6$
19. $36y^4 + 9y^2$
20. $42x^3 + 49x$

Factoring larger polynomials with 3 or 4 terms works the same way.

Example 2: $4x^5 + 16x^4 + 12x^3 + 8x^2$

Step 1: Find the greatest common factor of the whole numbers. 4 can be divided evenly into 4, 16, 12, and 8; therefore, 4 is the greatest common factor.

Step 2: Find the greatest common factor of the variables. x^5, x^4, x^3, and x^2 can be divided by x^2, the lowest power of x in each term.

$$4x^5 + 16x^4 + 12x^3 + 8x^2 = 4x^2\left(x^3 + 4x^2 + 3x + 2\right)$$

Factor each of the following polynomials.

1. $5a^3 + 15a^2 + 20a$
2. $18y^4 + 6y^3$ [illegible] $24y^2$
3. $12x^5 +$ [illegible] $x^3 + x^2$
4. [illegible] b^4 [illegible] $3b^3 + 15b^2$
5. $14c^3 + 28c^2 + 7c$
6. $15b^4 - 5b^2 + 20b$
7. [illegible] $+ 3t^2 -$ [illegible] t
8. $8a^3 - 4a^2 + 12a$
9. $16b^5 - 12b^4$ [illegible] $0b^2$
10. $20x^4 + 16x^3 - 24x^2 + 28x$
11. $40b^7 + 30b^5 - 50b^3$
12. $20y^4 - 15y^3 + 30y^2$
13. $4m^5 + 8m^4 + 12m^3 + 6m^2$
14. $16x^5 + 20x^4 - 12x^3 + 24x^2$
15. $18y^4 + 21y^3 - 9y^2$
16. $3n^5 + 9n^3 + 12n^2 + 15n$
17. $4d^6 - 8d^2 + 2d$
18. $10w^2 + 4w + 2$
19. $6t^3 - 3t^2 + 9t$
20. $25p^5 - 10p^3 - 5p^2$
21. $18x^4 + 9x^2 - 36x$
22. $6b^4 - 12b^2 - 6b$
23. $y^3 + 3y^2 - 9y$
24. [illegible] $- 2x^4 + 4x^2$

Example 3: Find the greatest common factor of $4a^3b^2 - 6a^2b^2 + 2a^4b^3$

Step 1: The greatest common factor of the whole numbers is 2.

$$4a^3b^2 - 6a^2b^2 + 2a^4b^3 = 2(2a^3b^2 - 3a^2b^2 + a^4b^3)$$

Step 2: Find the lowest power of each variable that is in each term. Factor them out of each term. The lowest power of a is a^2. The lowest power of b is b^2.

$$4a^3b^2 - 6a^2b^2 + 2a^4b^3 = 2a^2b^2(2a - 3 + a^2b)$$

Factor each of the following polynomials.

1. $3a^2b^2 - 6a^3b^4 + 9a^2b^3$
2. $12x^4y^3 + 18x^2y^{[illegible]} - 24x^3y^3$
3. $20x^2y$ [illegible] $5x^3y^3$
4. $12x^2y - 20x^2y^2 + 16xy^2$
5. $8a^3b + 12a^2b + 20a^2b^3$
6. $36c^4 + 42c^3 + 24c^2 - 18c$
7. $14m^3n^4 - 28m^3n^2 + 42m^2n^3$
8. $16x^4y^2 - 24x^3y^2 + 12x^2y^2 - 8xy^2$
9. $32c^3d^4 - 56c^2d^3 + 64c^3d^2$
10. $21a^4b^3 + 27a^2b^3 + 15a^3b^2$
11. $4w^3t^2 + 6w^2t - 8wt^2$
12. $5pw^3 - 2p^2q^2 - 9p^3q$
13. $49x^3t^3 + 7xt^2 - 14xt^3$
14. $9cd^4 - 3d^4 - 6c^2d^3$
15. $12a^2b^3 - 14ab + 10ab^2$
16. $25x^4 + 10x - 20x^2$
17. $bx^3 - b^2x^2 + b^3x$
18. $4k^3a^2 + 22ka + 16k^2a^2$
19. $33w^4y^2 - 9w^2y^2 + 24w^2y^2$
20. $18x^3 + 9x^5 + 27x^2$

10.1 Factor By Grouping

Not all polynomials have a common factor in each term. In this case they may sometimes be factored by grouping.

Example 4: Factor $ab + 4a + 2b + 8$

Step 1: Factor an a from the first two terms and a 2 from the last two terms.

$a\,(b + 4) + 2\,(b + 4)$

Now the polynomial has two terms, $a\,(b + 4)$ and $2\,(b + 4)$. Notice that $(b + 4)$ is a factor of each term.

Step 2: Factor out the common factor of each term:

$ab + 4a + 2b + 8 = (b + 4)\,(a + 2)$.

Check: Multiply using the FOIL method to check.

$(b + 4)\,(a + 2) = ab + 4a + 2b + 8$

Factor the following polynomials by grouping.

1. $xy + 4x + 2y + 8$
2. $cd + 5c + 4d + 20$
3. $xy - 4x + 6y - 24$
4. $ab + 6a + 3b + 18$
5. $ab + 3a - 5b - 15$
6. $xy - 2x + 6x - 12$
7. $cd + 4c + 4d + 16$
8. $mn - 5m + 3n - 15$
9. $ab + 4a + 3b + 12$
10. $xy + 7x - 4y - 28$
11. $ab - 2a + 8b - 16$
12. $cd + 4c - 5d - 20$
13. $mn + 6m - 2n - 12$
14. $xy - 9x - 3y + 27$
15. $bc - 3b + 5c - 15$
16. $ab + a + 7b + 7$
17. $xy + 4y + 2y + 8$
18. $cd + 9c - d - 9$
19. $ab + 2a - 7b - 14$
20. $xy - 6x - 2y + 12$
21. $wz + 6z - 4w - 24$

10.2 Factoring Trinomials

In the chapter on polynomials, you multiplied binomials (two terms) together, and the answer was a trinomial (three terms).

For example, $(x + 6)(x - 5) = x^2 + x - 30$

Now, you need to practice factoring a trinomial into two binomials.

Example 5: Factor $x^2 + 6x + 8$

Step 1: When the trinomial is in descending order as in the example above, you need to find a pair of numbers whose sum equals the number in the second term, while their product equals the third term. In the above example, find the pair of numbers that has a sum of 6 and a product of 8.

$\_\_\_\_ + \_\_\_\_ = 6$ and $\_\_\_\_ \times \_\_\_\_ = 8$

The pair of numbers that satisfy both equations is 4 and 2.

Step 2: Use the pair of numbers in the binomials.

The factors of $x^2 + 6x + 8$ are $(x + 4)(x + 2)$

Check: To check, use the FOIL method.
$(x + 4)(x + 2) = x^2 + 4x + 2x + 8 = x^2 + 6x + 8$

Notice, when the second term and the third term of the trinomial are both positive, both numbers in the solution are positive.

Example 6: Factor $x^2 - x - 6$ Find the pair of numbers where:

the sum is -1 and the product is -6

$\_\_\_\_ + \_\_\_\_ = -1$ and $\_\_\_\_ \times \_\_\_\_ = -6$

The pair of numbers that satisfies both equations is 2 and -3.
The factors of $x^2 - x - 6$ are $(x + 2)(x - 3)$

Notice, if the second term and the third term are negative, one number in the solution pair is positive, and the other number is negative.

Example 7: Factor $x^2 - 7x + 12$ Find the pair of numbers where:

the sum is -7 and the product is 12

$\_\_\_\_\_\_ + \_\_\_\_\_\_ = -7$ and $\_\_\_\_\_\_ \times \_\_\_\_\_\_ = 12$

The pair of numbers that satisfies both equations is -3 and -4
The factors of $x^2 - 7x + 12$ are $(x - 3)(x - 4)$.

Notice, if the second term of a trinomial is negative and the third term is positive, both numbers in the solution are negative.

Find the factors of the following trinomials.

1. $x^2 - x - 2$
2. $y^2 + y - 6$
3. $w^2 + 3w - 4$
4. $t^2 + 5t + 6$
5. $x^2 + 2x - 8$
6. $k^2 - 4k + 3$
7. $t^2 + 3t - 10$
8. $x^2 - 3x - 4$
9. $y^2 - 5y + 6$
10. $y^2 + y - 20$
11. $a^2 - a - 6$
12. $b^2 - 4b - 5$
13. $c^2 - 5c - 14$
14. $c^2 - c - 12$
15. $d^2 + d - 6$
16. $x^2 - 3x - 28$
17. $y^2 + 3y - 18$
18. $a^2 - 9a + 20$
19. $b^2 - 2b - 15$
20. $c^2 + 7c - 8$
21. $t^2 - 11t + 30$
22. $w^2 + 13w + 36$
23. $m^2 - 2m - 48$
24. $y^2 + 14y + 49$
25. $x^2 + 7x + 10$
26. $a^2 - 7a + 6$
27. $d^2 - 6d - 27$

10.3 More Factoring Trinomials

Sometimes a trinomial has a greatest common factor which must be factored out first.

Example 8: Factor $4x^2 + 8x - 32$

Step 1: Begin by factoring out the greatest common factor, 4.

$4\left(x^2 + 2x - 8\right)$

Step 2: Factor by finding a pair of numbers whose sum is 2 and product is -8. 4 and -2 will work, so

$4\left(x^2 + 2x - 8\right) = 4\left(x + 4\right)\left(x - 2\right)$

Check: Multiply to check $4\left(x + 4\right)\left(x - 2\right) = 4x^2 + 8x - 32$

Factor the following trinomials. Be sure to factor out the greatest common factor first.

1. $2x^2 + 6x + 4$
2. $3y^2 - 9y + 6$
3. $2a^2 + 2a - 12$
4. $4b^2 + 28b + 40$
5. $3y^2 - 6y - 9$
6. $10x^2 + 10x - 200$
7. $5c^2 - 10c - 40$
8. $6d^2 + 30d - 36$
9. $4x^2 + 8x - 60$
10. $6a^2 - 18a - 24$
11. $5b^2 + 40b + 75$
12. $3c^2 - 6c - 24$
13. $2x^2 - 18x + 28$
14. $4y^2 - 20y + 16$
15. $7a^2 - 7a - 42$
16. $6b^2 - 18b - 60$
17. $11d^2 + 66d + 88$
18. $3x^2 - 24x + 45$

10.4 Factoring More Trinomials

Some trinomials have a whole number in front of the first term that cannot be factored out of the trinomial. The trinomial can still be factored.

Example 9: Factor $2x^2 + 5x - 3$.

Step 1: To get a product of $2x^2$, one factor must begin with $2x$ and the other with x.

$(2x \quad)(x \quad)$

Step 2: Now think: What two numbers give a product of -3? The two possibilities are 3 and -1 or -3 and 1. We know they could be in any order so there are 4 possible arrangements.

$(2x + 3)(x - 1)$
$(2x - 3)(x + 1)$
$(2x + 1)(x - 3)$
$(2x - 1)(x + 3)$

Step 3: Multiply each possible answer until you find the arrangement of the numbers that works. Multiply the outside terms and the inside terms and add them together to see which one will equal $5x$.

$(2x + 3)(x - 1) = 2x^2 + x - 3$
$(2x - 3)(x + 1) = 2x^2 - x - 3$
$(2x + 1)(x - 3) = 2x^2 - 5 - 3$
$\boxed{(2x - 1)(x + 3) = 2x^2 + 5x - 3}$ ← This arrangement works, therefore:

The factors of $2x^2 + 5x - 3$ are $(2x - 1)(x + 3)$

Alternative: You can do some of the multiplying in your head. For the above example, ask yourself the following question: What two numbers give a product of -3 and give a sum of 5 (the whole number in the second term) when one number is first multiplied by 2 (the whole number in front of the first term)? The pair of numbers, -1 and 3, have a product of -3 and a sum of 5 when the 3 is first multiplied by 2. Therefore, the 3 will go opposite the factor with the $2x$ so that when the terms are multiplied, you get -5.

You can use this method to at least narrow down the possible pairs of numbers when you have several from which to choose.

Factor the following trinomials.

1. $3y^2 + 14y + 8$
2. $5a^2 + 24a - 5$
3. $7b^2 + 30b + 8$
4. $2c^2 - 9c + 9$
5. $2y^2 - 7$[illegible] $- 15$
6. [illegible] $+ 4x + 1$
7. $7y^2 + 13y - 2$
8. $11a^2 + 35a + 6$
9. $5y^2 + 17y - 12$
10. $3a^2 +$ [illegible] $- 7$
11. $2a^2 + 3a - 20$
12. $5b^2 - 13b - 6$
13. $3y^2 - 17$[illegible] $+ 36$
14. $2x^2 - 17x + 36$
15. $11x^2 - 29x - 12$
16. $5c^2 + 2c - 16$
17. $7y^2 - 30y + 27$
18. [illegible]$x^2 - 3x - 20$
19. $5b^2 + 24b - 5$
20. $7d^2 + 18d + 8$
21. $3x^2 - 20x + 25$
22. $2a^2 - 7a - 4$
23. $5m^2 + 12m + 4$
24. $9y^2 -$ [illegible]$y - 4$
25. $2b^2 - 13b + 18$
26. $7x^2 + 31x$ [illegible] 20
27. $3c^2$ [illegible] $2c - 21$

10.5 Factoring Trinomials with Two Variables

Some trinomials have two variables with exponents. You can still factor these trinomials.

Example 10: Factor $x^2 + 5xy + 6y^2$

Step 1: Notice there is an x^2 in the first term and a y^2 in the last term. When you see two different terms that are squared, you know there has to be an x and a y in each factor:

$(x \quad y)(x \quad y)$

Step 2: Now think: What are two numbers whose sum is 5 and product is 6? You see that 3 and 2 will work. Put 3 and 2 in the factors:

$(x + 3y)(x + 2y)$

Check: Multiply to check. $(x + 3y)(x + 2y) = x^2 + 3xy + 2xy + 6y^2 = x^2 + 5xy + 6y^2$

Factor the following trinomials.

1. $a^2 + 6ab + 8b^2$
2. $x^2 + 3xy - 4y^2$
3. $c^2 - 2cd - 15d^2$
4. $g^2 + 7gh + 10h^2$
5. $a^2 - 5ab + 6b^2$
6. $c^2 - cd - 30d^2$
7. $x^2 + 5xy - 24y^2$
8. $a^2 - 4ab + 4b^2$
9. $c^2 - 11cd + 30d^2$
10. $x^2 - 6xy + 8y^2$
11. $g^2 - gh - 42h^2$
12. $a^2 - ab - 20b^2$
13. $x^2 + 12xy + 32y^2$
14. $c^2 - 3cd - 40d^2$
15. $x^2 + 6xy - 27y^2$
16. $a^2 - 2ab - 48b^2$
17. $c^2 - 3cd - 28d^2$
18. $x^2 + xy - 6y^2$

10.6 Factoring the Difference of Two Squares

The product of a term and itself is called a **perfect square**.

> 25 is a perfect square because $5 \times 5 = 25$
> 49 is a perfect square because $7 \times 7 = 49$

Any variable with an even exponent is a perfect square.

> y^2 is a perfect square because $y \times y = y^2$
> y^4 is a perfect square because $y^2 \times y^2 = y^4$

When two terms that are both perfect squares are subtracted, factoring those terms is very easy. To factor the difference of perfect squares, you use the square root of each term, a plus sign in the first factor, and a minus sign in the second factor.

Example 11: Factor $4x^2 - 9$

This example has two terms which are both perfect squares, and the terms are subtracted.

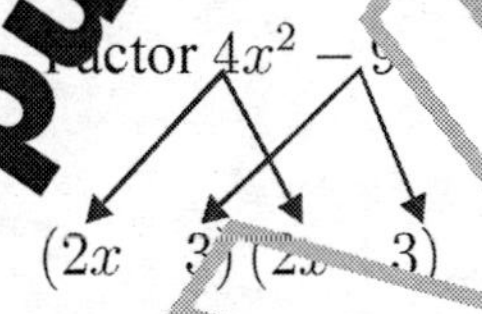

Step 1: $(2x \quad 3)(2x \quad 3)$

Find the square root of each term.
Use the square roots in each of the factors.

Step 2: $(2x + 3)(2x - 3)$

Use a plus sign in one factor and a minus sign in the other factor.

Check: Multiply to check $(2x + 3)(2x - 3) = 4x^2 - 6x + 6x - 9 = 4x^2 - 9$

The inner and outer terms add to zero.

Example 12: Factor $81y^4 - 1$

Step 1: $(9y^2 + 1)(9y^2 - 1)$

Factor like the example above.
Notice the second factor is also the difference of two perfect squares.

Step 2: $(9y^2 + 1)(3y + 1)(3y - 1)$

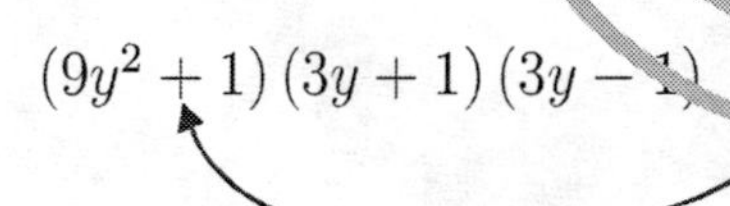

Factor the second term further.
Note: You cannot factor the sum of two perfect squares.

Check: Multiply in reverse to check your answer.
$(9y^2 + 1)(3y + 1)(3y - 1) = (9y^2 + 1)(9y^2 - 3y + 3y - 1) =$
$(9y^2 + 1)(9y^2 - 1) = 81y^4 + 9y^2 - 9y^2 - 1 = 81y^4 - 1$

Factor the following differences of perfect squares.

1. $64x^2 - 49$	11. $c^4 - 36$	21. $16x^{[illegible]} - 16$
2. $4y^4 - 25$	12. [illegible] $- 25$	22. $b^2 - 25$
3. $9a^4 - 4$	13. $25x^2 - 4$	23. $c^{[illegible]} - 144$
4. $25c^4 - 9$	14. $9x^4 - 64$	24. $9y^2 - 4$
5. $64y^2 - 9$	15. $49x^2 - 100$	25. $81x^4 - 16$
6. $x^4 - 16$	16. $16x^{[illegible]} - 81$	26. [illegible]$b^2 - 36$
7. 49[illegible]$- 4$	17. $9y^4 - 1$	27. $9w^2 - 9$
8. [illegible]$4d^2 - 25$	18. $49c^2 - 25$	28. $64c^{[illegible]} - 25$
9. $9a^2 - 16$	19. $25d^2 - 64$	29. $49y^2 - 121$
10. $100y^4 - 49$	20. $36a^4 - 49$	30. $x^6 - 9$

10.7 Simplifying Algebraic Ratios

Sometimes algebraic expressions are written as ratios. We will use what we learned so far in this chapter to factor the terms in the numerator and the denominator when possible, then simplify the ratio.

Example 13: Simplify $\dfrac{c^2 - 25}{c^2 + 5c}$

Step 1: The numerator is the difference of two perfect squares, so it can be easily factored as in the previous section. Use the square root of each of the terms in the parentheses, with a plus sign in one and a minus sign in the other.
$c^2 - 25 = (c - 5)(c + 5)$

Step 2: Find the greatest common factor in the denominator and factor it out. In this case, it is the variable c.
$c^2 + 5c = c(c + 5)$

Step 3: Simplify $\dfrac{c^2 - 25}{c^2 + 5c} = \dfrac{(c - 5)(c + 5)}{c(c + 5)} = \dfrac{c - 5}{c}$

Find the following algebraic ratios. Check for perfect squares and common factors.

1. $\dfrac{25x^2 - 4}{5x^2 - 2x}$
2. $\dfrac{64c^2 - 25}{8c^2 + 5c}$
3. $\dfrac{36a^2 - 49}{6a^2 - 7a}$
4. $\dfrac{x^2 - 9}{x^2 + 3x}$
5. $\dfrac{9a^2 - 16}{3a^2 - 4a}$
6. $\dfrac{16x^2 - 81}{4x^2 - 9x}$
7. $\dfrac{49x^2 - 100}{7x^2 + 10x}$
8. $\dfrac{x^4 - 16}{x^2 + 2x}$
9. $\dfrac{4y^2 - 36}{2y^2 + 6y}$
10. $\dfrac{81y^4 - 16}{9y^2 + 4}$
11. $\dfrac{25x^4 - 225}{5x^2 + 15}$
12. $\dfrac{3y^3 + 9}{y^9 - 9}$

Chapter 10 Review

Factor the following polynomials completely.

1. $8x - 18$

2. $6x^2 - 18x$

3. $16b^3 + 8b$

4. $15a^3 + 40$

5. $20y^6 - 12y^4$

6. $5a - 15a^2$

7. $4y^2 - 36$

8. $25a^4 - 49b^2$

9. $3ax + 3ay + 4x + 4y$

10. $ax - 2x + ay - 2y$

11. $2bx + 2x - 2by - 2y$

12. $2b^2 - 2b - 12$

13. $yx^3 + 14x - 3x^2 - 6$

14. $3a^3 + 4a^2 + 9a + 12$

15. $27y^2 + 42y - 5$

16. $12b^2 + 25b - 7$

17. $c^2 + cd - 20d^2$

18. $x^2 - 4xy - 21y^2$

19. $6y^2 + 30y + 36$

20. $2b^2 + 6b - 20$

21. $16b^4 - 81d^4$

22. $9w^2 - 54w - 63$

23. $m^2p^2 - 5mp + 2m^2p - 10m$

24. $12x^2 + 27x$

25. $2xy - 36 + 8y - 9x$

26. $2a^4 - 32$

27. $21c^2 + 41c + 10$

28. $x^2 - y + xy - x$

29. $2b^3 - 24 + 16b - 3b^2$

30. $5 - 2a - 25a^2 + 10a^3$

Chapter 11
Solving Quadratic Equations

In the previous chapter, we factored polynomials such as $y^2 - 4y - 5$ into two factors:

$$y^2 - 4y - 5 = (y + 1)(y - 5)$$

In this chapter, we learn that any equation that can be put in the form $ax^2 + bx + c = 0$ is a quadratic equation if a, b, and c are real numbers and $a \neq 0$. $ax^2 + bx + c = 0$ is the standard form of a quadratic equation. To solve these equations, follow the steps below.

Example 1: Solve $y^2 - 4y - 5 = 0$

Step 1: Factor the left side of the equation.

$$\begin{aligned} y^2 - 4y - 5 &= 0 \\ (y + 1)(y - 5) &= 0 \end{aligned}$$

Step 2: If the product of these two factors equals zero, then the two factors individually must be equal to zero. Therefore, to solve, we set each factor equal to zero.

$$\begin{array}{rcl} (y + 1) &=& 0 \\ -1 && -1 \\ \hline y &=& -1 \end{array} \qquad \begin{array}{rcl} (y - 5) &=& 0 \\ +5 && +5 \\ \hline y &=& 5 \end{array}$$

The equation has two solutions: $y = -1$ and $y = 5$

Check: To check, substitute each solution into the original equation.

When $y = -1$, the equation becomes:

$$\begin{aligned} (-1)^2 - (4)(-1) - 5 &= 0 \\ 1 + 4 - 5 &= 0 \\ 0 &= 0 \end{aligned}$$

When $y = 5$, the equation becomes:

$$\begin{aligned} 5^2 - (4)(5) - 5 &= 0 \\ 25 - 20 - 5 &= 0 \\ 0 &= 0 \end{aligned}$$

Both solutions produce true statements.
The solution set for the equation is $\{$[illegible]$, -5\}$

Solve each of the following quadratic equations by factoring and setting each factor equal to zero. Check by substituting answers back in the original equation.

1. $x^2 + x - 6 = 0$
2. $y^2 - 2y - 8 = 0$
3. $a^2 + 2a - 15 = 0$
4. $y^2 - 5y + 4 = 0$
5. $b^2 - 9b + 14 = 0$
6. $x^2 - 3x - 4 = 0$
7. $y^2 + y - 20 = 0$
8. $d^2 + 6d + 8 = 0$
9. $y^2 - 7y + 12 = 0$
10. $x^2 - 3x - 28 = 0$
11. $a^2 - 5a + 6 = 0$
12. $b^2 + 3b - 10 = 0$
13. $a^2 + 7a - 8 = 0$
14. $c^2 + 3c + 2 = 0$
15. $x^2 - x - 42 = 0$
16. $a^2 + a - 6 = 0$
17. $b^2 + 7b + 12 = 0$
18. $y^2 + 2y - 15 = 0$
19. $a^2 - 3a - 10 = 0$
20. $d^2 + 10d + 16 = 0$
21. $x^2 - 4x - 12 = 0$

Quadratic equations that have a whole number and a variable in the first term are solved the same way as the previous page. Factor the trinomial, and set each factor equal to zero to find the solution set.

Example 2: Solve $2x^2 + 3x - 2 = 0$

$(2x - 1)(x + 2) = 0$

Set each factor equal to zero and solve:

$$\begin{aligned} 2x - 1 &= 0 \\ +1 &\quad +1 \\ \frac{2x}{2} &= \frac{1}{2} \\ x &= \frac{1}{2} \end{aligned} \qquad \begin{aligned} x + 2 &= 0 \\ -2 &\quad -2 \\ x &= -2 \end{aligned}$$

The solution set is $\left\{\frac{1}{2}, -2\right\}$

Solve the following quadratic equations.

22. $3y^2 + 4y - 32 = 0$
23. $5c^2 - 2c - 16 = 0$
24. $7d^2 + 18d + 8 = 0$
25. $3a^2 - 10a - 8 = 0$
26. $11x^2 - 31x - 6 = 0$
27. $5b^2 + 17b + 6 = 0$
28. $3x^2 - 11x - 20 = 0$
29. $5a^2 + 47a - 30 = 0$
30. $2c^2 - 5c - 25 = 0$
31. $2y^2 + 11y - 21 = 0$
32. $5a^2 + 23a - 42 = 0$
33. $3d^2 + 11d - 20 = 0$
34. $3x^2 - 10x + 8 = 0$
35. $7b^2 + 23b - 20 = 0$
36. $9a^2$ [illegible] $58a + 24 = 0$
37. $4c^2 - 25c - 21 = 0$
38. $8d^2 + 53d + 30 = 0$
39. $4y^2 + 37a - 15 = 0$
40. $8a^2 + 37a - 15 = 0$
41. $3x^2 - 41x + 26 = 0$
42. $8b^2 + 2b - 3 = 0$

11.1 Solving the Difference of Two Squares

To solve the difference of two squares, first factor. Then set each factor equal to zero.

Example 3: $25x^2 - 36 = 0$

Step 1: Factor the left hand side of the equation.

$$25x^2 - 36 = 0$$
$$(5x + 6)(5x - 6) = 0$$

Step 2: Set each factor equal to zero and solve.

$$\begin{aligned} 5x + 6 &= 0 \\ -6 \quad &-6 \\ \hline \frac{5x}{5} &= \frac{-6}{5} \\ x &= -\frac{6}{5} \end{aligned} \qquad \begin{aligned} 5x - 6 &= 0 \\ +6 \quad &+6 \\ \hline \frac{5x}{5} &= \frac{6}{5} \\ x &= \frac{6}{5} \end{aligned}$$

Check: Substitute each solution in the equation to check.

for $x = -\frac{6}{5}$:

$25x^2 - 36 = 0$

$25\left(-\frac{6}{5}\right)\left(-\frac{6}{5}\right) - 36 = 0 \longleftarrow$ Substitute $-\frac{6}{5}$ for x.

$25\left(\frac{36}{25}\right) - 36 = 0 \longleftarrow$ Cancel the 25's.

$36 - 36 = 0 \longleftarrow$ A true statement. $x = -\frac{6}{5}$ is a solution.

for $x = \frac{6}{5}$:

$25x^2 - 36 = 0$

$25\left(\frac{6}{5}\right)\left(\frac{6}{5}\right) - 36 = 0 \longleftarrow$ Substitute $\frac{6}{5}$ for x.

$25\left(\frac{36}{25}\right) - 36 = 0 \longleftarrow$ Cancel the 25's.

$36 - 36 = 0 \longleftarrow$ A true statement. $x = \frac{6}{5}$ is a solution.

The solution set is $\left\{\frac{-6}{5}, \frac{6}{5}\right\}$.

Find the solution sets for the following.

1. $25a^2 - 16 = 0$
2. $c^2 - 36 = 0$
3. $9x^2 - 64 = 0$
4. $100y^2 - 49 = 0$
5. $4b^2 - 81 = 0$
6. $[illegible]^2 - 25 = 0$
7. $9x^2 - 1 = 0$
8. $16[illegible]^2 - 9 = 0$
9. $36[illegible]^2 - [illegible] = 0$
10. $36y^2 - 25 = 0$
11. $d^2 - 16 = 0$
12. $64[illegible]^2 - 9 = 0$
13. $81a^2 - 4 = 0$
14. $64y^2 - 25 = 0$
15. $4c^2 - 49 = 0$
16. $x^2 - 81 = 0$
17. $49b^2 - 9 = 0$
18. $a^2 - 64 = 0$
19. $x^2 - 1 = 0$
20. $4y^2 - 9 = 0$
21. $t^2 - 100 = 0$
22. $16[illegible]^2 - 81 = 0$
23. $a^2 - 4 = 0$
24. $36b^2 - 16 = 0$

11.2 Solving Perfect Squares

When the square root of a constant, variable, or polynomial results in a constant, variable, or polynomial without irrational numbers, the expression is a **perfect square**. Some examples are 49, x^2, and $(x-2)^2$.

Example 4: Solve the perfect square for x. $(x-5)^2 = 0$

Step 1: Take the square root of both sides.
$\sqrt{(x-5)^2} = \sqrt{0}$
$(x-5) = 0$

Step 2: Solve the equation.
$(x-5) = 0$
$x - 5 + 5 = 0 + 5$
$x = 5$

Example 5: Solve the perfect square for x. $(x-5)^2 = 64$

Step 1: Take the square root of both sides.
$\sqrt{(x-5)^2} = \sqrt{64}$
$(x-5) = \pm 8$
$(x-5) = 8$ and $(x-5) = -8$

Step 2: Solve the two equations.
$(x-5) = 8$ and $(x-5) = -8$
$x - 5 + 5 = 8 + 5$ and $x - 5 + 5 = -8 + 5$
$x = 13$ and $x = -3$

Solve the perfect square for x.

1. $(x-2)^2 = 0$
2. $(x+1)^2 = 0$
3. $(x+11)^2 = 0$
4. $(x-4)^2 = 0$
5. $(x-1)^2 = 0$
6. $(x+8)^2 = 0$
7. $(x+3)^2 = 4$
8. $(x-5)^2 = 16$
9. $(x-10)^2 = 100$
10. $(x+9)^2 = 9$
11. $(x-4.5)^2 = 25$
12. $(x+7)^2 = 36$
13. $(x +$ [illegible]$)^2 = 49$
14. $(x+1)^2 = 4$
15. $(x+8.9)^2 = 49$
16. $(x-6)^2 = 81$
17. $(x-12)^2 = 121$
18. $(x+2.5)^2 = 64$

11.3 Completing the Square

"Completing the Square" is another way of factoring a quadratic equation. To complete the square, convert the equation into a perfect square.

Example 6: Solve $x^2 - 10x + 9 = 0$ by completing the square.

Completing the square:

Step 1: The first step is to get the constant on the other side of the equation. Subtract 9 from both sides:
$x^2 - 10x + 9 - 9 = -9$
$x^2 - 10x = -9$

Step 2: Determine the coefficient of the x. The coefficient in this example is 10. Divide the coefficient by 2 and square the result.
$(10 \div 2)^2 = 5^2 = 25$

Step 3: Add the resulting value, 25, to both sides:
$x^2 - 10x + 25 = -9 + 25$
$x^2 - 10x + 25 = 16$

Step 4: Now factor the $x^2 - 10x + 25$ into a perfect square:
$(x - 5)^2 = 16$

Solving the perfect square:

Step 5: Take the square root of both sides.
$\sqrt{(x - 5)^2} = \sqrt{16}$
$(x - 5) = \pm 4$
$(x - 5) = 4$ and $(x - 5) = -4$

Step 6: Solve the two equations.

$(x - 5) = 4$	and	$(x - 5) = -4$
$x - 5 + 5 = 4 + 5$	and	$x - 5 + 5 = -4 + 5$
$x = 9$	and	$x = 1$

Solve for x by completing the square.

1. $x^2 + 2x - 3 = 0$
2. $x^2 - 8x + 7 = 0$
3. $x^2 + 6x - 7 = 0$
4. $x^2 - 16x - 36 = 0$
5. $x^2 - 14x + 49 = 0$
6. $x^2 - 4x = 0$
7. $x^2 + 12x + 27 = 0$
8. $x^2 + 2x - 24 = 0$
9. $x^2 + 12x - 85 = 0$
10. $x^2 - 8x + 15 = 0$
11. $x^2 - 16x + 60 = 0$
12. $x^2 - 8x - 48 = 0$
13. $x^2 + 24x + 44 = 0$
14. $x^2 + 6x + 5 = 0$
15. $x^2 - 11x + 5.25 = 0$

11.4 Using the Quadratic Formula

You may be asked on the math portion of the SAT test to use the quadratic formula to solve an algebra problem known as a **quadratic equation**. The equation should be in the form $ax^2+bx+c=0$.

Example 7: Using the quadratic formula, find x in the following equation: $x^2 - 8x = -7$.

Step 1: Make sure the equation is set equal to 0.

$x^2 - 8x + 7 = -7 + 7$
$x^2 - 8x + 7 = 0$

The quadratic formula, $\frac{-b \pm \sqrt{b^2 - 4ac}}{2a}$ will be given to you on your formula sheet with your test.

Step 2: In the formula, a is the number x^2 is multiplied by, b is the number x is multiplied by and c is the last term of the equation. For the equation in the example, $x^2 - 8x + 7$, $a = 1$, $b = -8$, and $c = 7$. When we look at the formula we notice a $\pm$ sign. This means that there will be two solutions to the equation, one when we use the plus sign and one when we use the minus sign. Substituting the numbers from the problem into the formula, we have:

$$\frac{8 + \sqrt{8^2 - (4)(1)(7)}}{2(1)} = 7 \quad \text{or} \quad \frac{8 - \sqrt{8^2 - (4)(1)(7)}}{2(1)} = 1$$

The solutions are $\{7, 1\}$

For each of the following equations, use the quadratic formula to find two solutions.

1. $x^2 + x - 6 = 0$
2. $y^2 - 2y - 8 = 0$
3. $a^2 - 2a - 15 = 0$
4. $y^2 - 5y + 4 = 0$
5. $b^2 - 9b + 14 = 0$
6. $x^2 - 3x - 4 = 0$
7. $y^2 + y - 20 = 0$
8. $d^2 + 6d + 8 = 0$
9. $y^2 - 7y + 12 = 0$
10. $x^2 - 3x - 28 = 0$
11. $a^2 - 5a + 6 = 0$
12. $b^2 + 3b - 10 = 0$
13. $a^2 + 7a - 8 = 0$
14. $c^2 + 3c + 2 = 0$
15. $x^2 - x -$ [illegible] $= 0$
16. $a^2 +$ [illegible]$a - 6 = 0$
17. $b^2 + 7b + 12 = 0$
18. $y^2 + y - 12 = 0$
19. $a^2 - 3a - 10 = 0$
20. $d^2 + 10d + 16 = 0$
21. $x^2 - 4x - 12 = 0$

Chapter 11 Review

Factor and solve each of the following quadratic equations.

1. $16b^2 - 25 = 0$

2. $a^2 - a - 30 = 0$

3. $x^2 - x = 6$

4. $100x^2 - 49 = 0$

5. $81y^2$ [illegible]

6. $y^2 = 21 - 4y$

7. $y^2 - 7y + 8 = 16$

8. $6x^2 + x - 2 = 0$

9. $3y^2 + y - 2 = 0$

10. $b^2 - 2b - 8 = 0$

11. $4x^2 + 19x - 5 = 0$

12. $8x^2 = 6x + 2$

13. $2y^2 - 6y - 20 = 0$

14. $-6x^2 + 7x - 2 = 0$

15. $y^2 + 3y - 18 = 0$

Using the quadratic formula, find both solutions for the variable.

16. $x^2 + 10x - 11 = 0$

17. $y^2 - 14y + 40 = 0$

18. $b^2 + 9b + 18 = 0$

19. $y^2 - 12y - 13 = 0$

20. $a^2 - 8a - 48 = 0$

21. $x^2 + 2x - 63 = 0$

Chapter 12
Graphing and Writing Equations and Inequalities

12.1 Graphing Linear Equations

In addition to graphing ordered pairs, the Cartesian plane can be used to graph the solution set for an equation. Any equation with two variables that are both to the first power is called a **linear equation.** The graph of a linear equation will always be a straight line.

Example 1: Graph the solution set for $x + y = 7$.

Step 1: Make a list of some pairs of numbers that will work in the equation.

$x + y = 7$		
$4 + 3 = 7$	$(4, 3)$	ordered pair solutions
$-1 + 8 = 7$	$(-1, 8)$	
$5 + 2 = 7$	$(5, 2)$	
$0 + 7 = 7$	$(0, 7)$	

Step 2: Plot these points on a Cartesian plane.

Step 3: By passing a line through these points, we graph the solution set for $x + y = 7$. This means that every point on the line is a solution to the equation $x + y = 7$. For example, $(1, 6)$ is a solution, so the line passes through the point $(1, 6)$.

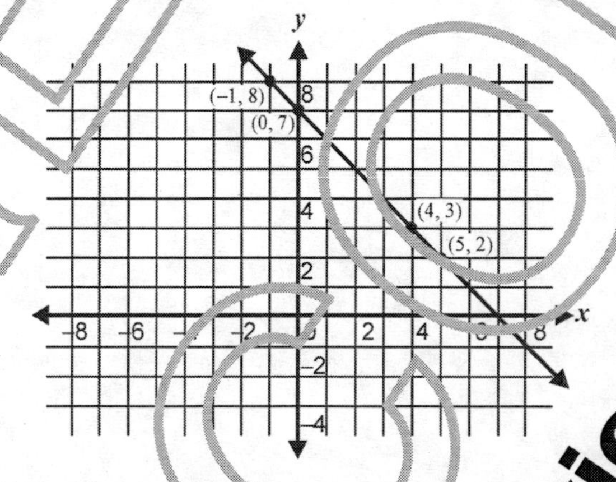

Make a table of solutions for each linear equation below. Then plot the ordered pair solutions on graph paper. Draw a line through the points. (If one of the points does not line up, you have made a mistake.)

1. $x + y = 6$
2. $y = x + 1$
3. $y = x - 2$
4. $x + 2 = y$
5. $x - 5 = y$
6. $x - y = 0$

Example 2: Graph the equation $y = 2x - 5$.

Step 1: This equation has 2 variables, both to the first power, so we know the graph will be a straight line. Substitute some numbers for x or y to find pairs of numbers that satisfy the equation. For the above equation, it will be easier to substitute values of x in order to find the corresponding value for y. Record the values for x and y in a table.

If x is 0, y would be -5
If x is 1, y would be -3
If x is 2, y would be -1
If x is 3, y would be 1

x	y
0	-5
1	-3
2	-1
3	1

Step 2: Graph the ordered pairs, and draw a line through the points.

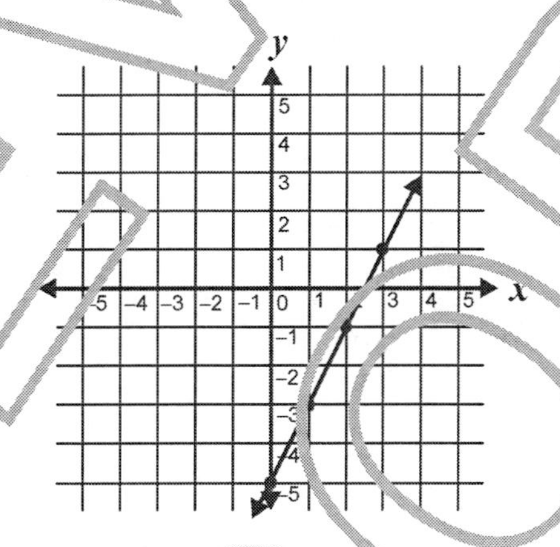

Find pairs of numbers that satisfy the equations below, and graph the line on graph paper.

1. $y = -2x + 2$
2. $2x - 2 = y$
3. $-x + 3 = y$
4. $y = x + 1$
5. $4x - 2 = y$
6. $y = 3x - 3$
7. $x = 4y - 3$
8. $2x = 3y + 1$
9. $x + 2y = 4$

12.2 Graphing Horizontal and Vertical Lines

The graph of some equations is a horizontal or a vertical line.

Example 3: $y = 3$

Step 1: Make a list of ordered pairs that satisfy the equation $y = 3$.

x	y
0	3
1	3
2	3
3	3

No matter what value of x you choose, y is always 3.

Step 2: Plot these points on an Cartesian plane, and draw a line through the points.

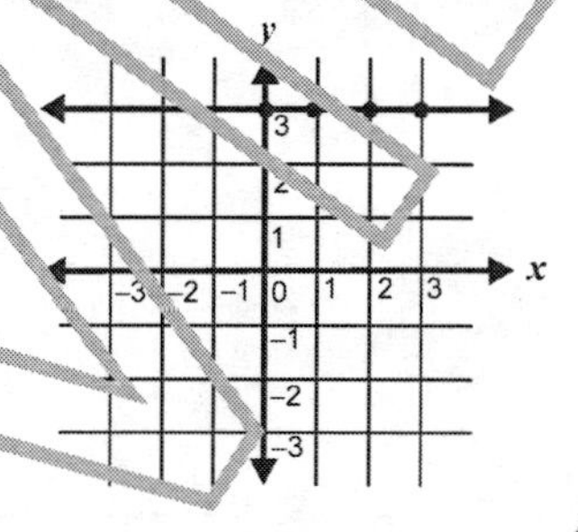

The graph is a horizontal line.

Example 4: $2x + 3 = 0$

Step 1: For these equations with only one variable, find what x equals first.

$2x + 3 = 0$

$2x = -3$

$x = \frac{-3}{2}$

Step 2: Using Example 3, find ordered pairs that satisfy the equation, plot the points, and graph the line.

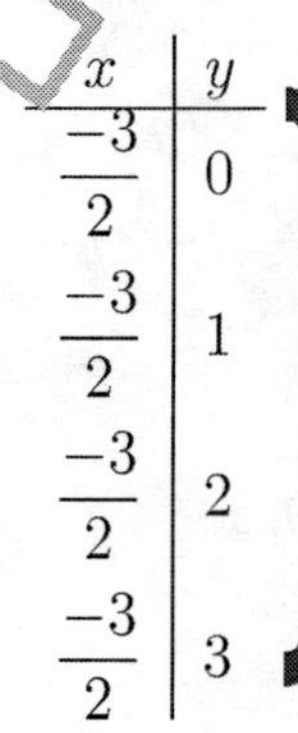

x	y
$\frac{-3}{2}$	0
$\frac{-3}{2}$	1
$\frac{-3}{2}$	2
$\frac{-3}{2}$	3

No matter which value of y you choose, the value of x does not change

The graph is a vertical line.

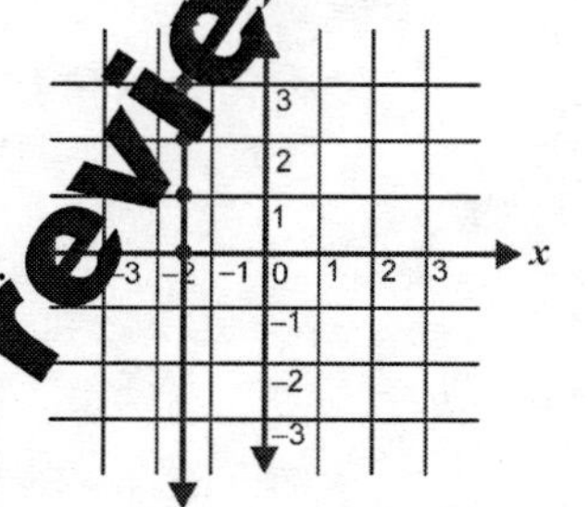

Find pairs of numbers that satisfy the equations below, and graph the line on graph paper.

1. $2y + 2 = 0$
2. $x = -4$
3. $3x = 3$
4. $y = 5$
5. $4x - 2 = 0$
6. $2x - 6 = 0$
7. $4y =$ [illegible]
8. $5x - 10 = 0$
9. [illegible] $+ 12 = 0$
10. $x + 1 = 0$
11. $2y - 8 = 0$
12. $3x = -9$
13. $x = -2$
14. $6y - 2 = 0$
15. $5x - 5 = 0$

12.3 Finding the Distance Between Two Points

Notice that a subscript added to the x and y identifies each ordered pair uniquely in the plane. For example, point 1 is identified as (x_1, y_1), point 2 as (x_2, y_2), and so on. This unique subscript identification allows us to calculate slope, distance, and midpoints of line segments in the plane using standard formulas like the distance formula. To find the distance between two points on a Cartesian plane, use the following formula:

$$d = \sqrt{(y_2 - y_1)^2 + (x_2 - x_1)^2}$$

Example 5: Find the distance between $(-2, 1)$ and $(3, -4)$.

Plugging the values from the ordered pairs into the formula, we find:

$d = \sqrt{(-4 - 1)^2 + [3 - (-2)]^2}$

$d = \sqrt{(-5)^2 + (5)^2}$

$d = \sqrt{25 + 25} = \sqrt{50}$

To simplify, we look for perfect squares that are a factor of 50. $50 = 25 \times 2$. Therefore,

$d = \sqrt{25} \times \sqrt{2} = 5\sqrt{2}$

Find the distance between the following pairs of points using the distance formula above.

1. $(6, -1)\ (5, 2)$
2. $(-4, 3)\ (2, -1)$
3. $(10, 2)\ (6, -1)$
4. $(-2, 5)\ (-4, 3)$
5. $(8, -2)\ (3, -9)$
6. $(2, -2)\ (8, 1)$
7. $(3, 1)\ (5, 5)$
8. $(-2, -1)\ (3, 4)$
9. $(5, -3)\ (-1, -5)$
10. $(6, 5)\ (3, -4)$
11. $(-1, 0)\ (-9, -8)$
12. $(-2, 0)\ (-6, 6)$
13. $(2, 4)\ (8, 10)$
14. $(-10, -5)\ (2, -7)$
15. $(-3, 6)\ (1, -1)$

12.4 Finding the Midpoint of a Line Segment

You can use the coordinates of the endpoints of a line segment to find the coordinates of the midpoint of the line segment. The formula to find the midpoint between two coordinates is:

$$\text{midpoint } M = \left(\frac{x_1 + x_2}{2}, \frac{y_1 + y_2}{2}\right)$$

Example 6: Find the midpoint of the line segment having endpoints at $(-3, -1)$ and $(4, 3)$.

Use the formula for the midpoint $M = \left(\frac{4 + (-3)}{2}, \frac{3 + (-1)}{2}\right)$

When we simplify each coordinate, we find the midpoint, M, is $\left(\frac{1}{2}, 1\right)$.

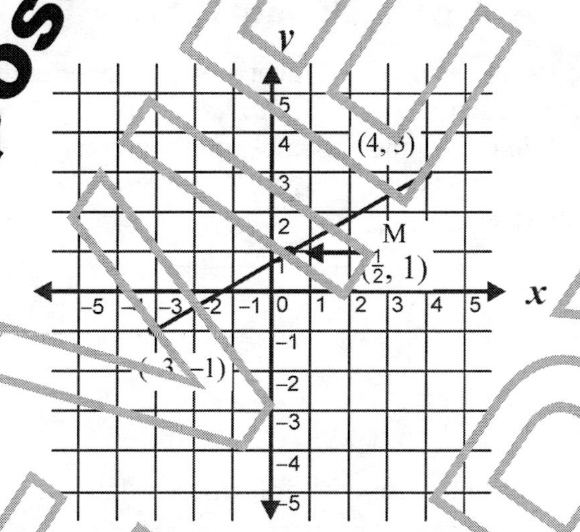

For each of the following pairs of points, find the coordinate of the midpoint, M, using the formula given above.

1. $(4, 5)\ (-6, 9)$
2. $(-3, 2)\ (-1, -2)$
3. $(3, 6)\ (9, 12)$
4. $(2, 5)\ (6, 9)$
5. $(8, 9)\ (6, 11)$
6. $(-4, 3)\ (8, 7)$
7. $(-1, -5)\ (-3, -11)$
8. $(4, 2)\ (-2, 8)$
9. $(4, 3)\ (-1, -5)$
10. $(-6, 2)\ (8, -8)$
11. $(-3, 9)\ (-9,$ [illegible]$)$
12. $(7, 8)\ ($[illegible]$)$
13. $($[illegible]$, 19)\ (2, 3)$
14. $(5, 4)\ (9, -2)$
15. $(-4, 6)\ (10, -2)$

12.5 Finding the Intercepts of a Line

The x-intercept is the point where the graph of a line crosses the x-axis. The y-intercept is the point where the graph of a line crosses the y-axis.

To find the x-intercept, set $y = 0$

To find the y-intercept, set $x = 0$

Example 7: Find the x- and y-intercepts of the line $6x + 2y = 18$

Step 1: To find the x-intercept, set $y = 0$.

$$\begin{aligned} 6x + 2(0) &= 18 \\ \frac{6x}{6} &= \frac{18}{6} \\ x &= 3 \end{aligned}$$

The x-intercept is at the point $(3, 0)$.
To find the y-intercept, set $x = 0$.

$$\begin{aligned} 6(0) + 2y &= 18 \\ \frac{2y}{2} &= \frac{18}{2} \\ y &= 9 \end{aligned}$$

The y-intercept is at the point $(0, 9)$.

You can now use the two intercepts to graph the line.

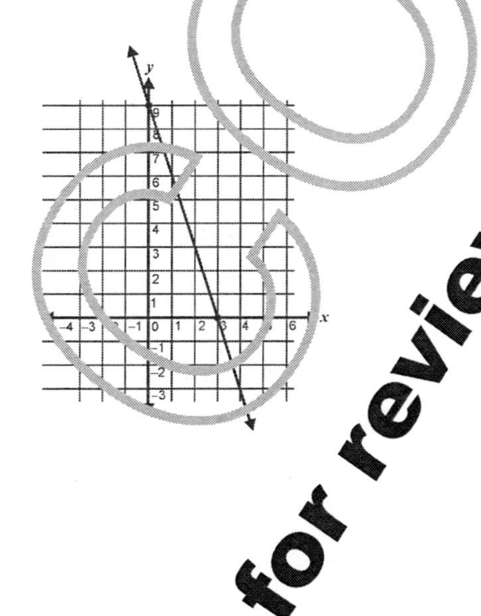

12.6 Understanding Slope

The slope of a line refers to how steep a line is. Slope is also defined as the rate of change. When we graph a line using ordered pairs, we can easily determine the slope. Slope is often represented by the letter m.

$$\text{The formula for slope of a line is: } m = \frac{y_2 - y_1}{x_2 - x_1} \text{ or } \frac{\text{rise}}{\text{run}}$$

Example 8: What is the slope of the following line that passes through the ordered pairs $(-4, -3)$ and $(1, 3)$?

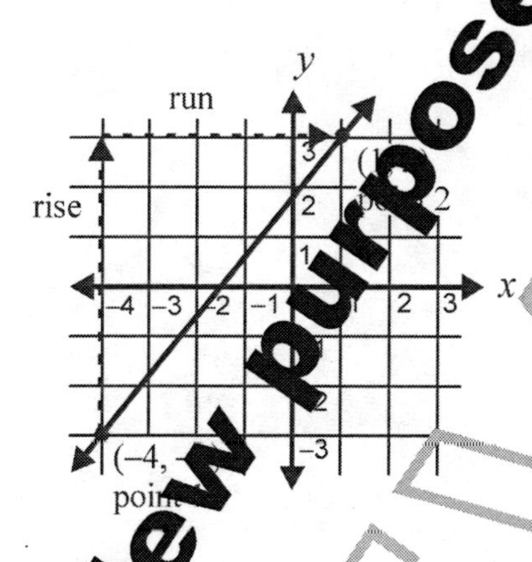

y_2 is 3, the y-coordinate of point 2.

y_1 is -3, the y-coordinate of point 1.

x_2 is 1, the x-coordinate of point 2.

x_1 is -4, the x-coordinate of point 1.

Use the formula for slope given above:

$$m = \frac{3 - (-3)}{1 - (-4)} = \frac{6}{5}$$

The slope is $\frac{6}{5}$. This shows us that we can go up 6 (rise) and over 5 to the right (run) to find another point on the line.

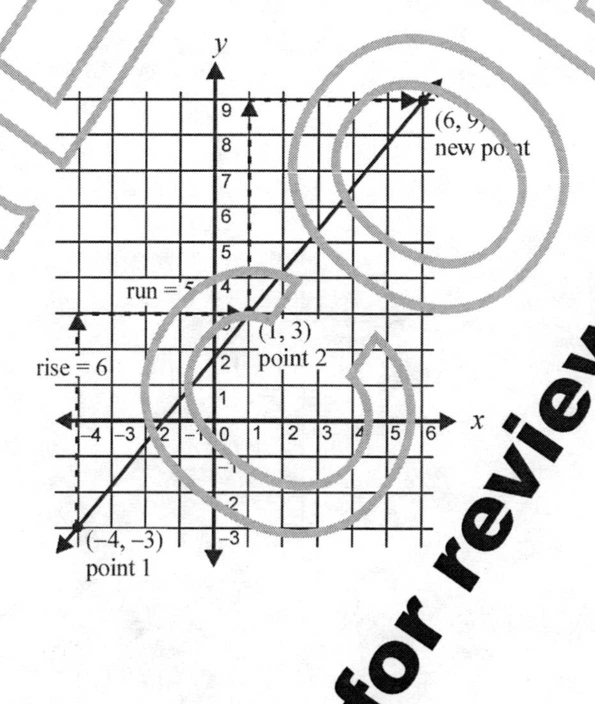

Example 9: Find the slope of a line through the points $(-2, 3)$ and $(1, -2)$. It doesn't matter which pair we choose for point 1 and point 2. The answer is the same.

Let point 1 be $(-2, 3)$
Let point 2 be $(1, -2)$

$$\text{slope} = \frac{(y_2 - y_1)}{(x_2 - x_1)} = \frac{-2 - 3}{1 - (-2)} = \frac{-5}{3}$$

When the slope is negative, the line will slant left. For this example, the line will go **down** 5 units and then over 3 units to the **right**.

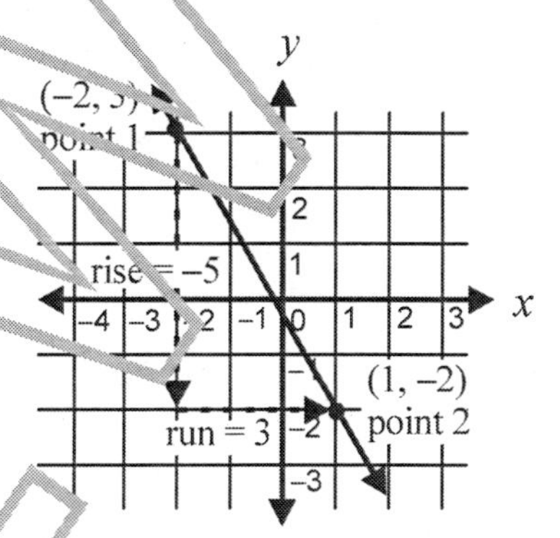

Example 8: What is the slope of a line that passes through $(1, 1)$ and $(3, 1)$?

$$\text{slope} = \frac{1 - 1}{3 - 1} = \frac{0}{2} = 0$$

When $y_2 - y_1 = 0$, the slope will equal 0, and the line will be horizontal.

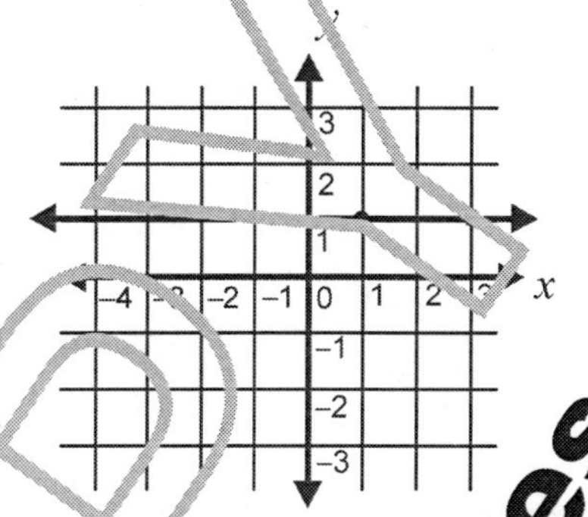

Example 9: What is the slope of a line that passes through $(2, 1)$ and $(2, -3)$?

$$\text{slope} = \frac{-3 - 1}{2 - 2} = \frac{4}{0} = \text{undefined}$$

When $x_2 - x_1 = 0$, the slope is undefined, and the line will be vertical.

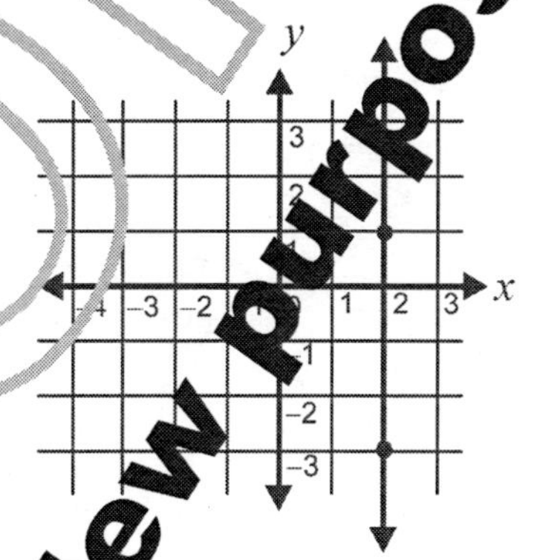

The following lines summarize what we know about slope.

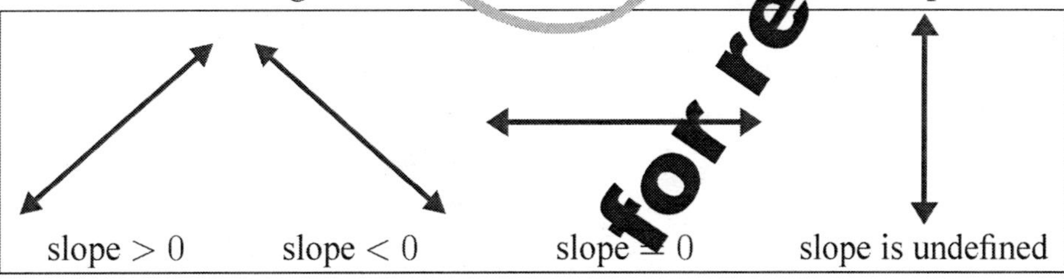

Find the slope of the line that goes through the following pairs of points. Then, using graph paper, graph the line through the two points, and label the rise and run. (See Examples 6–9).

1. $(2,3)\ (4,5)$
2. $(1,3)\ (2,5)$
3. $(-1,2)\ (4,1)$
4. $(1,-2)\ (4,-2)$
5. $(3,0)\ (3,4)$
6. $(3,2)\ (-1,8)$
7. $(4,3)\ (2,4)$
8. $(2,2)\ (1,0)$
9. $(3,4)\ (1,2)$
10. $(3,2)\ (3,6)$
11. $(6,-2)\ (3,-2)$
12. $(1,2)\ (3,4)$
13. $(-2,1)\ (-4,3)$
14. $(5,2)\ (4,-1)$
15. $(1,-3)\ (-2,4)$
16. $(2,-1)\ (3,5)$

12.7 Slope-Intercept Form of a Line

An equation that contains two variables, each to the first degree, is a **linear equation**. The graph for a linear equation is a straight line. To put a linear equation in slope-intercept form, solve the equation for y. This form of the equation shows the slope and the y-intercept. Slope-intercept form follows the pattern of $y = mx + b$. The "m" represents slope, and the "b" represents the y-intercept. The y-intercept is the point at which the line crosses the y-axis.

When the slope of a line is not 0, the graph of the equation shows a **direct variation** between y and x. When y increases, x increases in a certain proportion. The proportion stays constant. The constant is called the **slope** of the line.

Example 10: Put the equation $2x + 3y = 15$ in slope-intercept form. What is the slope of the line? What is the y-intercept? Graph the line.

Step 1: Solve for y

$$\begin{array}{rcrcl} 2x & + & 3y & = & 15 \\ -2x & & & & -2x \\ \hline \frac{3y}{3} & = & -\frac{2x}{3} & + & \frac{15}{3} \end{array}$$

slope-intercept form: $y = -\frac{2}{3}x + 5$

The slope is $-\frac{2}{3}$ and the y-intercept is 5.

Step 2: Knowing the slope and the y-intercept, we can graph the line.

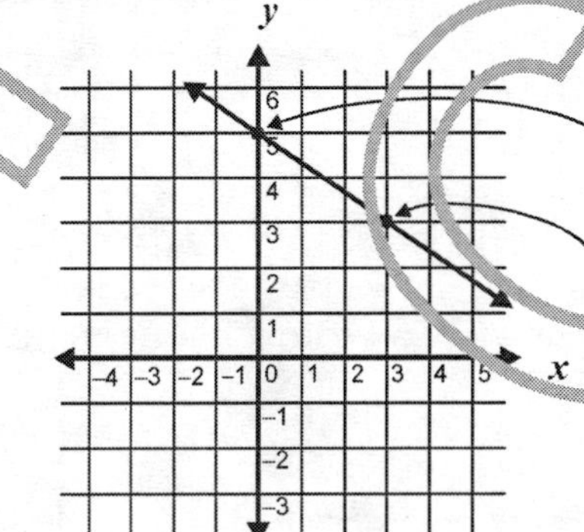

The y-intercept is 5, so the line passes through the point (0, 5) on the y-axis.

The slope is $-\frac{2}{3}$ so go down 2 and over 3 to get a second point.

Put each of the following equations in slope-intercept form by solving for y. On your graph paper, graph the line using the slope and y-intercept.

1. $4x - 5y = 5$	6. $8x - 5y = 10$	11. $3x - 2y = -6$	16. $4x + 2y = 8$
2. $2x + 4y = 16$	7. $-2x + y =$ [illegible]	12. $3x + 4y = 2$	17. $6x - y = 4$
3. $3x - 2y = 10$	8. $-4x +$ [illegible] $= 12$	13. $-x = 2 + 4y$	18. $-2x - 4y = 8$
4. $x + 3y = -12$	9. $-6x + 2y = 12$	14. $2x = 4y - 2$	19. $5x + 4y = 16$
5. $6x + 2y = 0$	10. $x - 5y = 5$	15. $6x - 3y = 9$	20. $6 = 2y - 3x$

12.8 Verify That a Point Lies on a Line

To know whether or not a point lies on a line, substitute the coordinates of the point into the formula for the line. If the point lies on the line, the equation will be true. If the point does not lie on the line, the equation will be false.

Example 11: Does the point $(5, 2)$ lie on the line given by the equation $x + y = 7$?

Solution: Substitute 5 for x and 2 for y in the equation. $5 + 2 = 7$. Since this is a true statement, the point $(5, 2)$ does lie on the line $x + y = 7$.

Example 12: Does the point $(0, 1)$ lie on the line given by the equation $5x + 4y = 16$?

Solution: Substitute 0 for x and 1 for y in the equation $5x + 4y = 16$. Does $5(0) + 4(1) = 16$? No, it equals 4, not 16. Therefore, the point $(0, 1)$ is not on the line given by the equation $5x + 4y = 16$.

For each point below, state whether or not it lies on the line given by the equation that follows the point coordinates.

1. $(2, 4)$ $6x - y = 8$	5. $(3, 7)$ $x - 5y = -32$	9. $(6, 8)$ $6x - y = 28$
2. $(1, 1)$ $6x - y = 5$	6. $(0, 5)$ $-6x - 5y = 3$	10. $(-2, 3)$ $x + 2y = 4$
3. $(3, 8)$ $-2x + y = 2$	7. $(2, 4)$ $4x + 2y = 16$	11. $(4, -1)$ $-x - 3y = -1$
4. $(9, 6)$ $-2x + y = 0$	8. $(9, 1)$ $3x - 2y =$ [illegible]	12. $(-1, -3)$ $2x + y = 1$

12.9 Graphing a Line Knowing a Point and Slope

If you are given a point of a line and the slope of a line, the line can be graphed.

Example 13: Given that line l has a slope of $\frac{4}{3}$ and contains the point $(2, -1)$, graph the line.

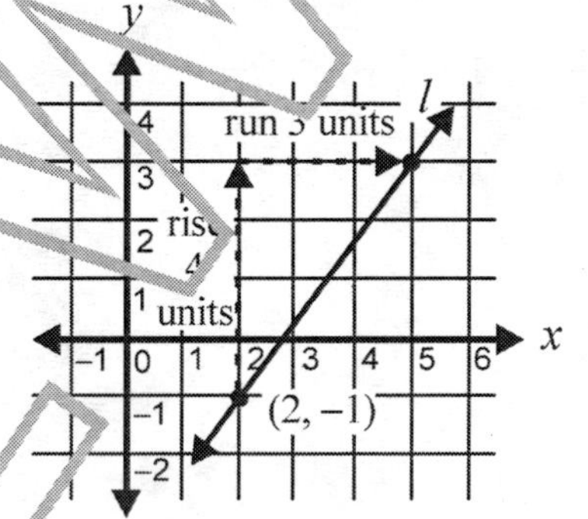

Plot and label the point $(2, -1)$ on a Cartesian plane.

The slope, $m = \frac{4}{3}$, so the rise is 4, and the run is 3. From the point $(2, -1)$, count 4 units up and 3 units to the right.

Draw the line through the two points.

Example 14: Given a line that has a slope of $-\frac{1}{4}$ and passes through the point $(-3, 2)$, graph the line.

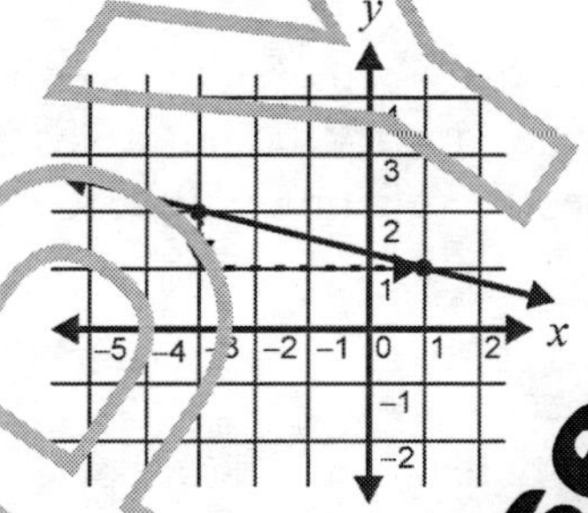

Plot the point $(-3, 2)$.

Since the slope is negative, go **down** 1 unit and over 4 units to get a second point.

Graph the line through the two points.

Graph a line on your own graph paper for each of the following problems. First plot the point. Then use the slope to find a second point. Draw the line formed from the point and the slope.

1. $(2, -2)$, $m = \frac{3}{4}$
2. $(3, -4)$, $m = \frac{1}{2}$
3. $(1, 3)$, $m = -\frac{1}{3}$
4. $(2, -4)$, $m = 1$
5. $(3, 0)$, $m = -\frac{1}{2}$
6. $(-2, 1)$, $m = \frac{4}{3}$
7. $(-4, -2)$, $m = \frac{1}{2}$
8. $(1, -4)$, $m = \frac{3}{4}$
9. $(2, -1)$, $m = -\frac{1}{2}$
10. $(5, -2)$, $m = \frac{1}{4}$
11. $(-2, -3)$, $m = \frac{2}{3}$
12. $(4, -1)$, $m = -\frac{1}{3}$
13. $(-1, 5)$, $m = \frac{2}{5}$
14. $(-2, 3)$, $m = \frac{3}{4}$
15. $(4, 4)$, $m = -$[illegible]
16. $(3, -3)$, $m = -\frac{3}{4}$
17. $(-2, 5)$, $m = \frac{1}{3}$
18. $(-2, -3)$, $m = -\frac{3}{4}$
19. $(4,$ [illegible]$)$, $m = \frac{2}{3}$
20. [illegible], $m = -\frac{1}{2}$

12.10 Finding the Equation of a Line Using Two Points or a Point and Slope

If you can find the slope of a line and know the coordinates of one point, you can write the equation for the line. You know the formula for the slope of a line is:

$$m = \frac{y_2 - y_1}{x_2 - x_1} \text{ or } \frac{y_2 - y_1}{x_2 - x_1} = m$$

Using algebra, you can see that if you multiply both sides of the equation by $x_2 - x_1$, you get:

$$y - y_1 = m(x - x_1) \longleftarrow \text{point-slope form of an equation}$$

Example 15: Write the equation of the line passing through the points $(-2, 3)$ and $(1, 5)$.

Step 1: First, find the slope of the line using the two points given.

$$m = \frac{y_2 - y_1}{x_2 - x_1} = \frac{5 - 3}{1 - (-2)} = \frac{2}{3}$$

Step 2: Pick one of the two points to use in the point-slope equation. For point $(-2, 3)$, we know $x_1 = -2$ and $y_1 = 3$, and we know $m = \frac{2}{3}$. Substitute these values into the point-slope form of the equation.

$$y - y_1 = m(x - x_1)$$

$$y - 3 = \frac{2}{3}[x - (-2)]$$

$$y - 3 = \frac{2}{3}x + \frac{4}{3}$$

$$y = \frac{2}{3}x + \frac{13}{3}$$

Use the point-slope formula to write an equation for each of the following lines.

1. $(1, -2), m = 2$
2. $(-3, 3), m = \frac{1}{3}$
3. $(4, 2), m = \frac{1}{4}$
4. $(5, 0), m = 1$
5. $(3, -4), m = \frac{1}{2}$
6. $(-1, -4)\ (2, -1)$
7. $(2, 1)\ (-1, -3)$
8. $(-2, 5)\ (-4, 3)$
9. $(-4, 3)\ (2, -1)$
10. $(3, 1)\ (5, 5)$
11. [illegible], $m = 2$
12. $(-1, 2), m = \frac{4}{3}$
13. $(2, -5), m = -2$
14. $(-1, 3), m = \frac{1}{3}$
15. $(0, -2), m = -\frac{3}{2}$

12.11 Graphing Inequalities

In the previous section, you would graph the equation $x = 3$ as:

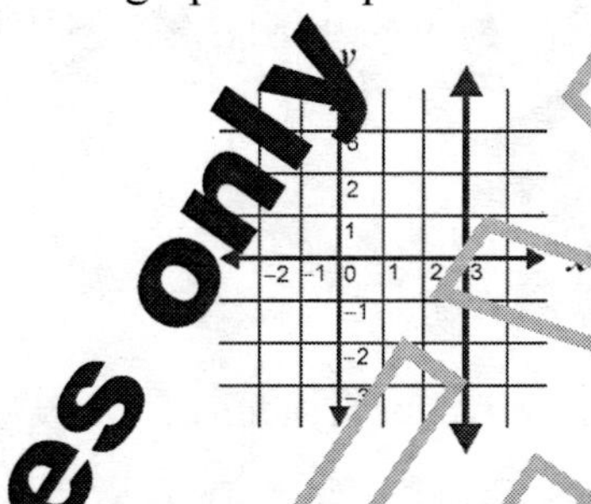

In this section, we graph inequalities such as $x > 3$ (read x is greater than 3). To show this, we use a broken line since the points on the line $x = 3$ are not included in the solution. We shade all points greater than 3.

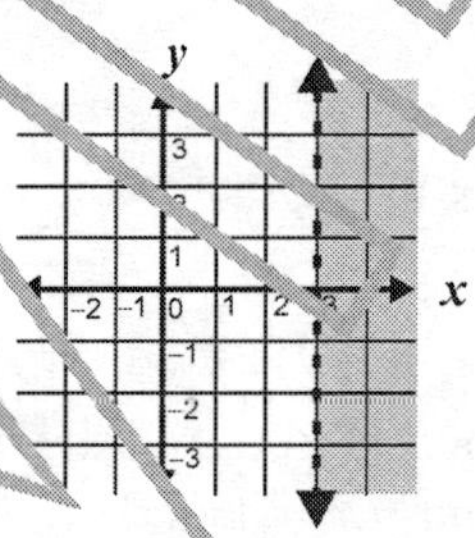

When we graph $x \geq 3$ (read x is greater than or equal to 3), we use a solid line because the points on the line $x = 3$ are included in the graph.

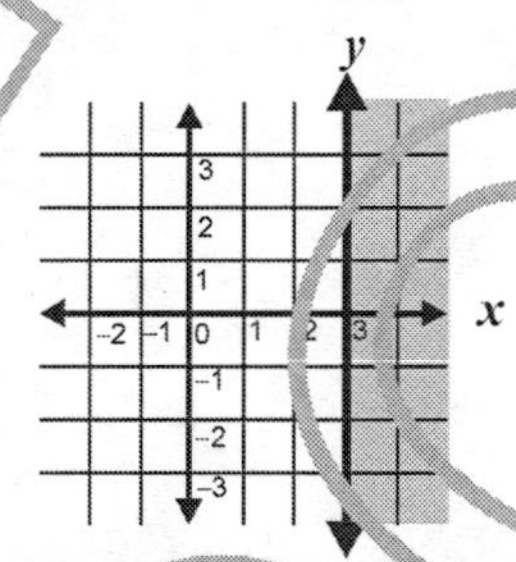

Graph the following inequalities on your own graph paper.

1. $y < 2$	7. $x > -3$	13. $x \leq 0$	19. $x \leq -2$
2. $x \geq 4$	8. $y \leq 3$	14. $y > -1$	20. $y < -2$
3. $y \geq 1$	9. $x \leq 5$	15. $y \leq 4$	21. $y \geq -4$
4. $x < -1$	10. $y > -5$	16. $x \geq 0$	22. $x \geq -1$
5. $y \geq -2$	11. $x \geq 3$	17. $y \geq 3$	23. $y \leq 5$
6. $x \leq -4$	12. $y < -1$	18. x [illegible] 4	24. $x < -3$

Example 16: Graph $x + y \geq 3$.

Step 1: First, we graph $x + y \geq 3$ by changing the inequality to an equality. Think of ordered pairs that will satisfy the equation $x + y = 3$. Then, plot the points, and draw the line. As shown below, this line divides the Cartesian plane into 2 half-planes, $x + y \geq 3$ and $x + y \leq 3$. One half plane is above the line, and the other is below the line.

x	y
2	1
0	3
3	0
4	−1

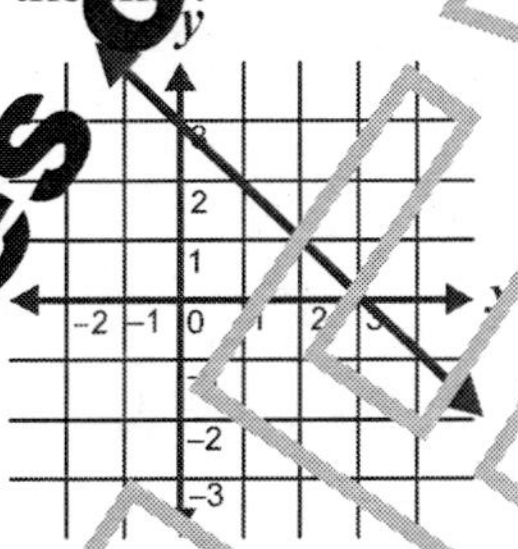

Step 2: To determine which side of the line to shade, first choose a test point. If the point you choose makes the inequality true, then the point is on the side you shade. If the point you choose does not make the inequality true, then shade the side that does not contain the test point.

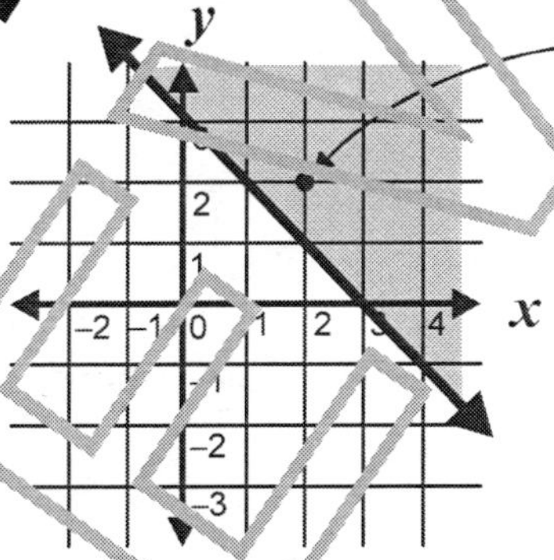

For our test point, let's choose $(2, 2)$. Substitute $(2, 2)$ into the inequality.

$x + y \geq 3$
$2 + 2 \geq 3$

$4 \geq 3$ is true, so shade the side that includes this point.

Use a solid line because of the $\geq$ sign.

Graph the following inequalities on your own graph paper.

1. $x + y \leq 4$
2. $x + y \geq 3$
3. $x \geq 5 - y$
4. $x \leq 1 + y$
5. $x - y \geq -2$
6. $x < y + 4$
7. $x + y < -1$
8. $x - y \leq 0$
9. $x \geq y + 2$
10. $x < -y + 1$
11. $-x + y > 1$
12. $-x - y < -2$

For more complex inequalities, it is easier to graph by first changing the inequality to an equality and then put the equation in slop-intercept form.

Example 17: Graph the inequality $2x + 4y \leq 8$.

Step 1: Change the inequality to an equality.
$2x + 4y = 8$

Step 2: Put the equation in slop-intercept form by solving the equation for y.

$$2x + 4y = 8$$

$$2x - 2x + 4y = -2x + 8 \quad \text{Subtract } 2x \text{ from both sides of the equation.}$$

$$4y = -2x + 8 \quad \text{Simplify.}$$

$$\frac{4y}{4} = \frac{-2x + 8}{4} \quad \text{Divide both sides by 4.}$$

$$y = \frac{-2x}{4} + \frac{8}{4} \quad \text{Find the lowest terms of the fractions.}$$

$$y = -\frac{1}{2}x + 2$$

Step 3: Graph the line. If the inequality is $<$ or $>$, use a dotted line. If the inequality is $\leq$ or $\geq$ use a solid line. For this example, we should use a solid line.

Step 4: Determine which side of the line to shade. Pick a point such as $(0, 0)$ to see if it is true in the inequality.

$2x + 4y \leq 8$, so substitute $(0, 0)$.
Is $0 + 0 \leq 8$? Yes, $0 \leq 8$, so shade the side of the line that includes the point $(0, 0)$.

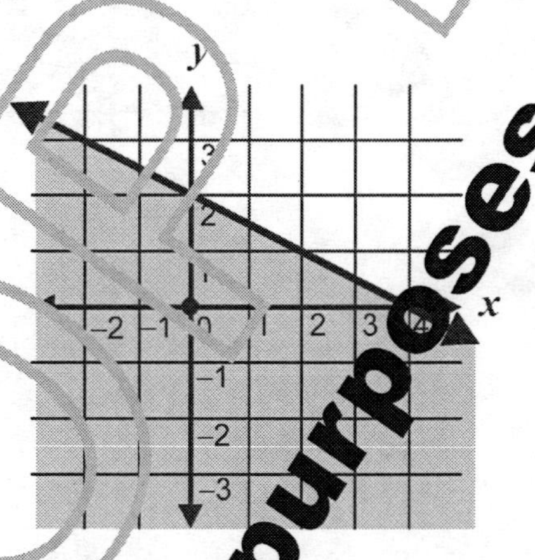

Graph the following inequalities on your own graph paper.

1. $2x + y \geq 1$
2. $3x - y \leq 3$
3. $x + 3y > 12$
4. $4x - 3y < 12$
5. $y \geq 3x + 1$
6. $x - 2y > -2$
7. $x \leq y + 4$
8. $x + y < -1$
9. $-4y \geq 2x + 1$
10. $x \leq 4y - 2$
11. $3x - y \geq 4$
12. $y \geq 2x - 5$

Chapter 12 Review

1. Graph the solution set for the linear equation: $x - 3 = y$.

2. Which of the following is not a solution of $3x = 5y - 1$?

(A) $(3, 2)$
(B) $(7, 4)$
(C) $(-\frac{1}{3}, 0)$
(D) $(-2, -1)$

3. $(-2, 1)$ is a solution for which of the following equations?

(A) $y + 2x = 4$
(B) $-2x - y = $ [illegible]
(C) $x + 2y = $ [illegible]4
(D) $2x - y = -5$

4. Graph the equation $2x - 4 = 0$.

5. What is the slope of the line that passes through the points $(5, 3)$ and $(6, 1)$?

6. What is the slope of the line that passes through the points $(-1, 4)$ and $(-5, -2)$?

7. What is the x-intercept for the following equation? $6x - y = 30$

8. What is the y-intercept for the following equation? $4x + 2y = 28$

9. Graph the equation $3y = 9$.

10. Write the following equation in slope-intercept form.
$3x = -2y + 4$

11. What is the slope of the line $y = -\frac{1}{2}x + 3$?

12. What is the x-intercept of the line $y = 5x + 6$?

13. What is the y-intercept of the line $y - \frac{2}{3}x + 3 = 0$?

14. Graph the line which has a slope of -2 and a y-intercept of -3.

15. Which of the following points does **not** lie on the line $y = 3x - 2$?

(A) $(0, -2)$
(B) $(1, 1)$
(C) $(-1, 5)$
(D) $(2, 4)$

16. Find the equation of the line which contains the point $(0, 2)$ and has a slope of $\frac{3}{4}$.

17. Which is the graph of $x - 3y = 6$?

(A)

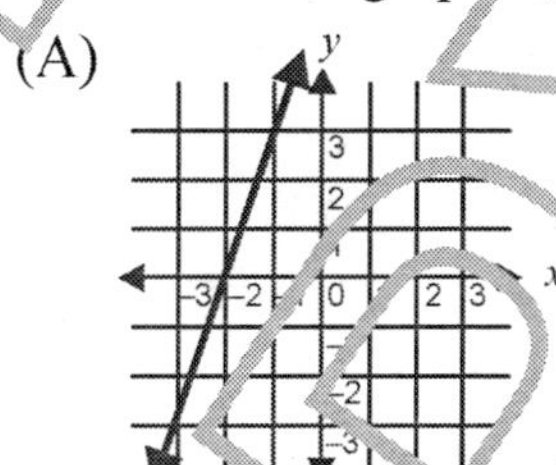

(B)

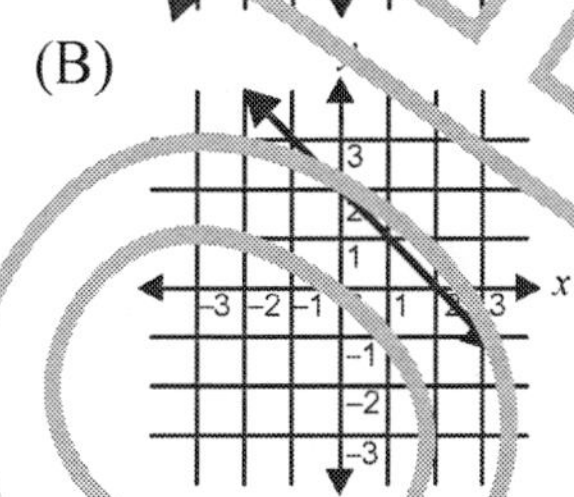

(C)

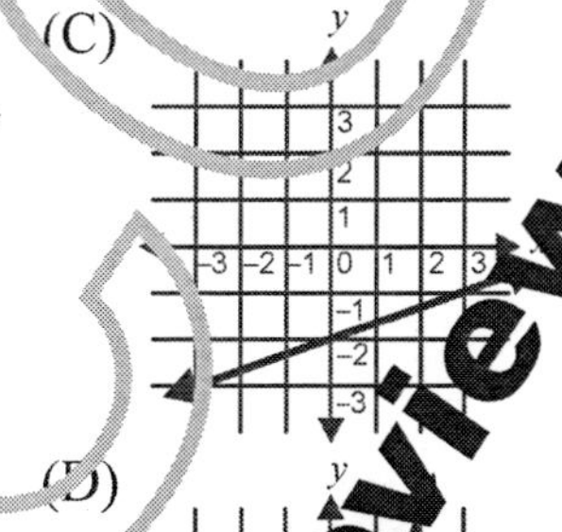

(D)

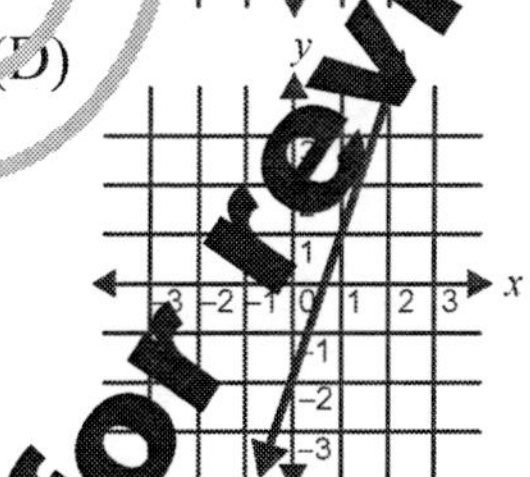

Graph the following inequalities on a Cartesian plane using your graph paper.

18. $x \geq 4$

19. $x \leq -2$

20. $5y > -10x + 5$

21. $y \leq 2$

22. $2x + y < 5$

23. $y - 2x \leq 3$

24. $y \geq x + 2$

25. $3 + y > x$

26. What is the distance between the points $(3, 3)$ and $(6, -1)$?

27. What is the distance between the two points $(-3, 0)$ and $(2, 5)$?

For questions 28 and 29, use the following formula to find the coordinates of the midpoint of the line segments with the given endpoints.

$$\text{midpoint} = \left(\frac{x_1 + x_2}{2}, \frac{y_1 + y_2}{2}\right)$$

28. $(6, 10)$ $(-4, 4)$

29. $(-1, -7)$ $(5, 3)$

Chapter 13
Applications of Graphs

13.1 Changing the Slope or Y-Intercept of a Line

When the slope and/or the y-intercept of a linear equation changes, the graph of the line will also change.

Example 1: Consider line l shown in Figure 1 at right. What happens to the graph of the line if the slope is changed to $\frac{4}{5}$?

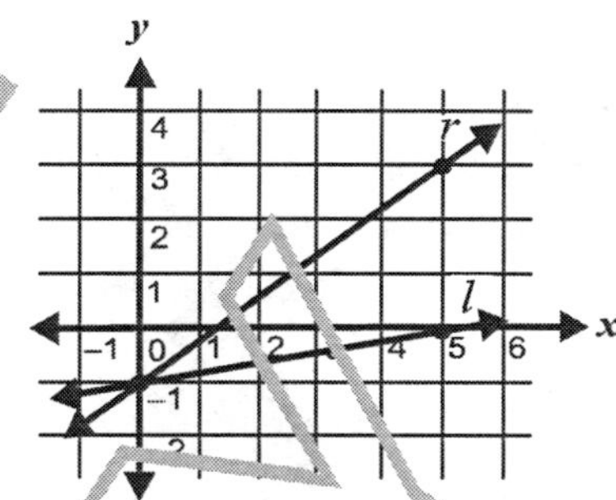

Figure 1

Determine the y-intercept of the line. For line l, it can easily be seen from the graph that the y-intercept is at the point $(0, -1)$.

Find the slope of the line using two points that the line goes through. $(0, -1)$ and $(5, 0)$.

$$m = \frac{y_2 - y_1}{x_2 - x_2} = \frac{0 - (-1)}{5 - 0} = \frac{1}{5}$$

Write the equation of line l in slope-intercept form:

$y = mx + b \quad \Longrightarrow \quad y = \frac{1}{5}x - 1$

Rewrite the equation of the line using a slope of $\frac{4}{5}$, and then graph the line. The equation of the new line is $y = \frac{4}{5}x - 1$.

The graph of the new line is labeled line r and is shown in Figure 1. A line with a slope of $\frac{4}{5}$ is steeper than a line with a slope of $\frac{1}{5}$.

Note: The greater the numerator, or "rise," of the slope, the steeper the line will be. The greater the denominator, or "run," of the slope, the flatter the line will be.

Example 2: Consider line l shown in Figure 2 at right. The equation of the line is $y = -\frac{1}{2}x + 3$. What happens to the graph of the line if the y-intercept is changed to -1?

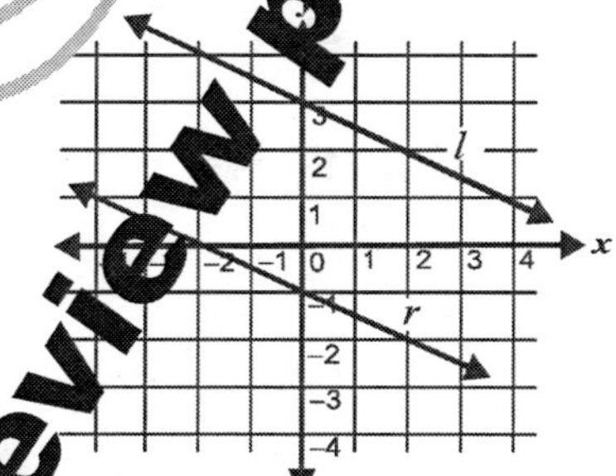

Figure 2

Rewrite the equation of the line replacing the y-intercept with -1. The equation of the new line is $y = -\frac{1}{2}x - 1$.

Graph the new line. Line r in Figure 2 is the graph of the equation $y = -\frac{1}{2}x - 1$. Since both lines l and r have the same slope, they are parallel. Line r, with a y-intercept of -1, sits below line l, with a y-intercept of 3.

Put each pair of the following equations in slope-intercept form. Write P if the lines are parallel and NP if the lines are not parallel.

1. $y = x + 1$
   $2y - 2x = 6$ _____
2. $2x + y = 6$
   $2x = 8 - y$ _____
3. $x + 5y = 0$
   $5y + 5 = x$ _____
4. $y = 3 + \frac{1}{3}x$
   $3y + x = -6$ _____
5. $x = 2y$
   $x = -2y + 14$ _____
6. $y = x + 2$
   $-y = x + 4$ _____
7. $y = 4 - \frac{1}{4}x$
   $3x + 4y = 4$ _____
8. $x + y = 5$
   $5 - y = 2x$ _____
9. $x - 4y = 0$
   $4y = x - 8$ _____

Consider the line (l) shown in each of the following graphs, and write the equation of the line in the space provided. Then, on the same graph, graph the line (r) for which the equation is given. Write how the slope and y-intercept of line l compare to the slope and y-intercept of line r for each graph.

10.

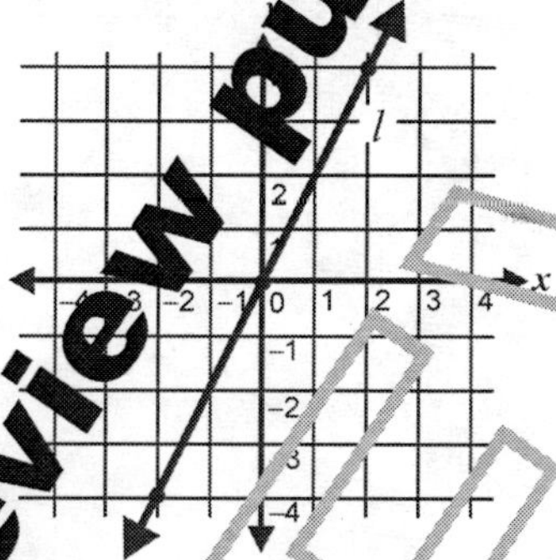

line l: _____
line r: $y = -2x$
slopes: _____
y-intercepts: _____

12.

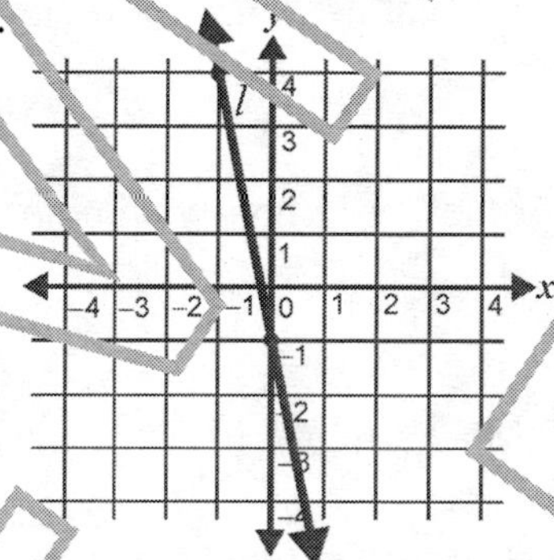

line l: _____
line r: $y = -3x - 1$
slopes: _____
y-intercepts: _____

14.

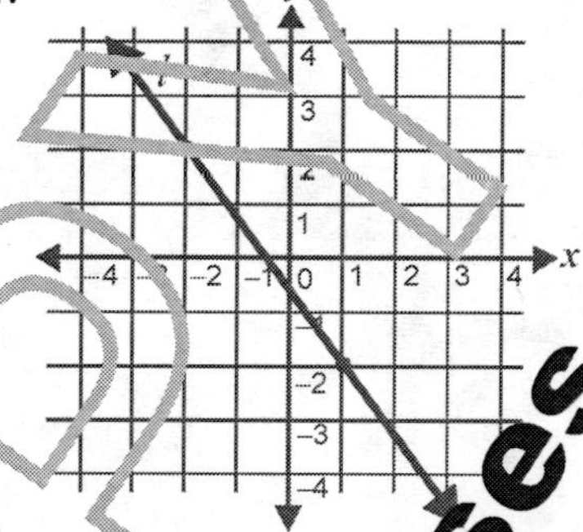

line l: _____
line r: $y = \frac{1}{4}x - $ _____
slopes: _____
y-intercepts: _____

11.

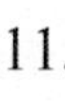

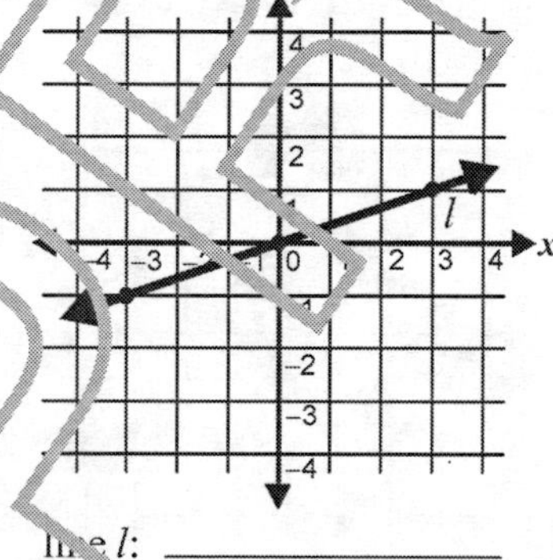

line l: _____
line r: $y = \frac{1}{3}x + 2$
slopes: _____
y-intercepts: _____

13.

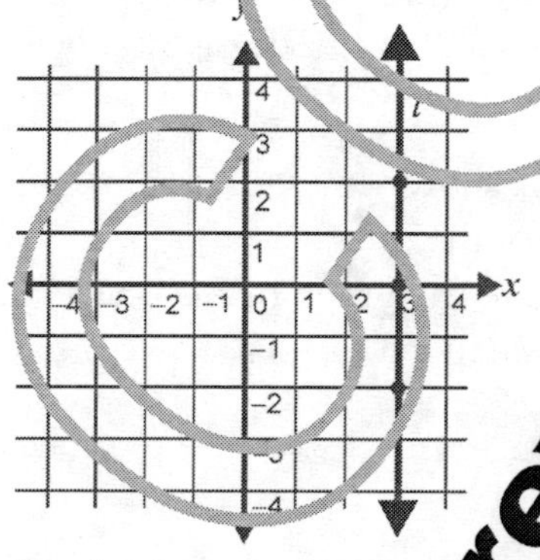

line l: _____
line r: $y = -3$
slopes: _____
y-intercepts: _____

15.

line l: _____
line r: $y = -\frac{1}{2}x - 3$
slopes: _____
y-intercepts: _____

13.2 Equations of Perpendicular Lines

Now that we know how to calculate the slope of lines using two points, we are going to learn how to calculate the slope of a line perpendicular to a given line, then find the equation of that perpendicular line. To find the slope of a line perpendicular to any given line, take the slope of the first line, m:

1. multiply the slope by -1
2. invert (or flip over) the slope

You now have the slope of a perpendicular line. Writing the equation for a line perpendicular to another line involves three steps:

1. find the slope of the perpendicular line
2. choose one point on the first line
3. use the point-slope form to write the equation

Example 3: The solid line on the graph below has a slope of $\frac{2}{3}$. Write the equation of a line perpendicular to the solid line.

Find the slope of the solid line. Multiply the slope by -1 and then find the inverse (flip it over).

$\frac{2}{3} \times -1 = -\frac{2}{3} \curvearrowright -\frac{3}{2}$

The slope of the perpendicular line, shown as a dotted line on the graph below, is $-\frac{3}{2}$.

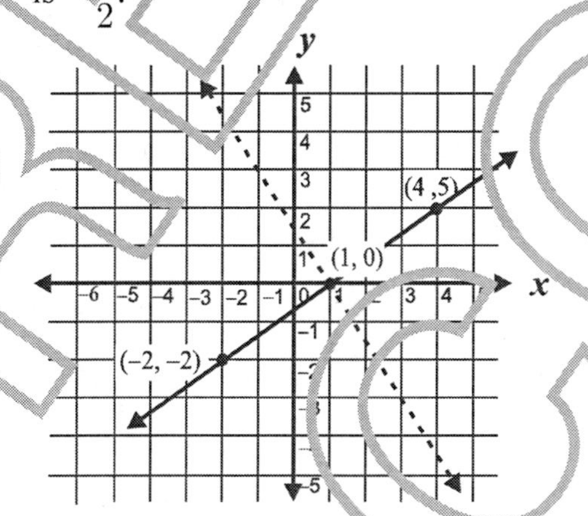

Step 2: Choose one point on the first line. We will use $(1, 0)$ in this example. The point $(-2, -2)$ or $(4, 5)$ could also be used.

Step 3: Use the point-slope formula, $(y - y_1) = m(x - x_1)$, to write the equation of the perpendicular line. Remember, we chose $(1, 0)$ as our point. So, $(y - 0) = -\frac{3}{2}(x - 1)$. Simplified, $y = -\frac{3}{2}x + \frac{3}{2}$.

Solve the following problems involving perpendicular lines.

1. Find the slope of the line perpendicular to the solid line shown at right, and draw the perpendicular as a dotted line. Use the point $(-1, 0)$ on the solid line and the calculated slope to find the equation of the perpendicular line.

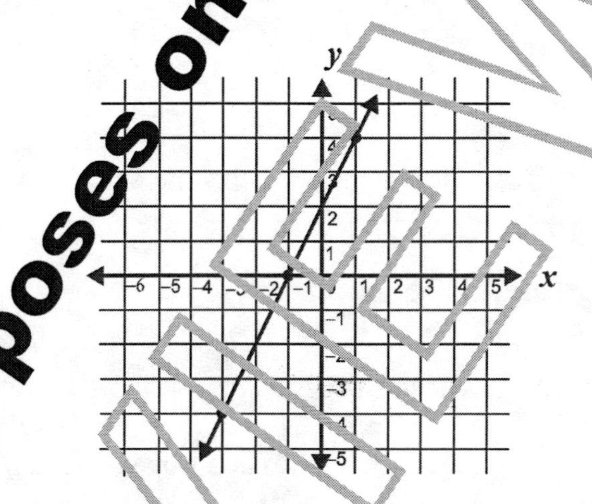

Find the equation of the perpendicular line using the point and slope given and the formula $(y - y_1) = m\,(x - x_1)$.

2. $(2, 1), 5$

3. $(3, 2), 2$

4. $(-2, 1), -3$

5. $(-4, 2), -\frac{1}{2}$

6. $(-1, 4), 1$

7. $(3, 3), \frac{2}{3}$

8. $(5, -1), -1$

9. $\left(\frac{1}{2}, \frac{3}{4}\right), 4$

10. $\left(\frac{2}{3}, \frac{3}{4}\right), -\frac{1}{6}$

11. $(7, -2), -\frac{1}{8}$

12. $(5, 6), \frac{2}{5}$

13. $(-3, -3), -\frac{7}{3}$

14. $\left(\frac{1}{4}, 4\right), \frac{1}{2}$

15. $(0, 6), -\frac{1}{9}$

13.3 Writing an Equation From Data

Data is often written in a two-column format. If the increases or decreases in the ordered pairs are at a constant rate, then a linear equation for the data can be found.

Example 4: Write an equation for the following set of data.

Dan set his car on cruise control and noted the distance he went every 5 minutes.

Minutes in operation (x)	Odometer reading (y)
5	28,490 miles
10	28,494 miles

Step 1: Write two order pairs in the form (minutes, distance) for Dan's driving, $(5, 28490)$ and $(10, 28494)$, and find the slope.

$$m = \frac{28494 - 28490}{10 - 5} = \frac{4}{5}$$

Step 2: Use the ordered pairs to write the equation in the form $y = mx + b$. Place the slope, m, that you found and one of the pairs of points as x_1 and y_1 in the following formula, $y - y_1 = m(x - x_1)$.

$$y - 28490 = \tfrac{4}{5}(x - 5)$$
$$y - 28490 = \tfrac{4}{5}x - 4$$
$$y - 28490 + 28490 = \tfrac{4}{5}x - 4 + 28490$$
$$y + 0 = \tfrac{4}{5}x + 28486$$
$$y = \tfrac{4}{5}x + 28486$$

Write an equation for each of the following sets of data, assuming the relationship is linear.

1. **Doug's Doughnut Shop**

Year in Business	Total Sales
1	$55,000
4	$85,000

2. **Gwen's Green Beans**

Days Growing	Height in Inches
2	5
6	12

3. **Jim's Depreciation of His Jet Ski**

Years	Value
1	$4,500
6	$2,500

4. **Stepping on the Brakes**

Seconds	MPH
2	51
5	18

13.4 Graphing Linear Data

Many types of data are related by a constant ratio. As you learned on the previous page, this type of data is linear. The slope of the line described by linear data is the ratio between the data. Plotting linear data with a constant ratio can be helpful in finding additional values.

Example 5: A department store prices socks per pair. Each pair of socks costs \$0.75. Plot pairs of socks versus price on a Cartesian plane.

Step 1: Since the price of the socks is constant, you know that one pair of socks costs \$0.75, 2 pairs of socks cost \$1.50, 3 pairs of socks cost \$2.25, and so on. Make a list of a few points.

Pair(s) x	Price y
1	**0.75**
2	**1.50**
3	**2.25**

Step 2: Plot these points on a Cartesian plane, and draw a straight line through the points.

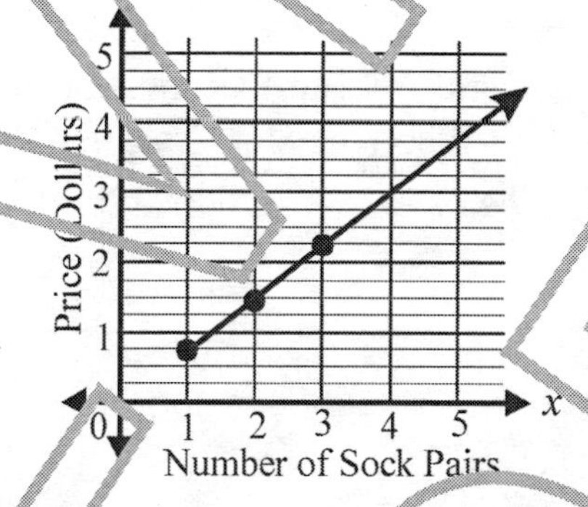

Example 6: What is the slope of the data in the example above? What does the slope describe?

Solution: You can determine the slope either by the graph or by the data points. For this data, the slope is .75. Remember, slope is rise/run. For every \$0.75 going up the y-axis, you go across one pair of socks on the x-axis. The slope describes the price per pair of socks.

Example 7: Use the graph created in the above example to answer the following questions. How much would 5 pairs of socks cost? How many pairs of socks could you purchase for \$3.00? Extending the line gives useful information about the price of additional pairs of socks.

Solution 1: The line that represents 5 pairs of socks intersects the data line at \$3.75 on the y-axis. Therefore, 5 pairs of socks would cost \$3.75.

Solution 2: The line representing the value of \$3.00 on the y-axis intersects the data line at 4 on the x-axis. Therefore, \$3.00 will buy exactly 4 pairs of socks.

Use the information given to make a line graph for each set of data, and answer the questions related to each graph.

1. The diameter of a circle versus the circumference of a circle is a constant ratio. Use the data given below to graph a line to fit the data. Extend the line, and use the graph to answer the next question.

Circle

Diameter	Circumference
4	12.56
5	15.70

2. Using the graph of the data in question 1, estimate the circumference of a circle that has a diameter of 3 inches.

3. If the circumference of a circle is 3 inches, about how long is the diameter?

4. What is the slope of the line you graphed in question 1?

5. What does the slope of the line in question 4 describe?

6. The length of a side of a square and the perimeter of a square are constant ratios to each other. Use the data below to graph this relationship.

Square

Length of side	Perimeter
2	8
3	12

7. Using the graph from question 6, what is the perimeter of a square with a side that measure 4 inches?

8. What is the slope of the line graphed in question 6?

9. Conversions are often constant ratios. For example, converting from pounds to ounces follows a constant ratio. Use the data below to graph a line that can be used to convert pounds to ounces.

Measurement Conversion

Pounds	Ounces
2	32
4	64

10. Use the graph from question 9 to convert [illegible]0 ounces to pounds.

11. What does the slope of the line graphs for question 9 represent?

12. Graph the data below, and create a line that shows the conversion from weeks to days.

Time

Weeks	Days
1	7
2	14

13. About how many days are in $2\frac{1}{2}$ weeks?

13.5 Identifying Graphs of Linear Equations

Match each equation below with the graph of the equation.

A.	$x = -4$	D.	$y = -4$	G.	$x - 2y = 6$
B.	$x = y$	E.	$4x + y = 4$	H.	$2x + 3y = 6$
C.	$-2x = y$	F.	$y = x - 3$	I.	$y = 3x + 2$

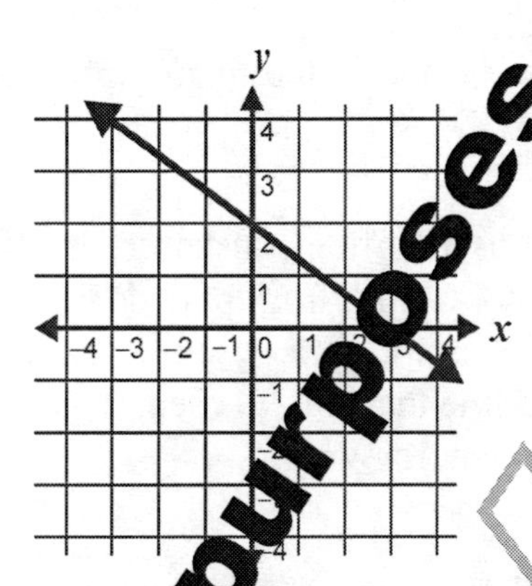

1. ________

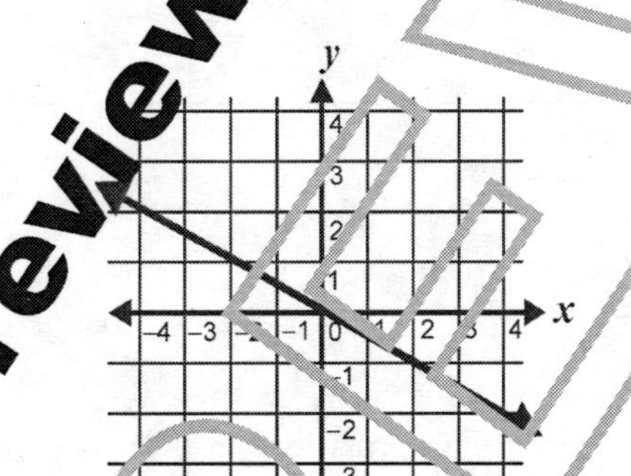

2. ________

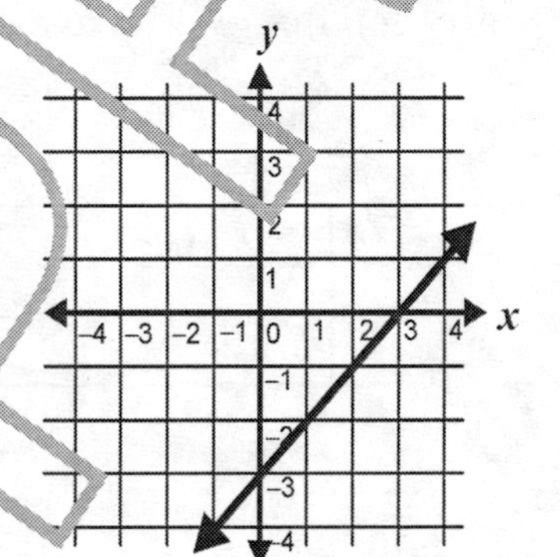

3. ________

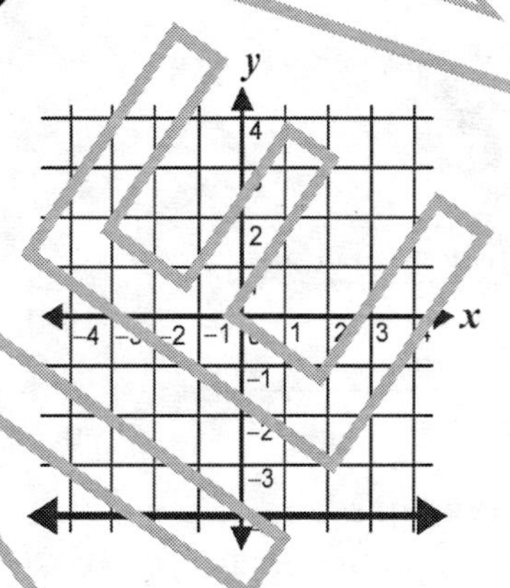

4. ________

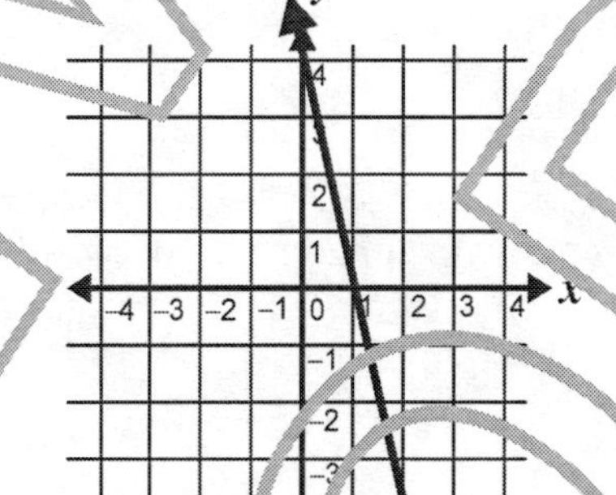

5. ________

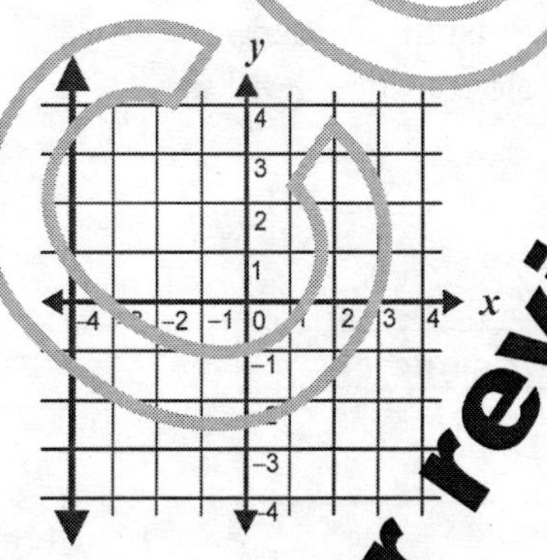

6. ________

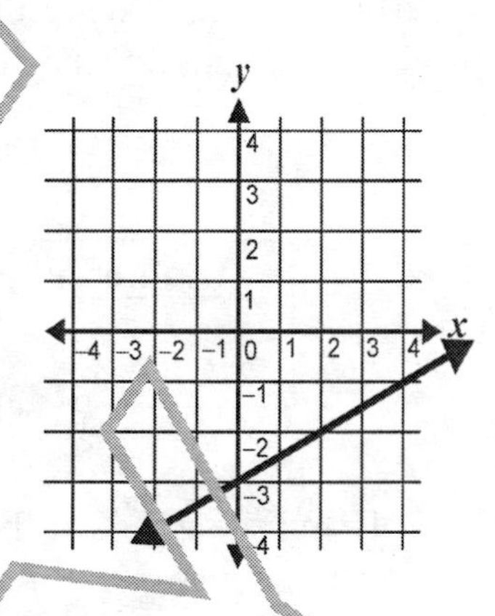

7. ________

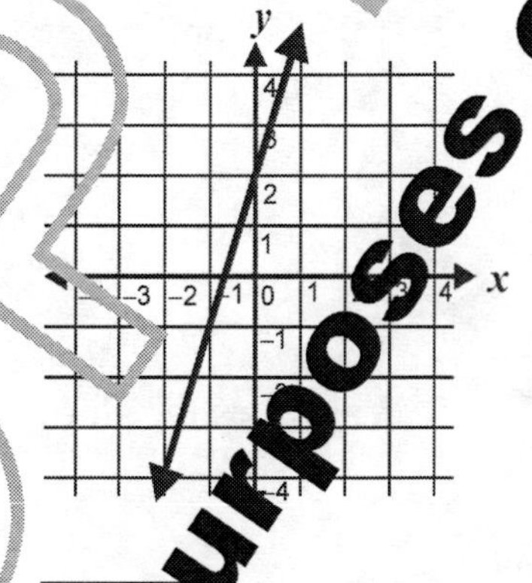

8. ________

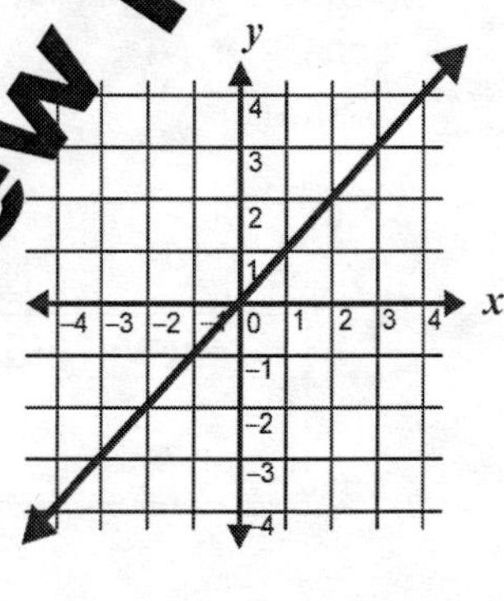

9. ________

13.6 Graphing Non-Linear Equations

Equations that you may encounter on the SAT exam will possibly involve variables which are squared (raised to the second power). The best way to find values for the x and y variables in an equation is to plug one number into x, and then find the corresponding value for y just as you did at the beginning of this chapter. Then, plot the points and draw a line through the points.

Example 8: Graph $y = x^2$.

Step 1: Make a table and find several values for x and y.

x	y
-2	4
-1	1
0	0
1	1
2	4

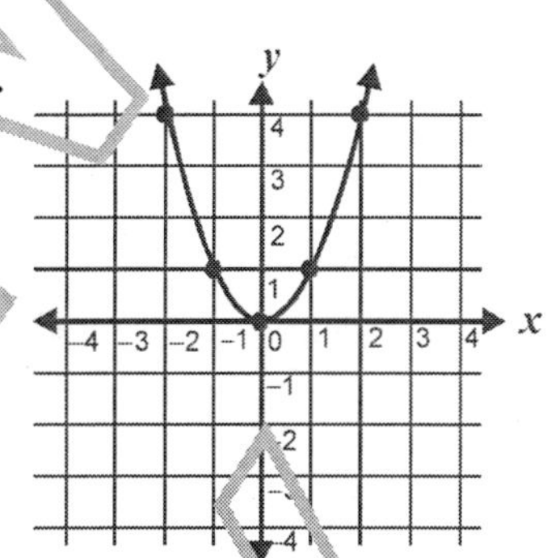

Step 2: Plot the points, and draw a curve through the points. Notice the shape of the curve. This type of curve is called a **parabola**. Equations with one squared term will be parabolas.

Example 9: Graph the equation $y = -2x^2 + 4$.

Step 1: Make a table and find several values for x and y.

x	y
-2	-4
-1	2
0	4
1	2
2	-4

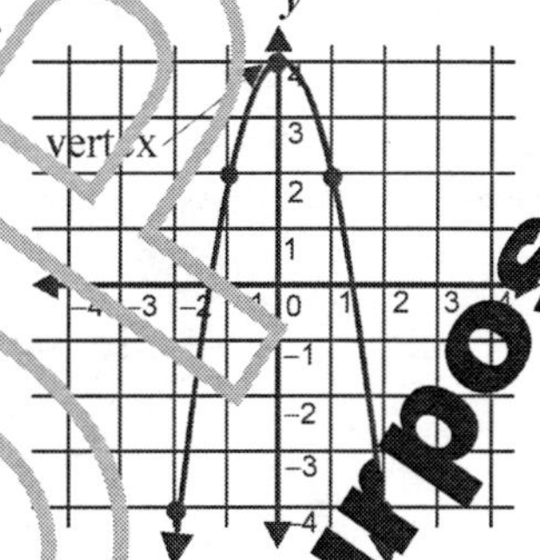

Step 2: Plot the points, and draw a curve through the points.

Note: In the equation $y = ax^2 + c$, changing the value of a will widen or narrow the parabola around the y-axis. If the value of a is a negative number, the parabola will be reflected across the x-axis (the vertex will be at the top of the parabola instead of at the bottom.) If $a = 0$, the graph will be a straight line, not a parabola. Changing the value of c will move the vertex of the parabola from the origin to a different point on the y-axis.

Graph the equations below on a Cartesian plane.

1. $y = 2x^2$
2. $y = 3 - x^2$
3. $y = x^2 - 2$
4. $y = -2x^2$
5. $y = x^2 + 3$
6. $y = -3x^2 + 2$
7. $y = 3x^2 - 5$
8. $y = x^2 + 1$
9. $y =$ [illegible] $- 6$
10. $y = -x^2$
11. $y = 2x^2 - 1$
12. $y = 2 - 2x^2$

13.7 Finding the Vertex of a Quadratic Equation

The vertex of a quadratic equation is the point on the parabola where the graph changes directions from increasing to decreasing or decreasing to increasing.

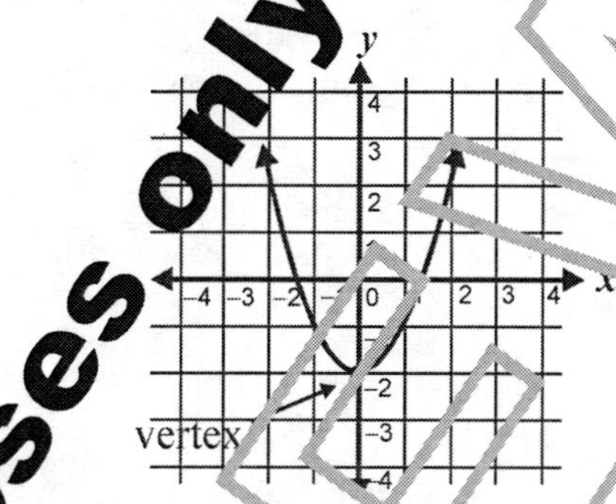

As a reminder, the quadratic equation is defined as $y = ax^2 + bx + c$, in which the coefficient a can never equal zero. If $a = 0$, then the equation will not have a x^2 term, which is what makes it a quadratic equation. The quadratic equation can also be written as a function of x by substituting $f(x)$ for y, such as $f(x) = ax^2 + bx + c$. To find the point of the vertex of the graph, you must use the formula below.

$$\text{vertex} = \left(-\frac{b}{2a}, f\left(-\frac{b}{2a}\right)\right)$$

where $f\left(-\frac{b}{2a}\right)$ is the quadratic equation evaluated at the value $-\frac{b}{2a}$. To do this, plug $-\frac{b}{2a}$ in for x.

Example 10: Find the point of the vertex of $f(x) = 2x^2 - 12x + 10$.

Step 1: First, find out what a and b equal. Since a is the coefficient of x^2, $a = 2$. b is the coefficient of x, so $b = -12$.

Step 2: Find the solution to $-\frac{b}{2a}$ by substituting the values of a and b from the equation into the expression.

$$-\frac{b}{2a} = -\left(\frac{-12}{2 \times 2}\right) = -\left(\frac{-12}{4}\right) = -(-3) = 3$$

Step 3: Find the solution to $f\left(-\frac{b}{2a}\right)$. We know that $-\frac{b}{2a} = 3$, so we need to find $f(3)$. To do this, we must substitute 3 into the quadratic equation for x.

$f(x) = 2x^2 - 12x + 10$

$f(3) = 2(3)^2 - 12(3) + 10 = 2(9) - 36 + 10 = 18 - 36 + 10 = -8$

The vertex equals $(3, -8)$.

Find the point of the vertex for the following equations.

1. $f(x) = x^2 + 6$
2. $f(x) = 2x^2 - 8$
3. $f(x) = x^2 + 10x - 4$
4. $f(x) = 2x^2 - 16x - 8$
5. $f(x) = x^2 + 4x - 5$
6. $f(x) = 3x^2 - 12x + 16$
7. $f(x) = x^2 - 25$
8. $f(x) = -x^2 + 6x - 12$
9. $f(x) = 4x^2 - 64x + 200$

13.8 Identifying Graphs of Real-World Situations

Real-world situations are sometimes modeled by graphs. Although an equation cannot be written for most of these graphs, interpreting these graphs provides valuable information. Situations may be represented on a graph as a function of time, length, temperature, etc.

The graph below depicts the temperature of a pond at different times of the day. Refer to the graph as you read through examples 1 and 2.

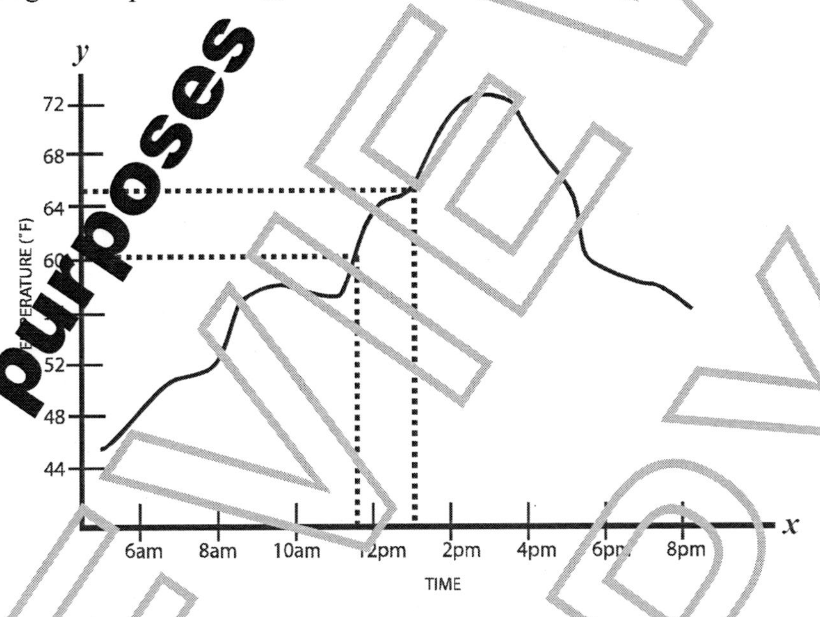

Example 11: If it is known that a specific breed of fish is most active in waters between 60°F and 65°F, what time of the day would this fish be the most active in this particular pond?

To find the answer, draw lines from the 60°F and 65°F points on the y-axis to the graph. Then, draw vertical lines from the graph to the x-axis. The time range between the two vertical lines on the x-axis indicates the time that the fish are most active. It can be determined from the graph that the fish are most active between 11:30 am and 1:00 PM.

Example 12: Describe the way the temperature of the pond acts as a function of time.

At 6 AM, the temperature of the pond is about 45°F. The temperature increases relatively steadily throughout the morning and early afternoon. The temperature peaks at 72°F, which is around 2:30 PM during the day. Afterwards, the temperature of the pond starts to decrease. The later it gets in the evening, the more the temperature of the water decreases. The graph shows that at 8 PM the temperature of the pond is about 57°F.

Use the graphs to answer the questions. Circle your answers.

The following graph depicts the number of articles of clothing as a function of time throughout the year. Use this graph for questions 1 and 2.

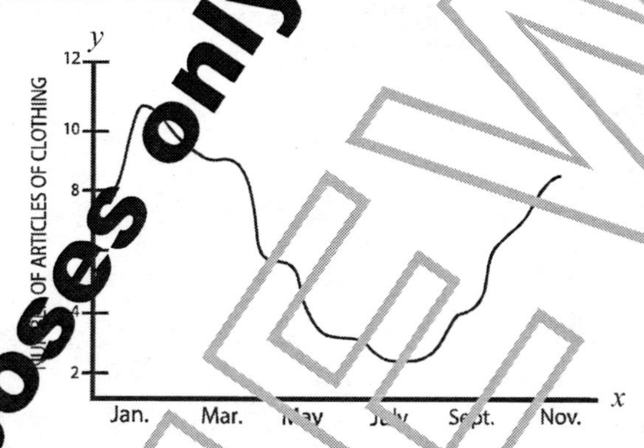

1. According to the graph, in what month are the most articles of clothing worn?
   (A) January
   (B) March
   (C) May
   (D) November

2. What is the average number of clothing a person wears in June?
   (A) 6
   (B) 3
   (C) 2
   (D) 5

The graph below depicts the efficiency of energy transfer as a function of distance in a certain element. Use the graph to answer question 3 and 4.

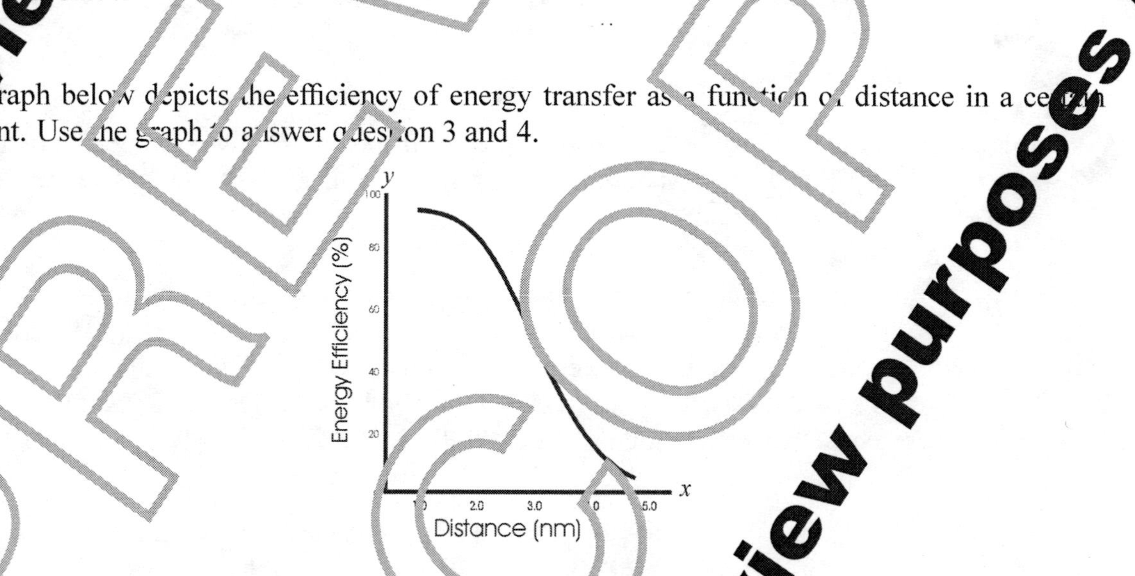

3. At what distance is the energy efficiency at 50%?
   (A) 1.0 nm
   (B) 2.0 nm
   (C) 3.5 nm
   (D) 3.0 nm

4. What is the energy efficiency at distance 2.5 nm?
   (A) 100%
   (B) 90%
   (C) [illegible]
   (D) [illegible]

Find the best non-linear graph to match each scenario.

1. Cathy begins her two-hour drive to her mother's house in her new sedan. She drives slowly through her city for thirty minutes to reach Interstate 95. After she enters the highway, she travels a constant 60–70 miles per hour for the next hour until she reaches her mother's exit. She then drives slowly down back roads to arrive at her mother's house.

2. Phillip is flying to Texas for a business meeting. When his flight leaves, the airplane increases its speed a great deal until it reaches about 500 miles per hour. After 20 minutes, the plane levels off for the last 45 minutes at 500 miles per hour. As the airplane nears the airport in Fort Worth, TX, it decreases its speed until it lands and reaches zero miles per hour.

3. Erica and her father like to build rockets for fun, and every Saturday they go to the park by their house to launch the rockets. Almost immediately after takeoff, the rocket reaches its greatest speed. Affected by gravity, it slows down until it reaches its peak height. It again speeds up as it descends to the ground.

4. Molly and her mother ride the train each time they go to the zoo. Molly knows that the train slows down twice so that the passengers can view the animals. Her favorite part of the ride, though, is when the train moves very quickly before it slows down to approach the station and come to a stop.

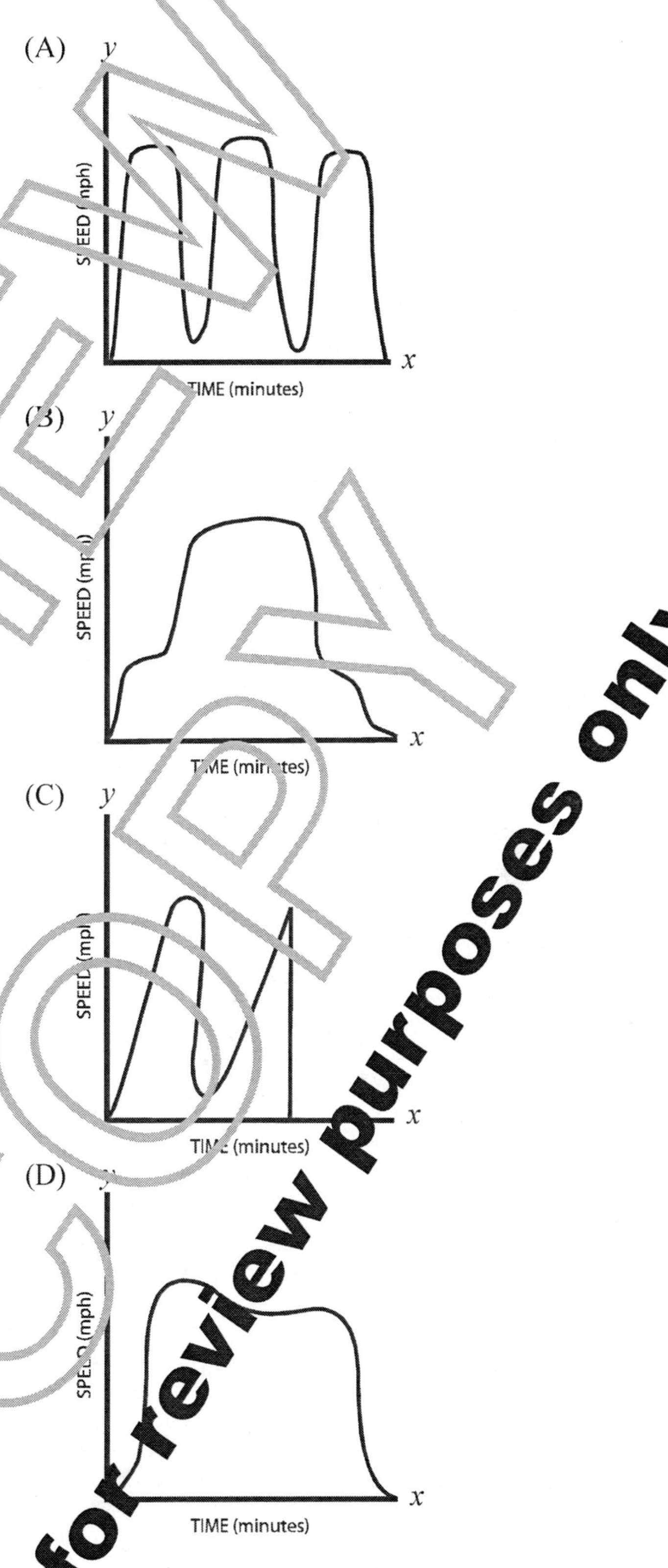

Chapter 13 Review

1. Paulo turned on the oven to preheat it. After one minute, the oven temperature was 200°. After 2 minutes, the oven temperature was 325°.

Oven Temperature

Minutes	Temperature
1	200°
2	325°

Assuming the oven temperature rose at a constant rate, write an equation that fits the data.

2. Write an equation that fits the data given below. Assume the data is linear.

Plumber Charges per Hour

Hours	Charge
1	\$170
2	\$220

3. Graph the equation $y = -\frac{1}{2}x^2 + 1$.

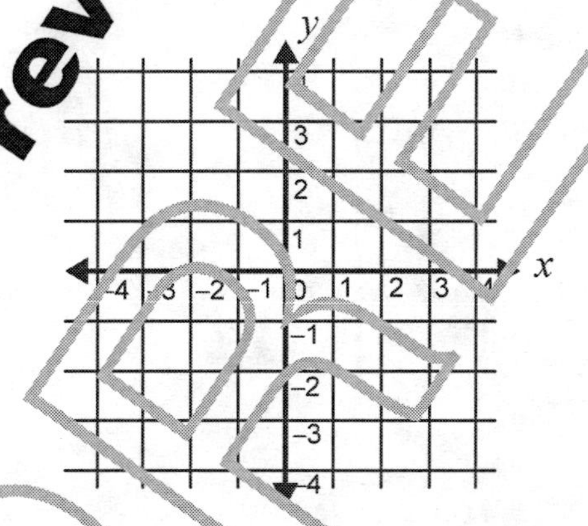

4. What happens to a graph of the line if the slope changes from 2 to −2?

(A) The graph will move down 4 spaces.
(B) The graph will slant downward towards the left instead of the right.
(C) The graph will flatten out to be more vertical.
(D) The graph will slant downward towards the right instead of the left.

5. What happens to a graph if the y-intercept changes from 4 to −2?

(A) The graph will move down 2 spaces.
(B) The graph will slant towards the left instead of the right.
(C) The graph will move down 6 spaces.
(D) The graph will move up 6 spaces.

6. The graph of the line $y = 3x - 1$ is shown below. On the same graph, draw the line $y = -\frac{1}{3}x - 1$.

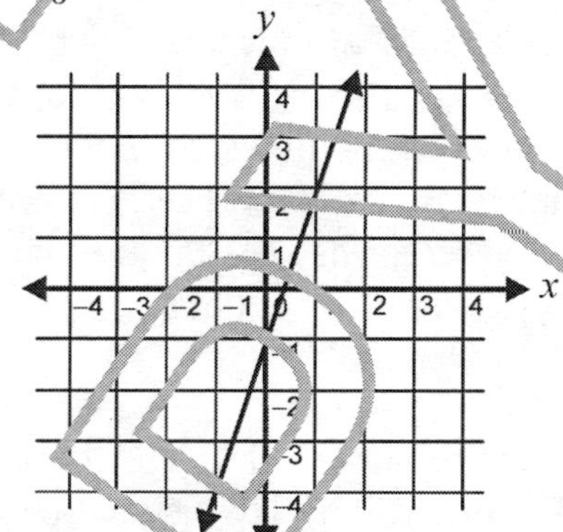

7. Which of the following statements is an accurate comparison of the lines $y = 3x - 1$ and $y = -\frac{1}{3}x - 1$?

(A) Only their y-intercepts are different.
(B) Only their slopes are different.
(C) Both their y-intercepts and their slopes are different.
(D) There is no difference between these two lines.

8. What is the name of the curve described by the equation $y = 2x^2 - 1$?

9. The data given below show conversions between miles per hour and kilometers per hour. Based on this data, graph a conversion line on the Cartesian plane below.

Speed	
MPH	**KPH**
5	8
10	16

KPH: 2, 4, 6, 8, 10, 12, 14, 16, 18, 20
MPH: 2, 4, 6, 8, 10, 12, 14

10. What would be the approximate conversion of 9 mph to kph?

11. What would be the approximate conversion of 15 kph to mph?

12. A bicyclist travels 12 mph downhill. Approximately how many kph is the bicyclist traveling?

13. Use the data given below to graph the interest rate versus the interest rate on $80.00 in one year.

$80.00 Principal	
Interest Rate	**Interest - 1 Year**
5%	$4.00
10%	$8.00

14. About how much interest would accrue in one year at an 8% interest rate?

15. What is the slope of the line describing interest versus interest rate?

16. What information does the slope give in problem 15?

17. Draw the graph of the following situation on the Cartesian plane provided. A girl rode her bicycle up a hill, then coasted down the other side of the hill on her bike. At the bottom she stopped.

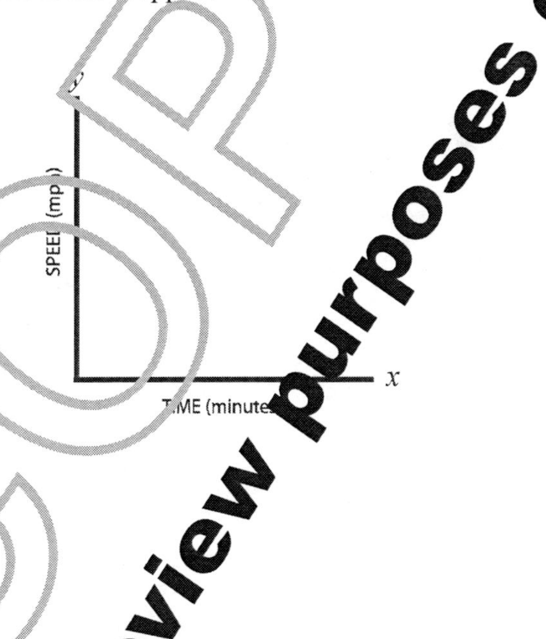

Chapter 14
Systems of Equations and Systems of Inequalities

Two linear equations considered at the same time are called a **system** of linear equations. The graph of a linear equation is a straight line. The graphs of two linear equations can show that the lines are **parallel, intersecting,** or **collinear**. Two lines that are **parallel** will never intersect and have no ordered pairs in common. If two lines are **intersecting**, they have one point in common, and in this chapter, you will learn to find the ordered pair for that point. If the graph of two linear equations is the same line, the lines are said to be **collinear**.

If you are given a system of two linear equations, and you put both equations in slope-intercept form, you can immediately tell if the graph of the lines will be **parallel, intersecting**, or **collinear**.

If two linear equations have the same slope and the same y-intercept, then they are both equations for the same line. They are called **collinear** or **coinciding** lines. A line is made up of an infinite number of points extending infinitely far in two directions. Therefore, collinear lines have an infinite number of points in common.

Example 1: $2x + 3y = -3$ **In slope intercept form:** $y = -\frac{2}{3}x - 1$

$4x + 6y = -6$ **In slope intercept form:** $y = -\frac{2}{3}x - 1$

The slope and y-intercept of both lines are the same.

If two linear equations have the same slope but different y-intercepts, they are **parallel** lines. Parallel lines never touch each other, so they have no points in common.

If two linear equations have different slopes, then they are intersecting lines and share exactly one point in common.

The chart below summarizes what we know about the graphs of two equations in slope-intercept form.

y-Intercepts	Slopes	Graphs	Number of Solutions
same	same	collinear	infinite
different	same	distinct parallel lines	none (they never touch)
same or different	different	intersecting lines	exactly one

For the pairs of equations below, put each equation in slope-intercept form, and tell whether the graphs of the lines will be collinear, parallel, or intersecting.

1. $x - y = -1$
   $-x + y = -1$

2. $x - 2y = 4$
   $-x + 2y = 6$

3. $y - 2 = x$
   $x + 2 = y$

4. $x = y - 1$
   $-x = y - 1$

5. $2x + 5y = 10$
   $4x + 10y = 20$

6. $x + y = 3$
   $x - y = 1$

7. $2y = 4y - 6$
   $-6x + y = 3$

8. $x + y = 5$
   $2x + 2y = 10$

9. $2x = 3y - 6$
   $4x = 6y - 6$

10. $2x - 2 = 2$
    $3y = -x + 5$

11. $x = -y$
    $x = 4 - y$

12. $2x = y$
    $x + y = 3$

13. $x = y + 1$
    $y = x + 1$

14. $x - 2y = 4$
    $-2x + 4y = -8$

15. $2x + 3y = 4$
    $-2x + 3y = -8$

16. $2x - 4y = 1$
    $-6x + 12y = 3$

17. $-3x + 4y = 1$
    $6x + 8y = 2$

18. $x + y = 2$
    $5x + 5y = 10$

19. $x + y = 4$
    $x - y = 4$

20. $y = -x + 3$
    $x - y = 1$

14.1 Finding Common Solutions for Intersecting Lines

When two lines intersect, they share exactly one point in common.

Example 2: $3x + 4y = 20$ and $4x + 2y = 12$

Put each equation in slope-intercept form.

$$\begin{aligned} 3x + 4y &= 20 \\ 4y &= -3x + 20 \\ y &= -\tfrac{3}{4}x + 5 \end{aligned} \qquad \begin{aligned} 2y - 4x &= 12 \\ 2y &= 4x + 12 \\ y &= 2x + 6 \end{aligned}$$

slope-intercept form

Straight lines with different slopes are **intersecting lines**. Look at the graphs of the lines on the same Cartesian plane.

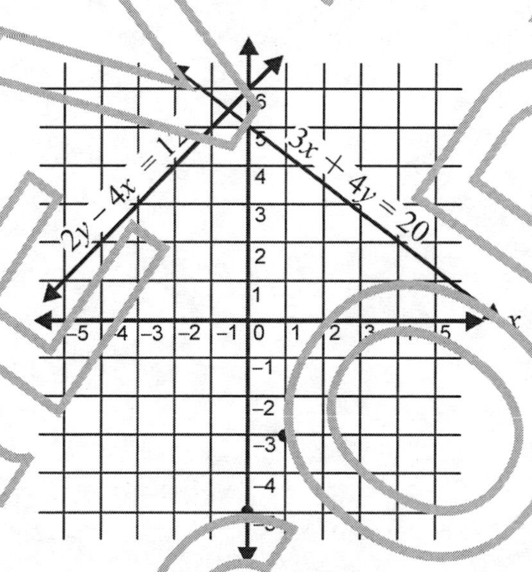

You can see from looking at the graph that the intersecting lines share one point in common. However, it is hard to tell from looking at the graph what the coordinates are for the point of intersection. To find the exact point of intersection, you can use the **substitution method** to solve the system of equations algebraically.

14.2 Solving Systems of Equations by Substitution

You can solve systems of equations by using the substitution method.

Example 3: Find the point of intersection of the following two equations:

Equation 1: $x - y = 3$

Equation 2: $2x + y = 9$

Step 1: Solve one of the equations for x or y. Let's choose to solve equation 1 for x.

Equation 1:
$$x - y = 3$$
$$x = y + 3$$

Step 2: Substitute the value of x from equation 1 in place of x in equation 2.

Equation 2:
$$2x + y = 9$$
$$2(y + 3) + y = 9$$
$$2y + 6 + y = 9$$
$$3y + 6 = 9$$
$$3y = 3$$
$$y = 1$$

Step 3: Substitute the solution for y back in equation 1 and solve for x.

Equation 1:
$$x - y = 3$$
$$x - 1 = 3$$
$$x = 4$$

Step 4: The solution set is $(4, 1)$. Substitute in one or both of the equations to check.

Equation 1:
$$x - y = 3$$
$$4 - 1 = 3$$
$$3 = 3$$

Equation 2:
$$2x + y = 9$$
$$2(4) + 1 = 9$$
$$8 + 1 = 9$$
$$9 = 9$$

The point $(4, 1)$ is common for both equations. This is the **point of intersection**.

For each of the following pairs of equations, find the point of intersection, the common solution, using the substitution method.

1. $x + 2y = 8$
   $2x - 3y = 2$
2. $x - y = -5$
   $x + y = 1$
3. $x - y = 4$
   $x + y = 2$
4. $x - y = -1$
   $x + y = 9$
5. $-x + y = 2$
   $x + y = 8$
6. $x + 4y = 10$
   $x + 5y = 12$
7. $2x + 3y = 2$
   $4x - 9y = -1$
8. $x + 3y = 5$
   $x - y = 1$
9. $-x = y - 1$
   $x = y - 1$
10. $x - 2y = 2$
    $2y + x = -2$
11. $5x + 2y = 1$
    $2x + 4y = 10$
12. $3x - y = 2$
    $5x + y = 6$
13. $2x + 3y = 3$
    $4x + 5y = 5$
14. $x - y = 1$
    $-x - y = 1$
15. $x = y + 3$
    $y = 3 - x$

14.3 Solving Systems of Equations by Adding or Subtracting

You can solve systems of equations algebraically by adding or subtracting an equation from another equation or system of equations.

Example 4: Find the point of intersection of the following two equations.
Equation 1: $x + y = 10$
Equation 2: $-x + 4y = 5$

Step 1: Eliminate one of the variables by adding the two equations together. Since the x has the same coefficient in each equation, but opposite signs, it will cancel nicely by adding.

$$\begin{aligned} x + y &= 10 \\ +(-x + 4y &= 5) && \text{Add each like term together.} \\ \hline 0 + 5y &= 15 && \text{Simplify.} \\ 5y &= 15 && \text{Divide both sides by 5.} \\ y &= 3 \end{aligned}$$

Step 2: Substitute the solution for y back into an equation, and solve for x.

Equation 1: $x + y = 10$ — Substitute 3 for y.
$x + 3 = 10$ — Subtract 3 from both sides.
$x = 7$

Step 3: The solution set is $(7, 3)$. To check, substitute the solution into both of the original equations.

Equation 1:
$$\begin{aligned} x + y &= 10 \\ 7 + 3 &= 10 \\ 10 &= 10 \end{aligned}$$

Equation 2:
$$\begin{aligned} -x + 4y &= 5 \\ -(7) + 4(3) &= 5 \\ -7 + 12 &= 5 \\ 5 &= 5 \end{aligned}$$

The point $(7, 3)$ is the point of intersection.

Example 5: Find the point of intersection of the following two equations:
Equation 1: $3x - 2y = -1$
Equation 2: $-4y = -x - 7$

Step 1: Put the variables on the same side of each equation. Take equation 2 out of y-intercept form.

$$\begin{aligned} -4y &= -x - 7 && \text{Add } x \text{ to both sides.} \\ x - 4y &= -x + x - 7 && \text{Simplify.} \\ x - 4y &= -7 \end{aligned}$$

Step 2: Add the two equations together to cancel one variable. Since each variable has the same sign and different coefficients, we have to multiply one equation by a negative number so one of the variables will cancel. Equation 1's y variable has a coefficient of 2, and if multiplied by -2 the y will have the same variable as the y in equation 2, but a different sign. This will cancel nicely when added.

$$\begin{aligned} -2(3x - 2y &= -1) && \text{Multiply by } -2. \\ -6x + 4y &= 2 \end{aligned}$$

Step 3: Add the two equations.

$-6x + 4y = 2$	
$+ (x - 4y = -7)$	Add equation 2 to equation 1.
$-5x + 0 = -5$	Simplify.
$-5x = -5$	Divide both sides by -5.
$x = 1$	

Step 4: Substitute the solution for x back into an equation and solve for y.

Equation 1:	$3x - 2y = -1$	Substitute 1 for x.
	$3(1) - 2y = -1$	Simplify.
	$3 - 2y = -1$	Subtract 3 from both sides.
	$3 - 3 - 2y = -1 - 3$	Simplify.
	$-2y = -4$	Divide both sides by -2.
	$y = 2$	

Step 5: The solution set is $(1, 2)$. To check, substitute the solution into both of the original equations.

Equation 1:		Equation 2:	
	$3x - 2y = -1$		$-4y = -x - 7$
	$3(1) - 2(2) = -1$		$-4(2) = -1 - 7$
	$3 - 4 = -1$		$-8 = -8$
	$-1 = -1$		

The point $(1, 2)$ is the point of intersection.

For each of the following pairs of equations, find the point of intersection by adding the equations together. Remember you might need to change the coefficients and/or signs of the variables before adding.

1. $x + 2y = 8$
   $-x - 3y = [illegible]$
2. $x - y = 5$
   $2x + y = 1$
3. $x - y = -1$
   $x + y = 9$
4. $3x - y = -1$
   $x + y = 13$
5. $-x + 4y = 2$
   $x + y = 8$
6. $x + 4y = 10$
   $x + 7y = 16$
7. $2x - y = 2$
   $4x - 9y = -3$
8. $x + 3y = 13$
   $5x - y = 1$
9. $-x = y - 1$
   $x = y - 1$
10. $x - y = 2$
    $2y + x = 5$
11. $5x + 2y = 1$
    $4x + 8y = [illegible]$
12. $3x - [illegible] = 14$
    $x - y = 6$
13. $2x + 3y = 3$
    $3x + 5y = 5$
14. $x - 4y = 6$
    $-x - y = -1$
15. $x = 2y + 3$
    $y = 3 - x$

14.4 Graphing Systems of Inequalities

Systems of inequalities are best solved graphically. Look at the following example.

Example 6: Sketch the solution set of the following systems of inequalities:

$y > -2x - 1$ and $y \le 3x$

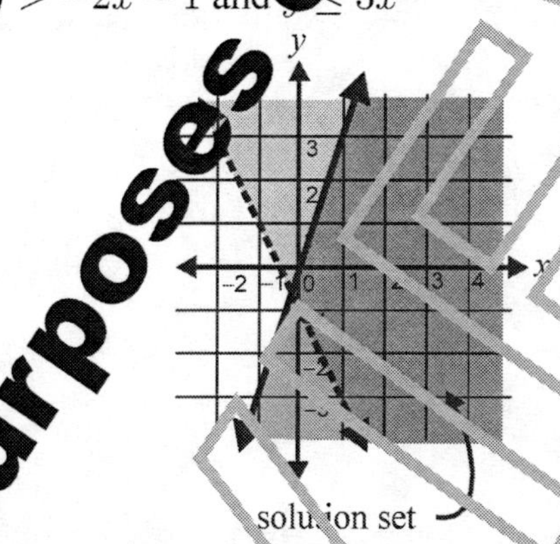

Step 1: Graph both inequalities on a Cartesian plane. Study the chapter on graphing inequalities if you need to review.

Step 2: Shade the portion of the graph that represents the solution set to each inequality just as you did in the chapter on graphing inequalities.

Step 3: Any shaded region that overlaps is the solution set of both inequalities.

Graph the following systems of inequalities on your own graph paper. Shade and identify the solution set for both inequalities.

1. $2x + 2y \ge -4$
   $3y < 2x + 6$

2. $7x + 7y \le 21$
   $8x < 6y - 24$

3. $9x + 12y < 36$
   $34x - 17y > 34$

4. $-11x - 22y \ge 44$
   $-4x + 2y \le 8$

5. $24x < 72 + 36y$
   $11x + 22y \le -33$

6. $15x - 60 < 30y$
   $20x + 10y < 40$

7. $-12x + 24y > -24$
   $10x < -5y + 15$

8. $y \ge 2x + 2$
   $y < -x - 3$

9. $3x + 4y \ge 12$
   $y > -3x + 2$

10. $-3x \le 6 + 2y$
    $y \ge -x - 2$

11. $2x - 2y \le 4$
    $3x + 3y \le -9$

12. $-x \ge -2y - 2$
    $-2x - 2y > 4$

14.5 Solving Word Problems with Systems of Equations

Certain word problems can be solved using systems of equations.

Example 7: In a game show, Andre earns 6 points for every right answer and loses 12 points for every wrong answer. He has answered correctly 12 times as many as he has missed. His final score was 120. How many times did he answer correctly?

Step 1: Let r = number of right answers.
Let w = number of wrong answers.

We know 2 sets of information that can be made into equations with 2 variables.

He earns 6 points for right answers and loses 12 points for wrong answers.

His wins and losses = 120

$$\begin{aligned} 6r - 12w &= 120 \\ 12w &= r \end{aligned}$$

12 times the number of wrong answers = the number of right answers.

Step 2: Substitute the value for r $(12w)$ in the first equation.

$$\begin{aligned} 6(12w) - 12w &= 120 \\ 72w - 12w &= 120 \\ 60w &= 120 \\ w &= 2 \end{aligned}$$

Step 3: Substitute the value for w back in the equation.

$$\begin{aligned} 6r - 12(2) &= 120 \\ 6r - 24 &= 120 \\ 6r &= 144 \\ r &= 24 \end{aligned}$$

Use systems of equations to solve the following word problems.

1. The sum of two numbers is 210 and their difference is 30. What are the two numbers?
2. The sum of two numbers is 126 and their difference is 42. What are the two numbers?
3. Kayla gets paid \$6.00 for raking leaves and \$8.00 for mowing the lawn for the neighbors around her subdivision. This year she mowed the lawns 12 times more than she raked leaves. In total, she made \$918.00 for doing both. How many times did she rake the leaves?
4. Prices for the movie were \$4.00 for children and \$8.00 for adults. The total amount of ticket sales was \$1,176. There were 172 tickets sold. How many adults and children bought tickets?
5. A farmer sells a dozen eggs at the market for \$2.00 and one of his bags of grain for \$5.00. He has sold 5 times as many bags of grain as he has dozens of eggs. By the end of the day, he has made \$243.00 worth of sales. How many bags of grain did he sell?
6. Every time Lauren does one of her chores, she gets 15 minutes to talk on the phone. When she does not perform one of her chores, she gets 20 minutes of phone time taken away. This week she has done her chores 5 times more than she has not performed her chores. In total, she has accumulated [illegible] minutes. How many times has Lauren not performed her chores?
7. The choir sold boxes of candy and teddy bears near Valentine's Day to raise money. They sold twice as many boxes of candy as they did teddy bears. Bears sold for \$8.00 each and candy sold for \$6.00. They collected \$380. How much of each item did they sell?
8. Mr. Marlow keeps ten and twenty dollar bills in his dresser drawer. He has 1 less than twice as many 10's as 20's. He has \$550 altogether. How many tens does he have?
9. Kosta was a contestant on a math quiz show. For every correct answer, Kosta received \$18.00. For every incorrect answer, Kosta lost \$24.00. Kosta answered the questions correctly twice as often as he answered the questions incorrectly. In total, Kosta won \$72.00. How many questions did Kosta answer incorrectly?
10. John Vasilovik works in landscaping. He gets paid \$50 for each house he pressure washes and \$20 for each lawn he mows. He gets 4 times more jobs for mowing lawns than for pressure washing houses. During a given month, John earned \$2,600. How many houses did John pressure wash?

14.6 Consecutive Integer Problems

	Examples:	Algebraic notation:
Consecutive integers follow each other in order	$1, 2, 3, 4$ $-3, -4, -5, -6$	$n, n+1, n+2, n+3$
Consecutive **even** integers:	$2, 4, 6, 8, 10$ $-12, -14, -16, -18$	$n, n+2, n+4, n+6$
Consecutive **odd** integers:	$3, 5, 7, 9$ $-5, -7, -9, -11$	$n, n+2, n+4, n+6$

Example 8: The sum of three consecutive odd integers is 63. Find the integers.

Step 1: Represent the three odd integers:
Let n = the first odd integer
$n + 2$ = the second odd integer
$n + 4$ = the third odd integer

Step 2: The sum of the integers is 63, so the algebraic equation is
$n + n + 2 + n + 4 = 63$. Solve for n.
$n = 19$

Solution: the first odd integer = 19
the second odd integer = 21
the third odd integer = 23

Check: Does $19 + 21 + 23 = 63$? Yes, it does.

Example 9: Find three consecutive odd integers such that the sum of the first and second is three less than the third.

Step 1: Represent the three odd integers just like above:
Let n = the first odd integer
$n + 2$ = the second odd integer
$n + 4$ = the third odd integer

Step 2: In this problem, the sum of the first and second integers is three less than the third integer, so the algebraic equation is written as follows:
$n + n + 2 = n + 4 - 3$
$n = -1$

Solution: the first odd integer = -1
the second odd integer = 1
the third odd integer = 3

Check: Is the sum of -1 and 1 three less than 3?
$-1 + 1 = 3 - 3$ or $0 = 0$ Yes, it is.

Solve the following problems.

1. Find three consecutive odd integers whose sum is 141.
2. Find three consecutive integers whose sum is -21.
3. The sum of three consecutive even integers is 48. What are the numbers?
4. Find two consecutive even integers such that six times the first equals five times the second.
5. Find two consecutive odd integers such that seven times the first equals five times the second.
6. Find two consecutive odd numbers whose sum is forty-four.

Chapter 14 Review

For each pair of equations below, tell whether the graphs of the lines will be collinear, parallel, or intersecting.

1. $y = 4x + 1$
   $y = 4x - 3$
2. $y - 4 = x$
   $2x + 8 = 2y$
3. x [illegible] $y = 5$
   x [illegible] $y = -1$
4. $2y - 3x = 6$
   $4y = 6x + 8$
5. $5y = 3x - 7$
   $4x - 3y = -7$
6. $2x - 2y = 2$
   $y - x = -1$

Find the common solution for each of the following pairs of equations.

7. $x - y = 2$
   $x + 4y = -3$
8. $x + y =$ [illegible]
   $x + 3y = 1$
9. $-4y = -2x + 4$
   $x = -2y$ [illegible] 2
10. $2x + 8y = 20$
    $5y = 12 - x$
11. $x = y - 3$
    $-x = y + 3$
12. $-2x + y = -3$
    x [illegible] $y = 9$

Find the point of intersection for each pair of equations by adding and/or subtracting the two equations.

13. $2x + y = 4$
    $3x - y = 6$
14. $x + 2y = 3$
    $x + 5y = 0$
15. $x + y = 1$
    $y = x + 7$
16. $2x + 4y = 5$
    $3x + 8y = 9$
17. $2x - 2y = 7$
    $3x - 5y = \frac{5}{2}$
18. $x - 3y = -2$
    $y = -\frac{1}{3}x + 4$

Graph the following systems of inequalities on your own graph paper. Identify the solution set to both inequalities.

19. $x + 2y \geq 2$
    $2x - y \leq 4$
20. $20x + 10y \leq 40$
    $3x + 2y \geq 6$
21. $6x + 8y \leq -24$
    $-4x + 8y \geq 16$
22. $14x - 7y \geq -28$
    $3x + 4y \leq$ [illegible] 12
23. $2y \geq$ [illegible] $+ 6$
    $2x - 4y \geq -4$
24. $9x - 6y \geq 18$
    [illegible]$y \geq 6x - 12$

Use systems of equations to solve the following word problems.

25. Chelsea Johnson is a bank teller. At the end of the day she had 85 \$5 and \$10 bills. They should total \$785. How many \$5's and \$10's should she have in her drawer?

26. Hargrove High School sold 227 tickets for their last basketball game. Adult tickets sold for \$5 and student tickets were \$2. How many adult tickets were sold if the ticket sales totalled \$574?

27. Every time Stephen walks the dog, he gets 30 minutes to play video or computer games. When he does not take out the dog on time, he gets a mess to clean up and loses 1 hour of video/computer game time. This week he has walked the dog on time 8 times more than he did not walk the dog on time. In total, he has accumulated 3 hours of video/computer time. How many times has Stephen not walked the dog on time?

28. On Friday, Rosa bought party hats and kazoos for her friend's birthday party. On Saturday she decided to purchase more when she found out more people were coming. How much did she pay for each party hat?

	Hats	Kazoos	Total Cost
Friday	15	20	\$15.00
Saturday	10	5	\$8.75

29. Timothy and Jesse went to purchase sports clothing they needed as soccer players. The table below shows what they bought and the amount they paid. What is the price of 1 soccer jersey?

	Soccer Jerseys	Tube Socks	Total Cost
Timothy	4	7	\$78.30
Jesse	3	5	\$57.60

30. Three consecutive integers have a sum of 240. Find the integers.

31. Find three consecutive even numbers whose sum is negative seventy-two.

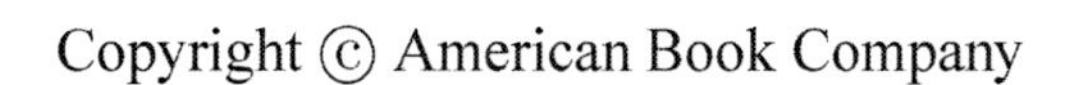

Chapter 15
Relations and Functions

15.1 Relations

A **relation** is a set of ordered pairs. The set of the first members of each ordered pair is called the **domain** of the relation. The set of the second members of each ordered pair is called the **range**.

Example 1: State the domain and range of the following relation:

$\{(2,4), (3,7), (4,9), (6,11)\}$

Solution: Domain: $\{2, 3, 4, 6\}$ the first member of each ordered pair

Range: $\{4, 7, 9, 11\}$ the second member of each ordered pair

State the domain and range for each relation.

1. $\{(2,5), (9,12), (3,8), (6,7)\}$

2. $\{(12,4), (3,4), (7,12), (26,19)\}$

3. $\{(4,3), (7,14), (16,34), (5,11)\}$

4. $\{(2,45), (33,43), (98,9), (43,61), (67,54)\}$

5. $\{(78,14), (29,67), (84,49), (16,18), (98,46)\}$

6. $\{(-8,16), (28,-7), (-4,-9), (16,-8), (-3,6)\}$

7. $\{(-7,-4), (-3,16), (-4,17), (-6,-8), (-8,12)\}$

8. $\{(-1,-2), (3,6), (-7,14), (-2,8), (-6,2)\}$

9. $\{(0,9), (-8,5), (3,12), (-8,-3), (7,18)\}$

10. $\{(58,14), (44,97), (74,32), (6,18), (63,44)\}$

When given an equation in two variables, the domain is the set of x values that satisfies the equation. The range is the set of y values that satisfies the equation.

Example 2: Find the range of the relation $3x = y + 2$ for the domain $\{-1, 0, 1, 2, 3\}$. Solve the equation for each value of x given. The result, the y values, will be the range.

Given:

x	y
-1	
0	
1	
2	
3	

Solution:

x	y
-1	-5
0	-2
1	1
2	4
3	7

The range is $\{-5, -2, 1, 4, 7\}$.

Find the range of each relation for the given domain.

	Relation	Domain	Range
1.	$y = 5x$	$\{1, 2, 3, 4\}$	
2.	$y = \|x\|$	$\{-3, -2, -1, 0, 1\}$	
3.	$y = 3x + 2$	$\{0, 1, 3, 4\}$	
4.	$y = -\|x\|$	$\{-2 - 1, 0, 1, 2\}$	
5.	$y = -2x + 1$	$\{0, 1, 3, 4\}$	
6.	$y = 10x - 2$	$\{-2 - 1, 0, 1, 2\}$	
7.	$y = 3\|x\| + 1$	$\{-2 - 1, 0, 1, 2\}$	
8.	$y - x = 0$	$\{1, 2, 3, 4\}$	
9.	$y - 2x = 0$	$\{1, 2, 3, 4\}$	
10.	$y = 3x - 1$	$\{0, 1, 3, 4\}$	
11.	$y = 4x + 2$	$\{0, 1, 3, 4\}$	
12.	$y = 2\|x\| - 1$	$\{-2 - 1, 0, 1, 2\}$	

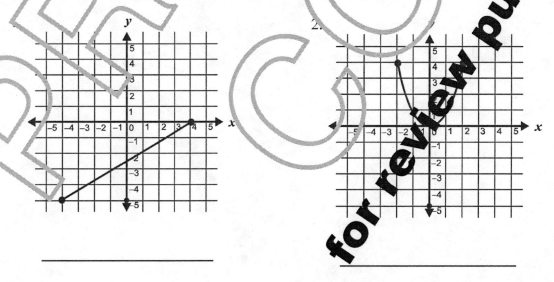

15.2 Determining Domain and Range From Graphs

The domain is all of the x values that lie on the function in the graph from the lowest x value to the highest x value. The range is all of the y values that lie on the function in the graph from the lowest y to the highest y.

Example 3: Find the domain and range of the graph.

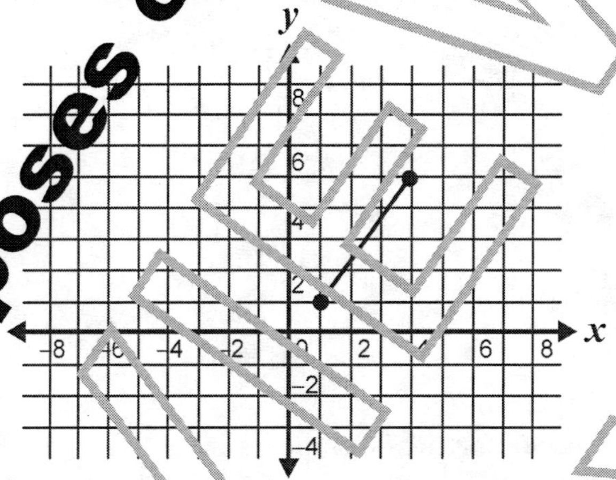

Step 1: First find the lowest x value depicted on the graph. In this case it is 1. Then find the highest x value depicted on the graph. The highest value of x on the graph is 4. The domain must contain all of the values between the lowest x value and the highest x value. The easiest way to write this is $1 \leq \text{Domain} \leq 4$ or $1 \leq x \leq 4$.

Step 2: Perform the same process for the range, but this time look at the lowest and highest y values. The answer is $1 \leq \text{Range} \leq 5$ or $1 \leq y \leq 5$.

Find the domain and range of each graph below. Write your answers on the line provided.

1.

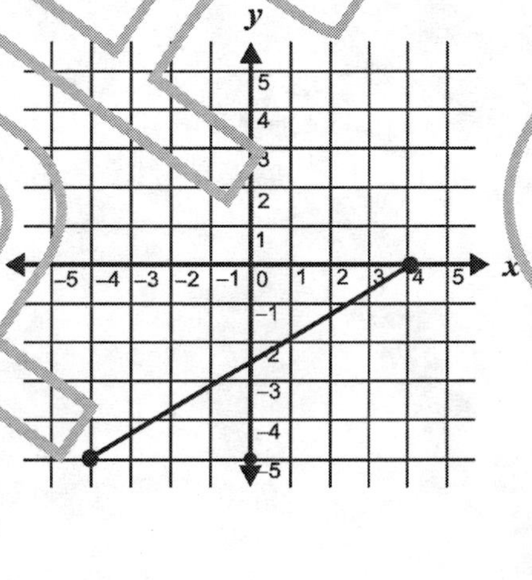

____

2.

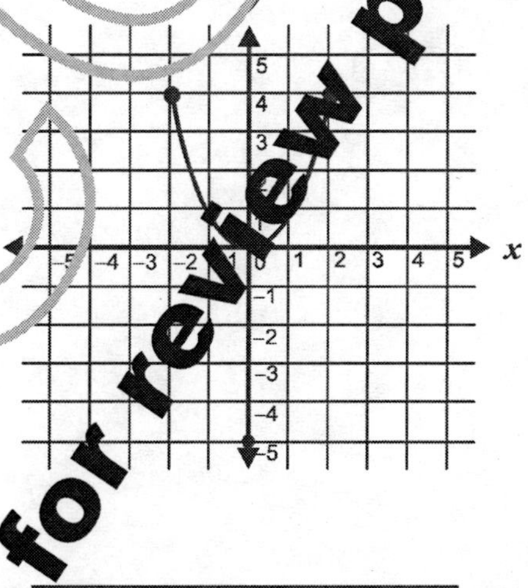

____

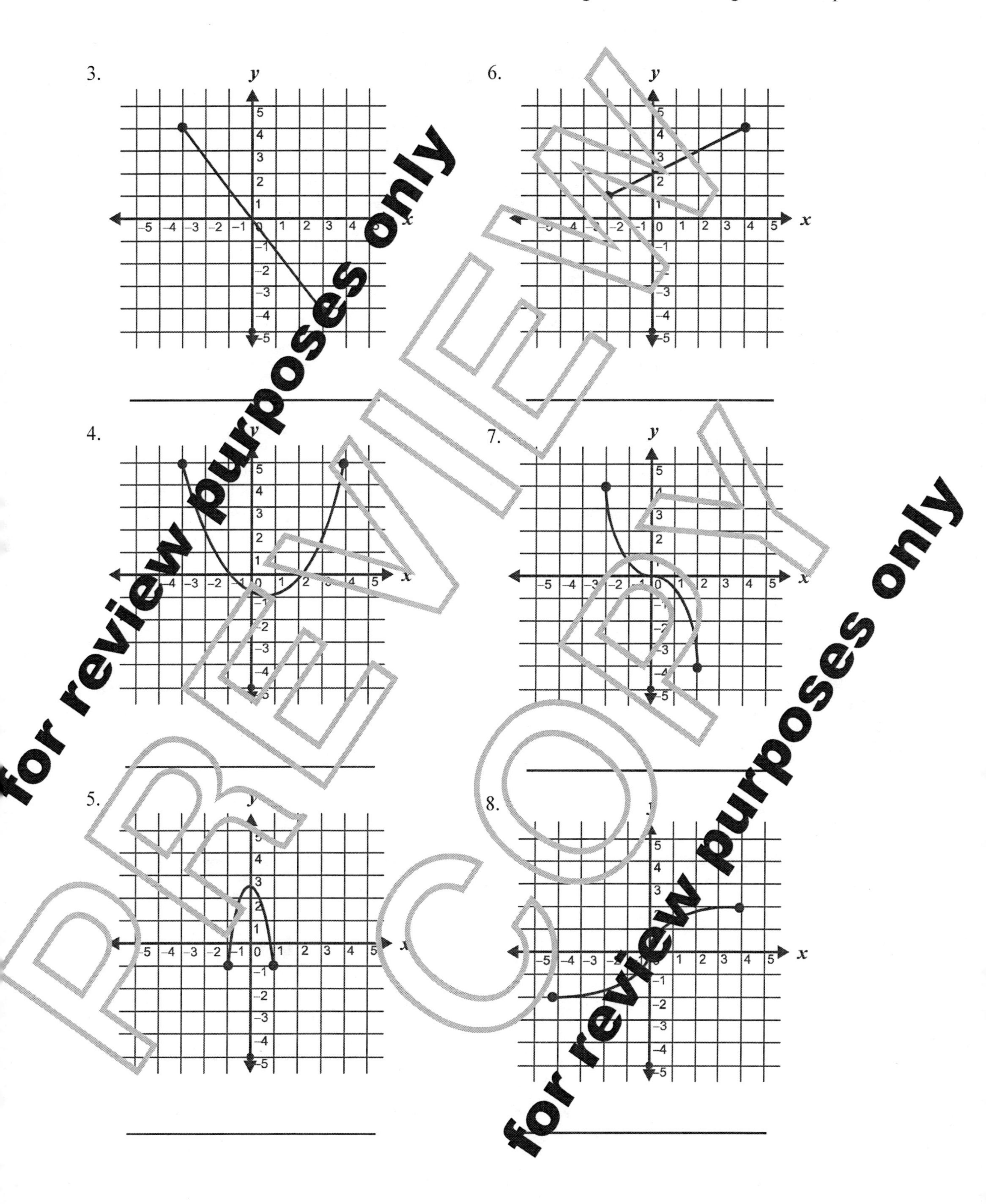
3.
6.
4.
7.
5.
8.
for review purposes only
PREVIEW COPY

15.3 Domain and Range of Quadratic Equations

The **domain** of a quadratic equation is the set of independent variables, or x values, over which the equation is defined. The **range** is the set of y values for which an equation given in two variables, x and y, is satisfied. A quadratic equation in the form of $y = x^2$ is represented by the following graph:

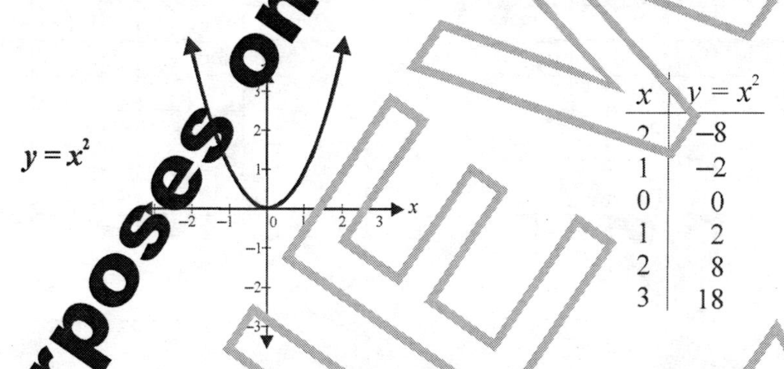

$y = x^2$
x
$y = x^2$
2 −8
1 −2
0 0
1 2
2 8
3 18

In this example the domain, or set of independent variables over which the equation is defined, will be all real numbers (positive and negative.) The range, however, will only include all **positive** real numbers. How would the graph be affected by multiplying x^2 by a constant, that is $y = ax^2$? If 'a' is a positive number greater than 1, the graph will be the same shape but will be taller and thinner. For example, let $a = 2$:

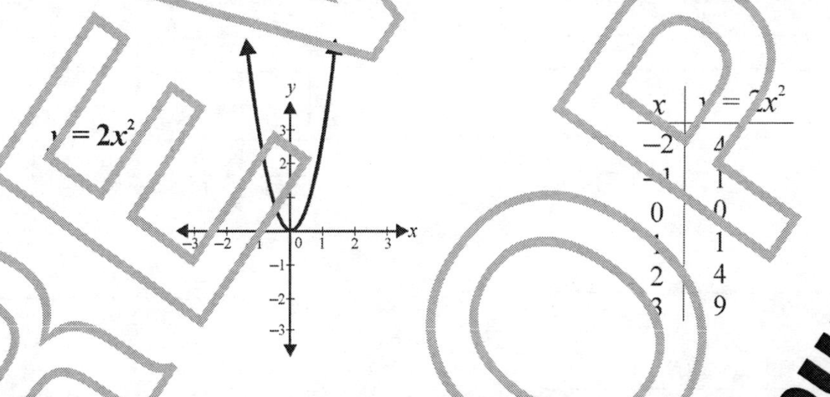

If 'a' is a negative number smaller than −1, the graph is the same shape (tall and thin), but is inverted. For example, let $a = -2$.

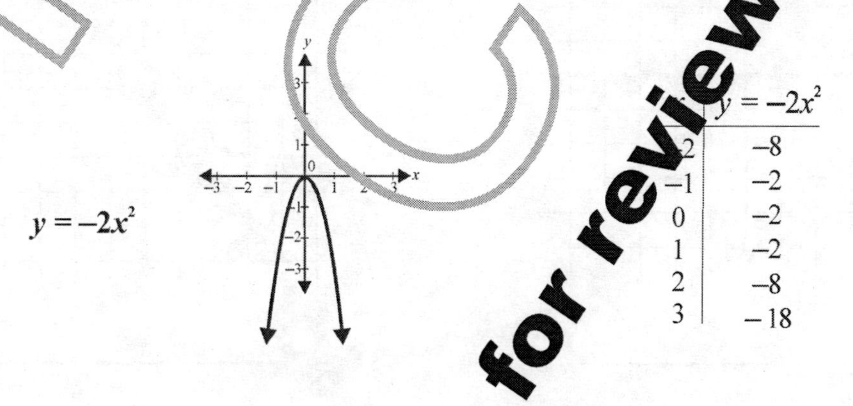

Using the same logic, you can see that when $0 < a < 1$ or $-1 < a < 0$, the graph will widen and flatten as shown in the figures below:

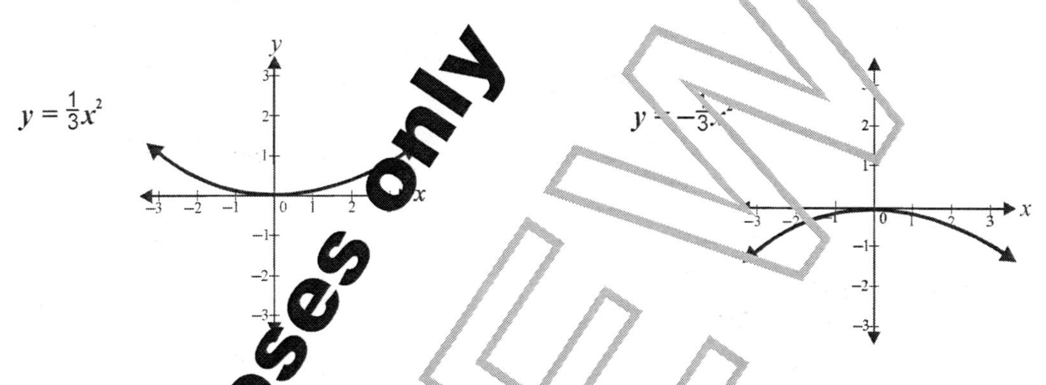

When a constant 'c' is added to the equation, then the graph is shifted up (or down) the y-axis by the constant amount 'c'.

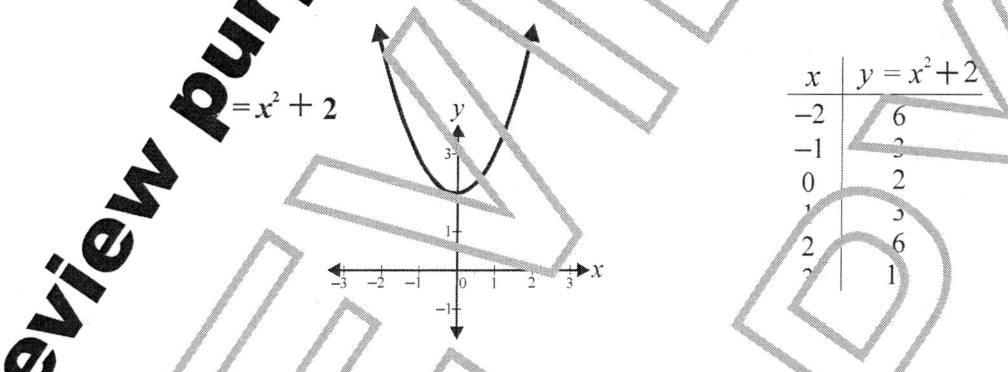

The magnitude of the constant 'a' determines the width of the curve, the sign of 'a' determines the orientation of the curve, and the constant 'c' determines the y-intercept.

Roots of the Quadratic Equation

When factoring quadratic equations, the answer is often left in the form of two factors in parentheses multiplied together. These factors are called the **roots** of the quadratic equation. For example, the quadratic equation $2x^2 - 11x + 12 = 0$ can be factored as $(2x - 3)(x - 4) = 0$. In this example, $(2x - 3)$ and $(x - 4)$ are the **roots** of the equation. To find the **solution** or **solution set** to the equation, each of these roots must be set equal to zero, and then solved for x. In this case, the solutions are:

$$\begin{aligned} 2x - 3 &= 0 \\ +3 \quad &\quad +3 \\ \hline \frac{2x}{2} &= \frac{3}{2} \\ x &= \frac{3}{2} \end{aligned} \qquad \text{and} \qquad \begin{aligned} x - 4 &= 0 \\ +4 \quad &\quad +4 \\ \hline x &= 4 \end{aligned}$$

The solution set $\left\{\frac{3}{2}, 4\right\}$ of the equation is derived from the roots of the equation. The solution(s) will satisfy the original equation when substituted and simplified.

Answer the following questions about the quadratic equation graphs.

1. Which of the following graphs has the largest value of a in $y = ax^2$?

A.

B.

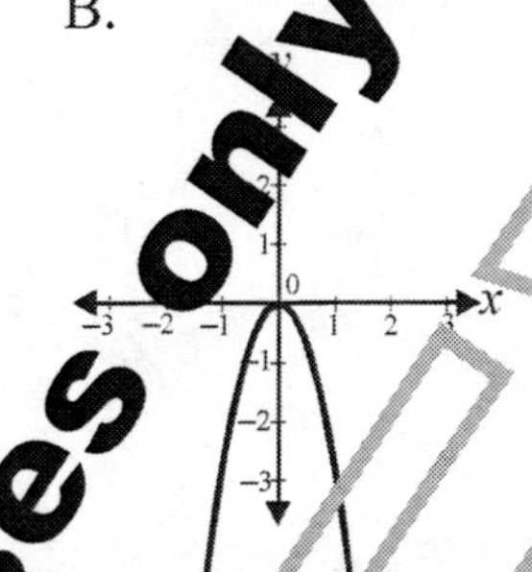

C.

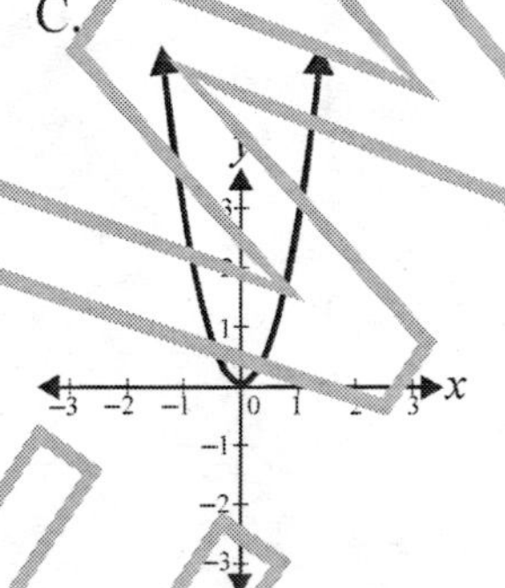

D.

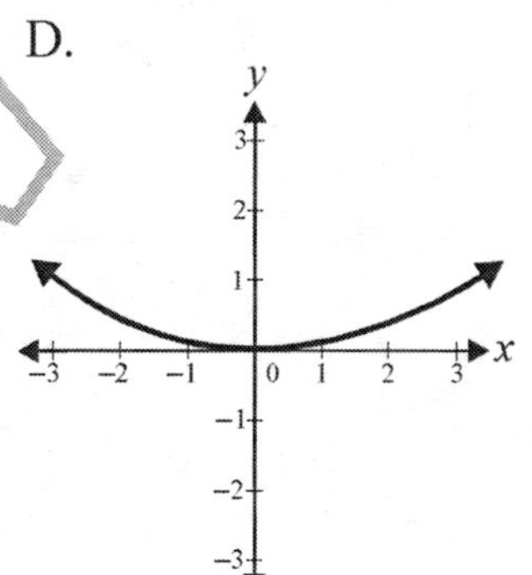

2. Fill in the tables for the following equations:

A.

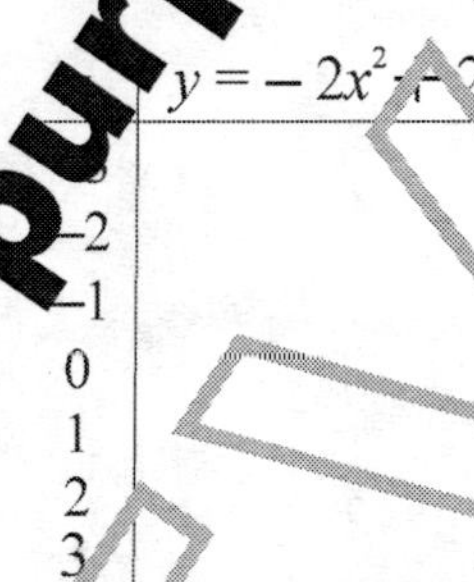

x	$y = -2x^2 + 2$
-3	
-2	
-1	
0	
1	
2	
3	

B.

x	$y = -\frac{1}{3}x^2 - 2$
-3	
-2	
-1	
0	
1	
2	
3	

What is the domain and range of each of these equations? What is the y-intercept in each of the equations?

3. Identify the domain, range, and y-intercept for the following equations. Constructing a table and graph on another sheet of paper would be helpful.

A. $y = x^2 + 4$ D. $y = -2x^2 - 2$ G. $\frac{y}{2} - 2 = 2x^2$

B. $y = 4x^2 - 3$ E. $y = \frac{1}{2}x^2 + 3$ H. $7 + y = x^2$

C. $y = -\frac{1}{5}x^2 - 5$ F. $y = \frac{1}{x^2}$ I. $4 = x^2 + y$

4. Factor the quadratic equation $6x^2 + x - 2 = 0$. Show the roots of the equation and solve for the solution set. Substitute one of the roots into the original equation, and verify that it does solve the equation.

15.4 Functions

Some relations are also **functions**. A relation is a function if **for every element in the domain, there is exactly one element in the range.** In other words, for each value for x there is only one unique value for y.

Example 4: $\{(2,4), (2,5), (3,4)\}$ is **NOT** a function because in the first pair, 2 is paired with 4, and in the second pair, 2 is paired with 5. The 2 can be paired with only one number to be a function. In this example, the x value of 2 has more than one value for y: 4 and 5.

Example 5: $\{(1,2), (3,2), (5,6)\}$ **IS** a function. Each first number is paired with only one second number. The 2 is repeated as a second number, but the relation remains a function.

Determine whether the ordered pairs of numbers below represent a function. Write "F" if it is a function. Write "NF" if it is not a function.

1. $\{(-1,1), (-3,3), (0,0), (2,2)\}$ ________
2. $\{(-4,-3), (-2,-3), (-1,-3), (2,-3)\}$ ________
3. $\{(3,-1), (2,0), (2,2), (5,3)\}$ ________
4. $\{(-3,3), (0,2), (1,1), (2,0)\}$ ________
5. $\{(-2,-5), (-2,-1), (-2,1), (-2,3)\}$ ________
6. $\{(0,2), (1,$ [illegible]$), (2,2), (4,3)\}$ ________
7. $\{(4,2), (3,3), (2,$ [illegible]$), (0,3)\}$ ________
8. $\{(-1,-1), (-2,-2), (3,-1), (3,2)\}$ ________
9. $\{(2,-2), (0,-2), (-2,0), (1,-3)\}$ ________
10. $\{(2,1), (3,2), (4,3), (5,-1)\}$ ________
11. $\{(-1,0), (2,1), (2,4), (-2,2)\}$ ________
12. $\{(1,4), (2,3), (0,2), (0,4)\}$ ________
13. $\{(0,0), (1,0), (2,0), (3,0)\}$ ________
14. $\{(-5,-1), (-3,-2), (-4,-9), (-7,-3)\}$ ________
15. $\{(8,-3), (-4,4), (8,0), (6,2)\}$ ________
16. $\{(7,-1), (4,3), (8,2), (2,8)\}$ ________
17. $\{(4,-3), (2,0), (5,3), (4,1)\}$ ________
18. $\{(2,-6), (7,3), (-3,4), (2,-3)\}$ ________
19. $\{(1,1), (3,-2), (4,16), (1,-5)\}$ ________
20. $\{(5,7), (3,8), (5,3), (6,9)\}$ ________

15.5 Function Notation

Function notation is used to represent relations which are functions. Some commonly used letters to represent functions include f, g, h, F, G, and H.

Example 6: $f(x) = 2x - 1$; find $f(-3)$

Step 1: Find $f(-3)$ means to replace x with -3 in the relation $2x - 1$.
$f(-3) = 2(-3) - 1$

Step 2: Solve $f(-3)$. $f(-3) = 2(-3) - 1 = -6 - 1 = -7$
$f(-3) = -7$

Example 7: $g(x) = 4 - 2x^2$; find $g(2)$

Step 1: Replace x with 2 in the relation $4 - 2x^2$.
$g(2) = 4 - 2(2)^2$

Step 2: Solve $g(2)$. $g(2) = 4 - 2(2)^2 = 4 - 2(4) = 4 - 8 = -4$
$g(2) = -4$

Find the solutions for each of the following.

1. $F(x) = 2 + 3x^2$; find $F(3)$
2. $f(x) = 4x + 6$; find $f(-4)$
3. $H(x) = 6 - 2x^2$; find $H(-1)$
4. $g(x) = -3x + 7$; find $g(-3)$
5. $f(x) = -5 + 4x$; find $F(7)$
6. $G(x) = 4x^2 + 4$; find $G(0)$
7. $f(x) = 7 - 6x$; find $f(-4)$
8. $h(x) = 2x^2 + 10$; find $h(5)$
9. $F(x) = 7 - 5x$; find $F(2)$
10. $f(x) = 4x^2 + 5$; find $f(-2)$

15.6 Recognizing Functions

Recall that a relation is a function with only one y value for every x value. We can depict functions in many ways including through graphs.

Example 8:

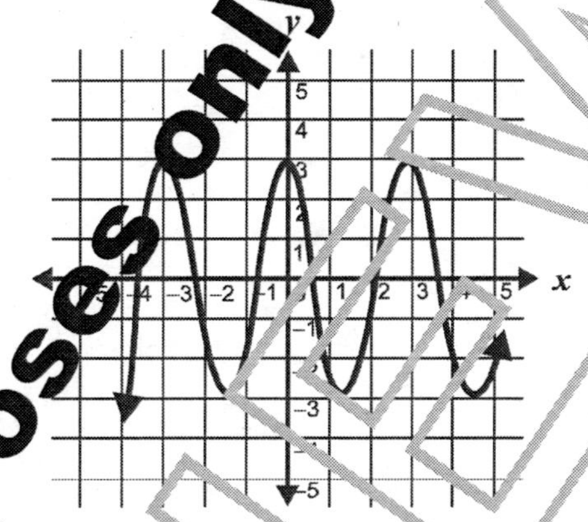

This graph **IS** a function because it has only one y value for each value of x.

Example 9:

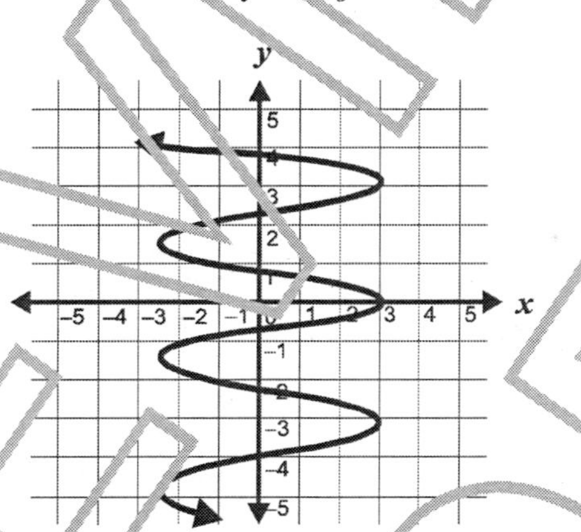

This graph is **NOT** a function because there is more than one y value for each value of x.

Hint: An easy way to determine a function from a graph is to do a vertical line test. First, draw a vertical line that crosses over the whole graph. If the line crosses the graph more than one time, then it is not a function. If it only crosses it once, it is a function. Take Example 9 above:

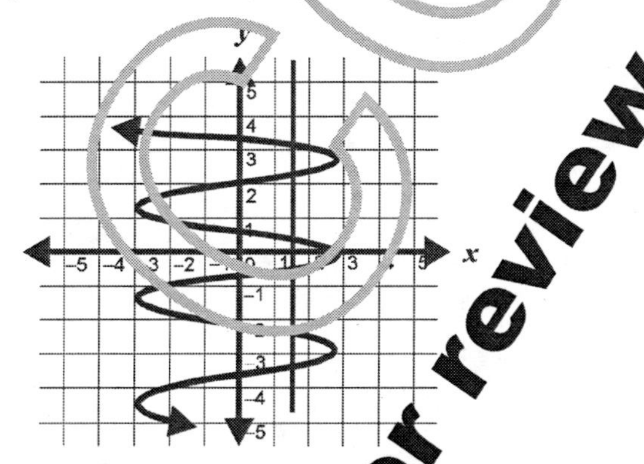

Since the vertical line passes over the graph six times, it is not a function.

Determine whether or not each of the following graphs is a function. If it is, write function on the line provided. If it is not a function, write NOT a function on the line provided.

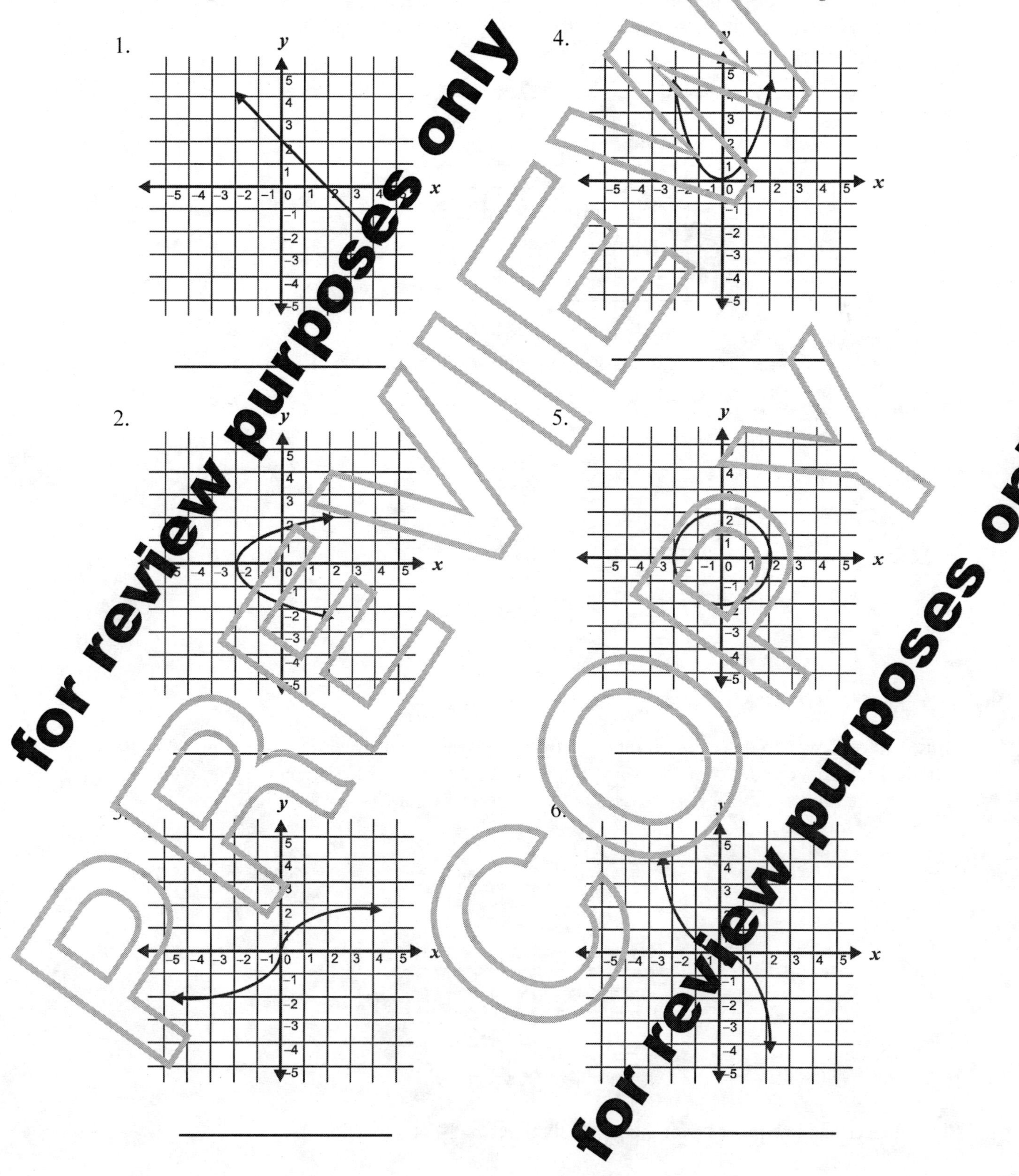

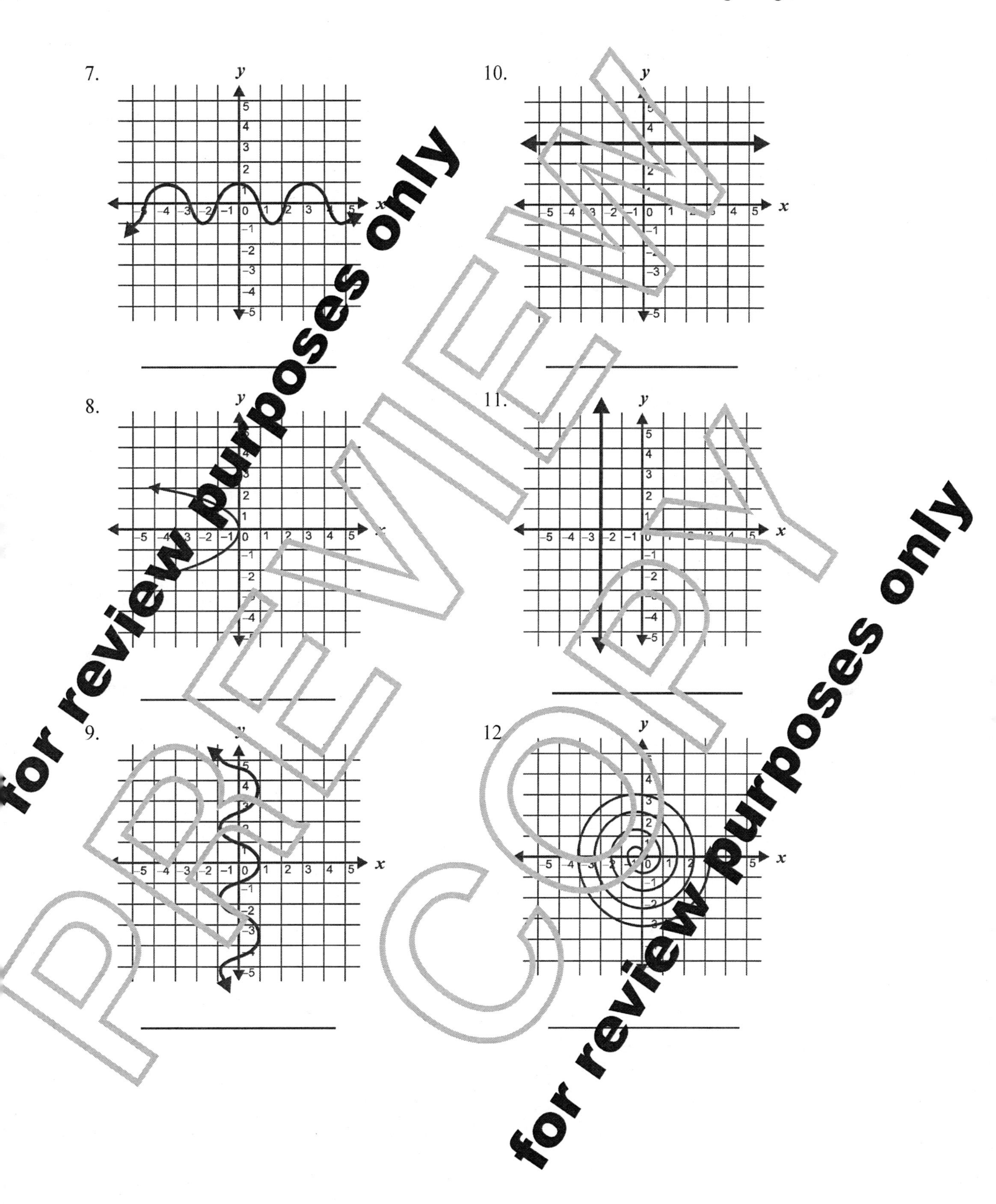
7.
10.
8.
11.
9.
12.
y
x
for review purposes only
PREVIEW COPY

15.7 Qualitative Behavior of Graphs

The qualitative behavior of graphs and function is determined by reading and understanding the graph of the function.

Example 10: The graph of the function $y = f(x)$ is shown below.

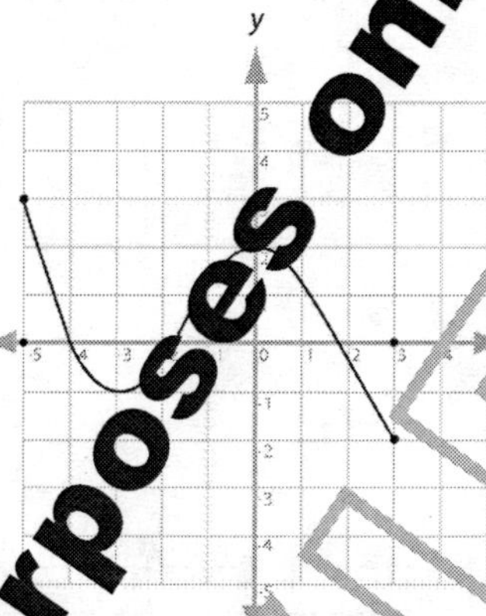

If $-5 \leq x \leq 3$ is the domain of the function, for how many values of x does
(A) $f(x) = 1$?
(B) $f(x) = 2$?

Step 1: Trace a line though the graph at the indicated value.

Step 2: Count the number of times the graph of the function intersects the traced line.

(A)

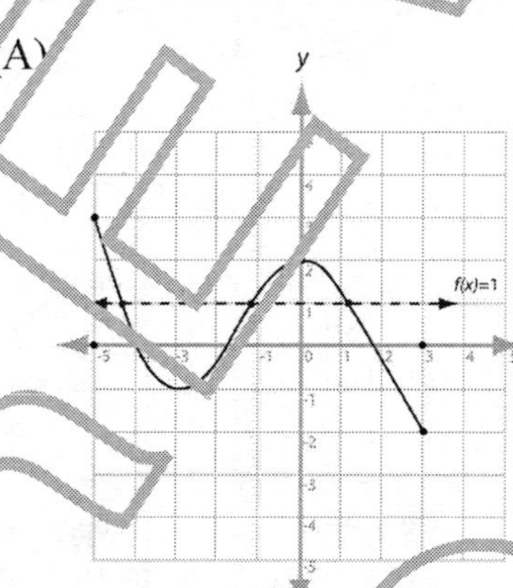

Notice how the line at $f(x) = 1$ goes through the graph of the function three times. This means that there are three values of x when $f(x) = 1$.

(B)

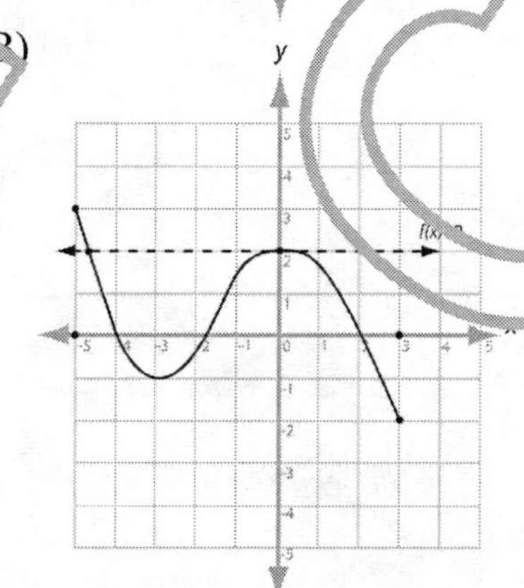

Notice how the line at $f(x) = 2$ goes through the graph of the function two times. This means that there are two values of x when $f(x) = 2$.

Use the following graph of $y = f(x)$ to answer questions 1–3.

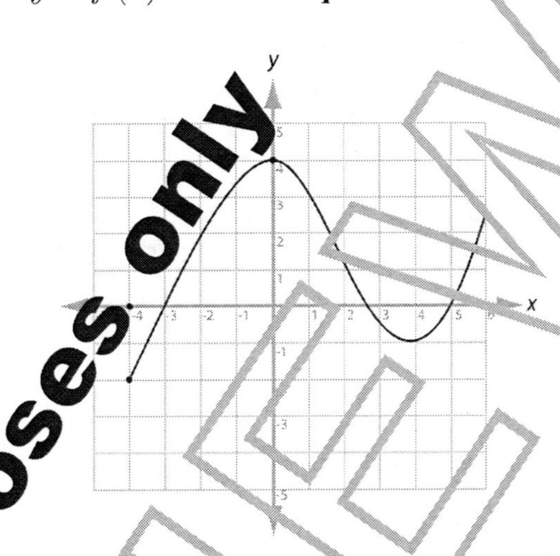

If $-4 \leq x \leq 6$ is the domain of the function, for how many values of x does

1. $f(x) = 3$?
2. $f(x) = 0$?
3. $f(x) = 7$?

Use the following graph of $y = f(x)$ to answer questions 4 and 5.

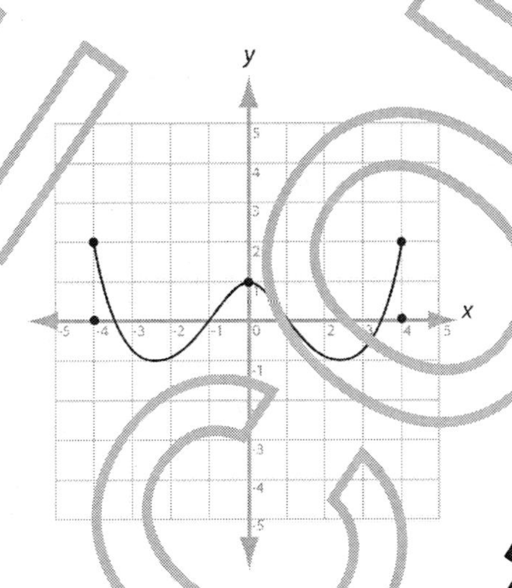

If $-4 \leq x \leq 4$ is the domain of the function, for how many values of x does

4. $f(x) = 0$?
5. $f(x) = 1$?

15.8 Relations That Can Be Represented by Functions

Real-life examples can be represented by functions. The most common functions are exponential growth and decay and half-life.

Example 11: Atlanta, GA has a population of about 410,000 people. The U.S. Census Bureau estimates that the population will double in 26 years. If the population continues at the same rate, what will the population be in
a) 10 years?
b) 50 years?

Step 1: Use the double growth equation $P = P_0(2^{t/d})$, where P = population at time t, P_0 = population at time $t = 0$, and d = double time.

Step 2: Determine the variable of each of the facts given in the problem. In this case, $P_0 = 410,000$ people, $d = 26$ years, and $t = 10$ years for part a and $t = 50$ years for part b.

Step 3: Plug all of the information into the given equation. Round to the nearest whole number.
a) $P = 410,000(2^{10/26}) = 410,000\,(1.3055) = 535,260$ people
b) $P = 410,000(2^{50/26}) = 410,000\,(3.7923) = 1,554,847$ people

Find the answers to the real-life problems by using the equations and variables given. Round your answer to the nearest whole numbers.

For questions 1 and 2 use the following half-life formula.

$A = A_0\left(\frac{1}{2}\right)^{t/h}$
A = amount at time t
A_0 = amount at time $t = 0$
h is the half-life

1. If you have 6,000 atoms of hydrogen (H), and hydrogen's half-life is 12.3 years, how many atoms will you have left after 7 years?

2. Chlorine (Cl) has a half-life of 55.5 minutes. If you start with 200 milligrams of chlorine, how many will be left after 5 hours?

For questions 3 and 4 use the double growth formula.

$P = P_0(2)^{t/d}$
P = amount at time t
P_0 = amount at time $t = 0$
d is the half-life

3. There are about 3,390,000 Girl Scouts in the United States. The Girl Scout Council says that there is a growth rate of 5.[illegible]10% per year, so they expect the Girl Scout population in the United States to double in [illegible] years. If the Girl Scout's organization expands as continuously as it has been, what will the population be
   (A) in 8 years?
   (B) next year?

4. Dr. Kellie noticed the bacteria growth in her laboratory. After observing the bacteria, she concluded that the double time of the bacteria is 40 minutes, and she started off with just 2,500 bacteria. Assuming this information is accurate and constant, how many bacteria will be in Dr. Kellie's lab
   (A) in [illegible] minutes?
   (B) after 3 hours?

For questions 5 and 6 use the compound interest formula.

$A = P\left(1 + \frac{r}{k}\right)^{kt}$
A = amount at time t
P = principle amount invested
k = how many times per year interest is compounded
r = rate

5. Lisa invested $1,000 into an account that pays 6% interest compounded monthly. If this account is for her newborn, how much will the account be worth on [illegible] 21st birthday, which is exactly 21 years from now?

6. Mr. Dumple wants to open up a savings account. He has looked at two different banks. Bank 1 is offering a rate of 5% compounded daily. Bank 2 is offering an account that has a rate of 8%, but is only compounded semi-yearly. Mr. Dumple puts $5,000 in an account and wants to take it out for his retirement in 10 years. Which bank will give him the most money back?

15.9 Exponential Growth and Decay

Many quantities experience exponential growth or decay under certain conditions. Examples include bacteria, populations, disease, money in a savings account that compounds interest, and radioisotopes. Exponential functions are those functions in which the independent variable is time, and time is an exponent (thus the name exponential function). For instance, the formula for growth of money in a savings account that compounds interest annually is:

$$A = P(1 + r)^t$$

where A is the value of the account after t years, P is the original amount of money in the account, and r is the annual interest rate.

Below are graphs of the general forms of exponential growth functions and exponential decay functions. Time is represented on the x-axis. Whatever is growing or decaying exponentially, such as population or money, is represented on the y-axis. Note that exponential function graphs are generally in Quadrant I since time and objects cannot be assigned negative values.

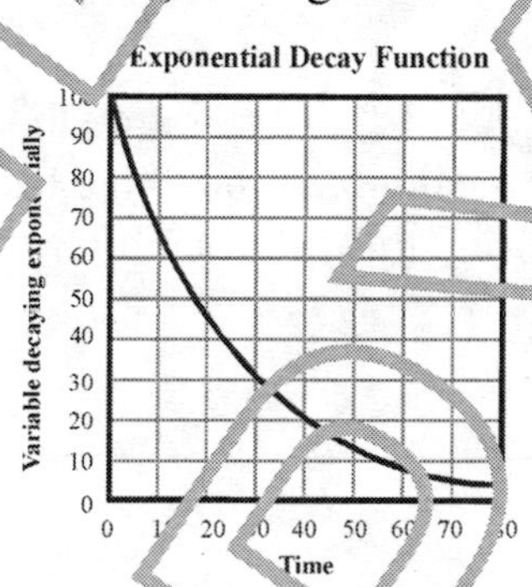

Example 12: Mason deposited \$2,000 into a savings account that pays an annual interest rate of 9% compounded annually. Using the formula $A = P(1 + r)^t$ determine the amount of money in the savings account after 1 year, 5 years, and 20 years. Using the calculated values, construct a graph.

Step 1: Consider the known values. $P = 2{,}000$, and $r = 0.09$. The problem will have to be worked three times where $t = 1$, $t = 5$, and $t = 20$. A is the amount being calculated.

Step 2:

$A = 2000(1 + 0.09)^1$ $A = 2000(1 + 0.09)^5$ $A = 2000(1 + 0.09)^{20}$

$A = 2000(1.09)^1$ $A = 2000(1.09)^5$ $A = 2000(1.09)^{20}$

$A = \$2,180$ $A = \$3,077.25$ $A = \$11,208.82$

Step 3: Use the calculated values to graph the function.

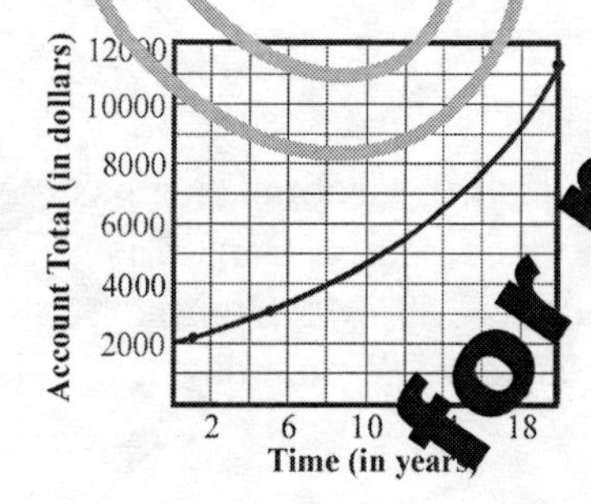

Fill in the tables for the following functions. On the line under each table, label the given function as an exponential growth function or an exponential decay function. Round your answers to two decimal places. For extra practice, graph the functions.

1. $F(t) = 15(1.01)^t$

t	$F(t)$
1	
2	
3	
4	

____

2. $S(t) = 350$ ([illegible]$)^t$

t	$S(t)$
1	
[illegible]	
5	
7	

____

3. $M(t) = 1000(1.04)^t$

t	$M(t)$
2	
4	
6	
8	

____

4. $B(t) = 2(2.50)^t$

t	$B(t)$
1	
2	
3	
4	

____

5. $C(t) =$ [illegible]$300(0.5)^t$

t	$C(t)$
5	
10	
15	
20	

____

6. $R(t) = 80\left(\frac{1}{3}\right)^t$

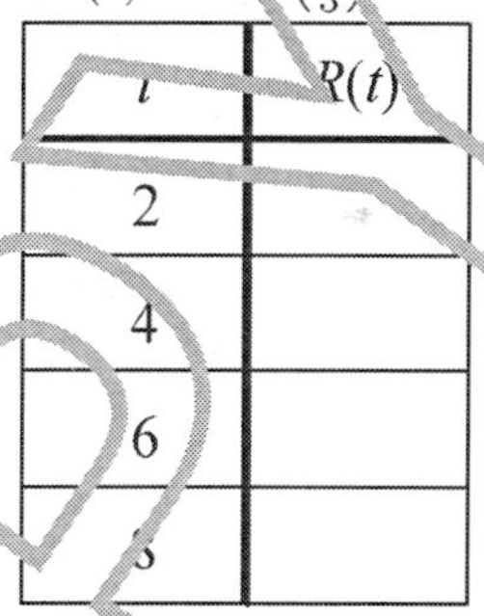

t	$R(t)$
2	
4	
6	
8	

____

Refer to the graph at right to answer questions 7–10.

7. Which town is experiencing exponential decay? growth?

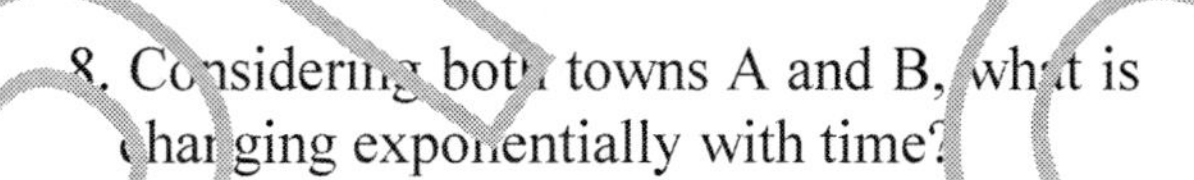

8. Considering both towns A and B, what is changing exponentially with time?

9. Why would it not make sense to draw the graph of town B below the x-axis?

10. In what year does the population of town B reach 3,000?

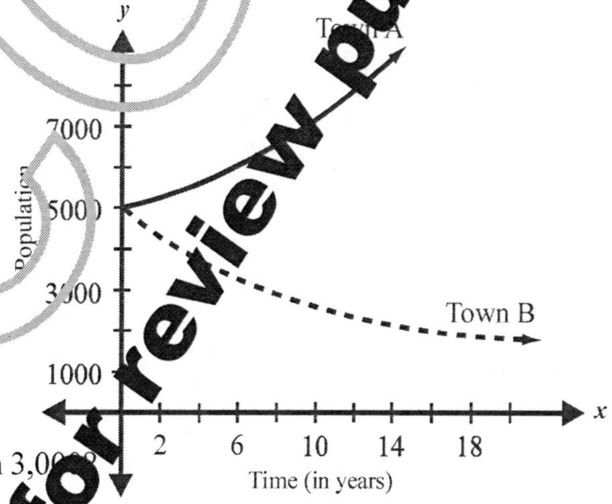

15.10 Piecewise Functions

A **piecewise function** is a function consisting of 2 or more formulas over a sequence of intervals. These **intervals** are defined by the possible values of x, also known as the domain of the function. The graph of a piecewise function consists of the graphs of each interval formula.

Example 13: $f(x) = \begin{cases} 3 & \text{if } 0 \leq x < 1 \\ 2 & \text{if } 1 \leq x < 2 \\ 1 & \text{if } 2 \leq x < 3 \end{cases}$

Graph $f(x)$.

Step 1: Graph each formula over the given interval.
For example, $f(x) = 3$ when the domain is $0 \leq x < 1$. This means that you would draw the graph $y = 3$ first. (Recall that this is a horizontal line that passes through the point $(0, 3)$.) After this, you would only draw $y = 3$ between the points $(0, 3)$ and $(1, 3)$ because of the domain. The graph cannot go outside of those points.
When $f(x) = 2$ and the domain is $1 \leq x < 2$, draw the graph $y = 2$ between the points $(1, 2)$ and $(2, 2)$.
When $f(x) = 3$ and the domain is $2 \leq x < 3$, draw the graph $y = 3$ between the points $(2, 3)$ and $(3, 3)$.

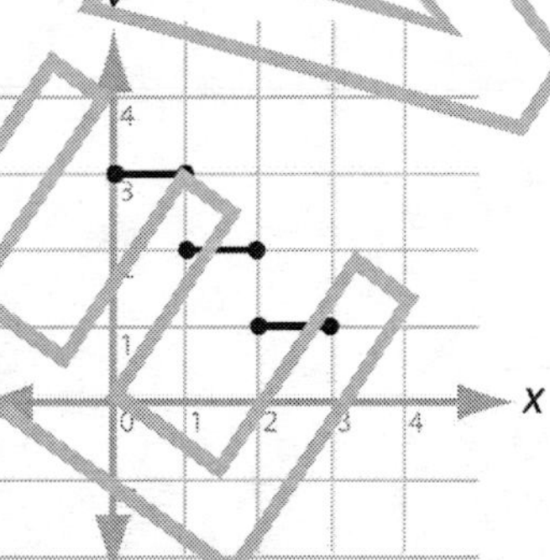

Step 2: Since the function cannot equal two y values for a x value (otherwise, it would not be a function), you must look at the inequalities in the domain. When the inequality is less than or equal to ($\leq$), you must draw the endpoint as a filled in circle. This shows that the function can equal to that point. For the strict inequalities ($<$), you must draw an endpoint with an open (not filled in) circle.

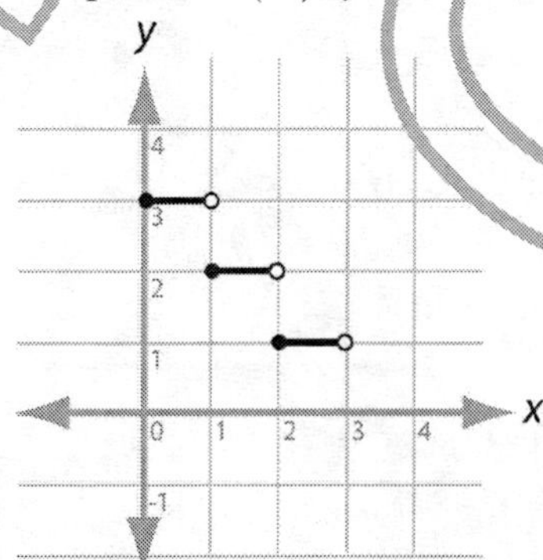

Example 14: $f(x) = \begin{cases} x^2 & \text{if } x \geq 2 \\ 3 - x & \text{if } x < 2 \end{cases}$

Find (A) $f(1)$, (B) $f(3)$, and (C) $f(2)$.

Step 1: Determine which interval of the domain includes the value of x.
(A) For $f(1)$, $x = 1$. Since 1 is less than 2, you would plug $x = 1$ into $3 - x$.
(B) For $f(3)$, $x = 3$. Since 3 is greater than 2, you would plug $x = 3$ into x^2.
(C) For $f(2)$, $x = 2$. Since 2 is equal to 2, you would plug $x = 2$ into x^2.

Step 2: Plug the value of x into the appropriate formula to solve for the value of $f(x)$.
(A) $f(x) = 3 - x$, so $f(1) = 3 - 1 = 2$
(B) $f(x) = x^2$, so $f(3) = (3)^2 = 9$
(C) $f(x) = x^2$, so $f(2) = (2)^2 = 4$

Graph each of the following functions.

1. $f(x) = \begin{cases} x & \text{if } x \geq 0 \\ -x & \text{if } x < 0 \end{cases}$

2. $f(x) = \begin{cases} 1 & \text{if } x < 1 \\ x^2 & \text{if } x \geq 1 \end{cases}$

3. $f(x) = \begin{cases} \sqrt{x} & \text{if } x \geq 2 \\ x^2 & \text{if } x < 2 \end{cases}$

4. $f(x) = \begin{cases} 2x + 3 & \text{if } x < 0 \\ 2x - 3 & \text{if } x \geq 0 \end{cases}$

5. $f(x) = \begin{cases} x^2 & \text{if } x < -1 \\ x & \text{if } -1 \leq x \leq 1 \\ -(x^2) & \text{if } x > 1 \end{cases}$

6. Phil's long distance phone service charges him 50 cents for the first 10 minutes and 10 cents for each minute afterwards. Graph the function that represents Phil's long distance phone service and find how much he would pay for
(A) a 5 minute call.
(B) a 10 minute call.
(C) a 15 minute call.

7. The tuition at State University is determined by the number of class hours a student takes. Tuition is $100 for the first three hours and doubles every 3 hours up to 12 hours. After 12 hours, tuition does not change. Graph the function that represents the tuition at State University and find the tuition for a student taking
(A) 6 class hours.
(B) 12 class hours.
(C) 15 class hours.

Chapter 15 Review

1. What is the domain of the following relation? $\{(-1,2),(2,5),(4,9),(6,11)\}$

2. What is the range of the following relation? $\{(0,-2),(-1,-4),(-2,6),(-3,-8)\}$

3. Find the range of the relation $y = 5x$ for the domain $\{0,1,2,3,4\}$.

4. Find the values of $M(y)$ of the relation $M(y) = 2(1.1)^y$ for the domain $\{2,3,4,5,6\}$.

5. Find the range of the following relation for the domain $\{0,2,6,8,10\}$. $B(t) = 600(0.75)^t$

6. Find the range of the relation $y = \dfrac{3(x-2)}{5}$ for the domain $\{-8,-3,7,12,17\}$.

7. Find the range of the relation $y = 10 - 2x$ for the domain $\{-8,-4,0,4,8\}$.

8. Find the range of the relation $y = \dfrac{4+x}{3}$ for the domain $\{-7,-1,2,5,8\}$.

For each of the following relations given in questions 9–13, write F if it is a function and NF if it is not a function.

9. $\{(1,2),(2,2),(3,2)\}$

10. $\{(-1,0),(0,1),(1,2),(2,3)\}$

11. $\{(2,1),(2,2),(2,3)\}$

12. $\{(1,7),(2,5),(3,6),(2,4)\}$

13. $\{(0,-1),(-1,-2),(-2,-3),(-3,-4)\}$

For questions 14–19, find the range of the following functions for the given value of the domain.

14. For $g(x) = 2x^2 - 4x$; find $g(-1)$

15. For $h(x) = 3x(x-4)$; find $h(3)$

16. For $f(n) = \dfrac{1}{n+3}$; find $f(4)$

17. For $G(n) = \dfrac{2-n}{2}$; find $G(8)$

18. For $H(x) = 2x(x-1)$; find $H(4)$

19. For $f(x) = 7x^2 + 3x - 2$; find $f(2)$

Chapter 16
Series, Sequences, and Algorithms

16.1 Number Patterns

In each of the examples below, there is a sequence of numbers that follows a pattern. Think of the sequence of numbers like the output for a function. You must find the pattern (or function) that holds true for each number in the sequence. Once you determine the pattern, you can find the next number in the sequence or any number in the sequence.

	Sequence	Pattern	Next Number	20th number in the sequence
Example 1:	$3, 4, 5, 6, 7$	$n+2$	8	22

In number patterns, the sequence is the output. The input can be the set of whole numbers starting with 1. But, you must determine the "rule" or pattern. Look at the table below.

input	sequence
1 →	3
2 →	4
3 →	5
4 →	6
5 →	7

What pattern or "rule" can you come up with that gives you the first number in the sequence, 3, when you input 1? $n+2$ will work because when $n = 1$, the first number in the sequence $= 3$. Does this pattern hold true for the rest of the numbers in the sequence? Yes, it does. When $n = 2$, the second number in the sequence $= 4$. When $n = 3$, the third number in the sequence $= 5$, and so on. Therefore $n+2$ is the pattern. Even without knowing the algebraic form of the pattern, you could figure out that 8 is the next pattern in the sequence. To find the 20th number in the pattern, use $n = 20$ to get 22.

	Sequence	Pattern	Next Number	20th number in the sequence
Example 2:	$1, 4, 9, 16, 25$	n^2	36	400
Example 3:	$-2, -4, -6, -8, -10$	$-2n$	-12	-40

Find the pattern and the next number in each of the sequences below.

	Sequence	Pattern	Next Number	20th number in the sequence
1.	$-2, -1, 0, 1, 2$			
2.	$5, 6, 7, 8, 9$			
3.	$3, 7, 9, 11, 15, 19$			
4.	$-3, -6, -9, -12, -15$			
5.	$3, 5, 7, 9, 11$			
6.	$2, 4, 8, 16, 32$			$1,048,576$
7.	$1, 8, 27, 64, 125$			
8.	$0, -1, -2, -3, -4$			
9.	$2, 5, 10, 17, 26$			
10.	$4, 6, 8, 10, 12$			

16.2 Geometric Patterns

Sometimes the rule that governs a pattern cannot be written as a simple mathematical formula. Nevertheless, inductive reasoning will lead to the next item in the sequence.

Example 4: Find the shape that comes next in the pattern below.

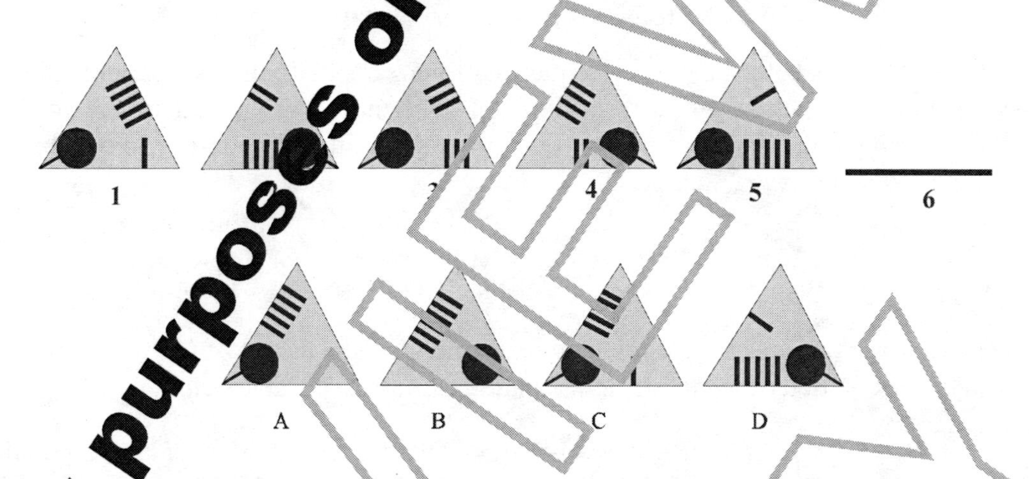

Step 1: Notice that the outline shape of all the objects in the pattern is a triangle. Next, recognize any distinguishing marks or shapes within the objects. The first object has a circle in the lower left-hand corner, a single line on the bottom side of the triangle, and five lines on the right-hand side of the triangle.

Step 2: Look at the rest of the objects within the pattern. The very next object in the sequence has a circle in the lower right-hand side, the third object has a circle in the lower left-hand side, and so on. Continuing the pattern, you can conclude that the final object in the pattern will have a circle in the right-hand side.

Step 3: The final step is to determine the pattern governing the number and position of the lines. We know that the circle will be on the lower right-hand side in the missing object, so let's concentrate on triangles 2 and 4 in our pattern that also have circles in the lower right-hand corner. The lines on the bottom decrease by two from triangle 2 to triangle 4. Continuing the pattern and decreasing by two, it can be determined that there should be zero lines on the bottom in figure 6, the missing object. Using the same logic, look at the lines on the side opposite the circle in triangles 2 and 4 in the sequence. These lines increase by 2. Continuing the pattern, the missing object should have 6 lines.

Step 4: The next triangle in the pattern has to have a circle in its lower right-hand side, no lines on the bottom side, and six lines on the side opposite the circle. The correct answer is B.

Find the object that comes next in each pattern. Circle your answer.

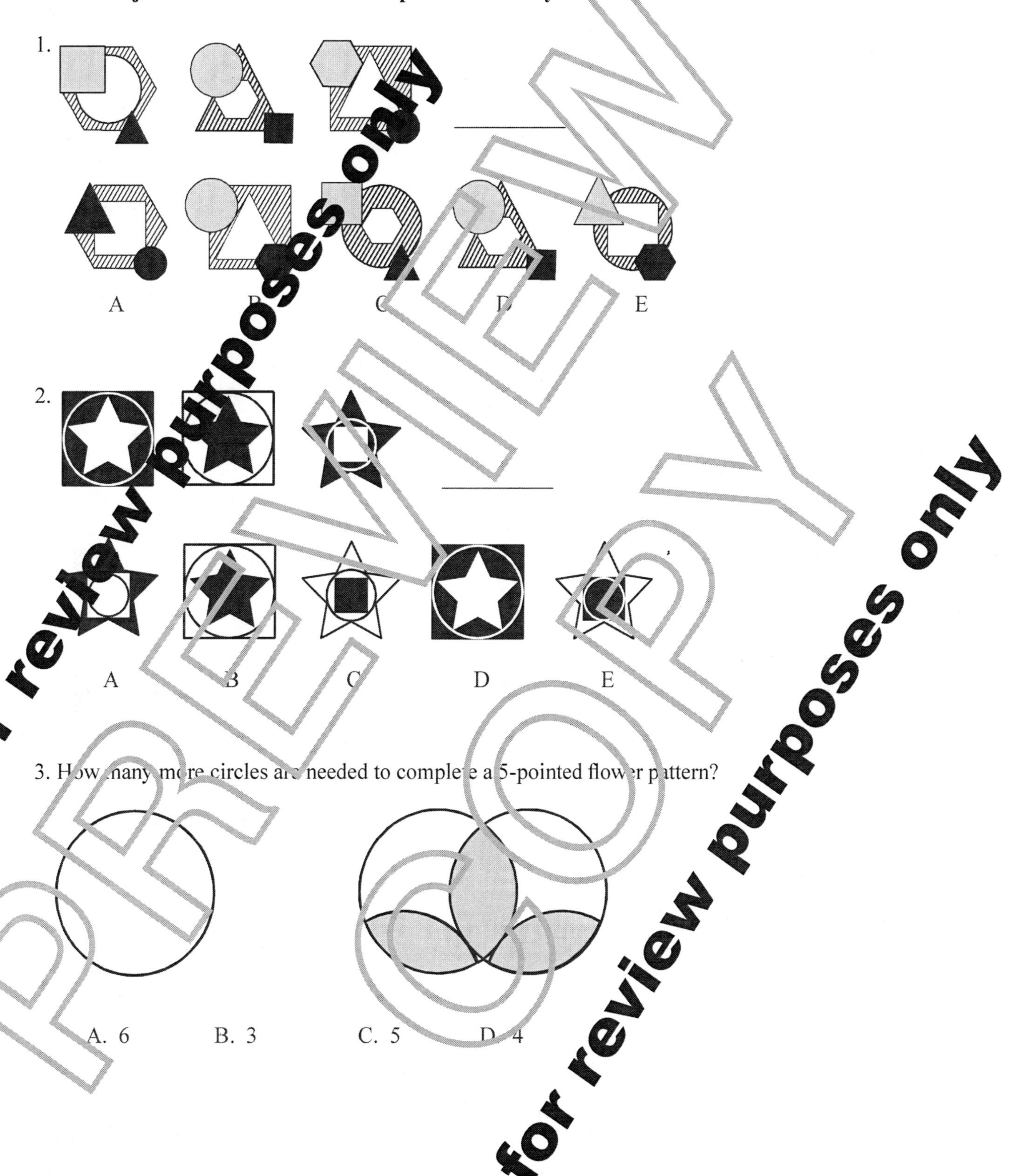

16.3 Limits of Series and Sequences

Limits are the extremes in a series or sequence. A limit may be either a minimum value or a maximum value. The limit is the value that the sequence or series approaches, but never actually reaches.

Example 5: Find the limit of the repeating decimal $0.\overline{99}$.

Answer: 0.999999... keeps repeating forever. It is an infinite decimal. $0.\overline{99}$ approaches 1, but never actually reaches it. The next closest number that follows $0.\overline{99}$ is 1. Therefore, the limit of $0.\overline{99}$ is 1.

Example 6: Find the limit of $\frac{1}{n}$ as n approaches infinity, denoted by the symbol ∞.

Step 1: The function must be evaluated as n approaches infinity, ∞. Since infinity has no boundaries, it can never be reached. We must plug values into the function, $\frac{1}{n}$ to find what it is approaching.

$n = 10$	$n = 100$	$n = 1000$	$n = 10000$	$n = 100000$	...
$\frac{1}{10} = 0.1$	$\frac{1}{100} = 0.01$	$\frac{1}{1000} = 0.001$	$\frac{1}{10000} = 0.0001$	$\frac{1}{100000} = 0.00001$	...

Step 2: Notice that the bigger n gets, the smaller the function gets. The smaller the function gets, the closer it gets to zero. The function can never equal zero because if $n = 0$, then the function will be undefined. Therefore, the limit (the value it gets closest to, but never reaches) is zero.

NOTE: All functions can be looked at graphically, and most of the time the limit can be determined by looking at the graph. The graph of a is shown below for both positive and negative values of n. Notice how the function gets closer and closer to zero, but will never reach it.

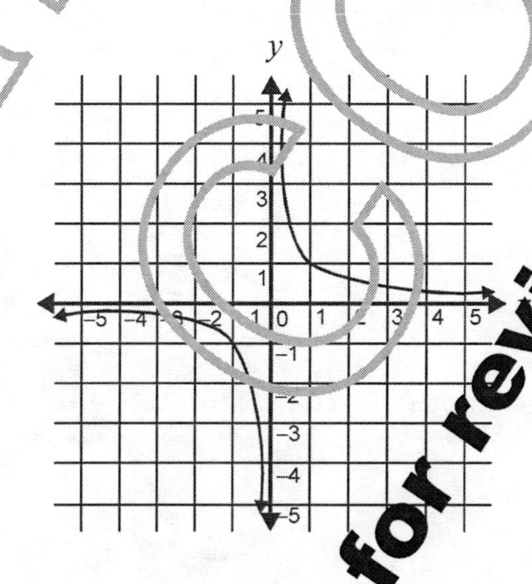

Example 7: Find the limit of the infinite sequence, 1, $\frac{1}{2}$, $\frac{1}{4}$, $\frac{1}{8}$, $\frac{1}{16}$, . . .

Answer: Each value in this sequence is cut in half to reach the next term. This sequence will go on infinitely because there are infinite amounts of numbers. It will continue to get smaller but will never reach zero. Thus, the limit is zero.

HINT: The limit of a series or sequence can be infinity, ∞, or negative infinity, $-\infty$. Negative infinity contains the negative numbers, but can never be reached.

Find the limit in each sequence.

1. 5, 6, 7, 8, 9, . . .
2. 32, 8, 2, $\frac{1}{2}$, $\frac{1}{8}$, . . .
3. −3, −6, −9, −12, −15, . . .
4. 1, $\frac{1}{2}$, $\frac{1}{3}$, $\frac{1}{4}$, $\frac{1}{5}$, . . .
5. 0, −1, −2, −3, −4, . . .
6. −1, $-\frac{1}{2}$, $-\frac{1}{3}$, $-\frac{1}{4}$, $-\frac{1}{5}$, . . .
7. $\frac{1}{3}$, $\frac{1}{9}$, $\frac{1}{27}$, $\frac{1}{81}$, $\frac{1}{243}$, . . .
8. 16, 8, 4, 2, 1, . . .
9. −1, 0, 1, 2, 3, . . .
10. 2, 5, 10, 17, 26, . . .
11. −2, −4, −8, −16, −32, . . .
12. −9, −3, −1, $-\frac{1}{3}$, $-\frac{1}{9}$, . . .
13. 1, 8, 27, 64, 125, . . .
14. 4, 6, 8, 10, 12, . . .
15. −625, −125, −25, −5, −1, . . .

Find the limit of each function.

16. $F(x) = \sqrt{x}$
17. $f(n) = \frac{1}{n+3}$
18. $g(x) = \frac{1}{x^2}$
19. $f(x) = 2x - 1$
20. $G(n) = \frac{1}{1-n}$
21. $F(x) = \frac{1}{\sqrt{x}}$
22. $f(x) = x^2 + 3$
23. $G(n) = \frac{1}{2n-1}$
24. $h(x) = \frac{1}{2x + 4}$

Find the limit of each series or sequence in the following word problems.

25. Sammy drops a tennis ball from the roof of his high school while Aubrey records the motion of the ball. Later, the video is analyzed, and Sammy and Aubrey realize that each time the tennis ball bounced, it would come back up to $\frac{2}{3}$ of its height from the previous bounce. Assuming the ball could bounce indefinitely, find the limit of the height of the bounces.

26. Sally's second grade class ordered a pizza for their Christmas party. The first student took $\frac{1}{2}$ of the pizza. The next student took $\frac{1}{2}$ of the remaining pizza. The third student took $\frac{1}{2}$ of the remaining pizza available, and so on until all of the students have had some pizza. What is the limit?

16.4 Summing Arithmetic and Geometric Series

A series is the sum of a sequence. Three examples of series are finite arithmetic series and finite and infinite geometric series. An arithmetic series is a series where the difference between each successive term is a fixed constant. A geometric series is a series where each term is being multiplied by the same value to get the next term. There are specific formulas for the sum of each type of series.

Series	Formula for the Sum
Finite Arithmetic Series	$S_n = \dfrac{n(t_1 + t_n)}{2}$
Finite Geometric Series	$S_n = \dfrac{t_1(1 - r^n)}{(1 - r)}$
Infinite Geometric Series	$S = \dfrac{t_1}{1 - r}$

The sum is denoted by S (S_n is the sum of the series to a specific term, n). t_1 is the first term in the series, and t_n is the nth term in the series. In the geometric series, r is the value being multiplied to each term to reach the next term in the series.

NOTE: In the infinite geometric series, if $|r| < 1$, then the sum of the series can be found. If $|r| > 1$, then the sum of the series cannot be found.

Example 1: Find the sum of the first 30 terms of the arithmetic series:
$3 + 4 + 5 + 6 + 7 + \ldots$

Step 1: First, find the pattern of the series. The pattern of this series is $n + 2$.

Step 2: Next find all the variables for the summing equation, $S_n = \dfrac{n(t_1 + t_n)}{2}$.
$t_1 = 3$
$n = 30$ (how many terms we will find the sum of)
$t_n = t_{30} = 30 + 2 = 32$

Step 3: Plug all the values into the equation: $S_{30} = \dfrac{30(3 + 32)}{2} = 525$.
The sum of the arithmetic series is 525.

Example 2: Find the sum of the first 5 terms of the finite geometric series:
$\frac{1}{3} + \frac{1}{9} + \frac{1}{27} + \frac{1}{81} + \frac{1}{243}$

Step 1: Find the value of r, by which each term is multiplied to get the next number in the series. r in this series is $\frac{1}{3}$.

Step 2: Find all the variables for the summing equation, $S_n = \dfrac{t_1(1 - r^n)}{(1 - r)}$
$t_1 = \frac{1}{3}$
$n = 5$
$r = \frac{1}{3}$

Step 3: Insert the values in the equations: $S_5 = \dfrac{\frac{1}{3}\left(1 - \left(\frac{1}{3}\right)^5\right)}{\left(1 - \frac{1}{3}\right)} = \dfrac{\frac{1}{3}\left(\frac{242}{243}\right)}{\left(\frac{2}{3}\right)} = \dfrac{121}{243}$

The sum of the geometric series is $\dfrac{121}{243} = 0.498$.

Example 10: Find the sum of the infinite geometric series:
$16 + 8 + 4 + 2 + 1 + \ldots$

Step 1: Find the value, r, by which each term is multiplied to get the next number in the series. r in this series is $\frac{1}{2}$.

Step 2: Find all the values for the equations, $S = \frac{t_1}{1 - r}$.
$t_1 = 16$
$r = \frac{1}{2}$

Step 3: Determine if $|r| < 1$ or if $|r| > 1$.
$|r| = \left|\frac{1}{2}\right| = \frac{1}{2} < 1$. Therefore, there is a sum for this series.

Step 4: Insert the values into the summing equation: $S = \frac{16}{1 - \frac{1}{2}} = 32$.
The sum of the infinite geometric series is 32.

Find the sum of the first 20 terms in the finite arithmetic series.

1. $5 + 6 + 7 + 8 + 9 + \ldots$
2. $-3 + -6 + -9 + -12 + -15 + \ldots$
3. $0 + -1 + -2 + -3 + -4 + \ldots$
4. $4 + 6 + 8 + 10 + 12 + \ldots$
5. $-1 + 0 + 1 + 2 + 3 + \ldots$
6. $-2 + -1 + 0 + 1 + 2 + \ldots$

Find the sum of the first 10 terms in the finite geometric series.

7. $1 + \frac{1}{2} + \frac{1}{4} + \frac{1}{8} + \frac{1}{16} + \ldots$
8. $25 + 5 + 1 + \frac{1}{5} + \ldots$
9. $100 + 10 + 1 + \frac{1}{10} + \frac{1}{100} + \ldots$
10. $32 + 8 + 2 + \frac{1}{2} + \frac{1}{8} + \ldots$
11. $-625 + -125 + -25 + -25 + -5 + \ldots$
12. $-9 + -3 + -1 + -\frac{1}{3} + -\frac{1}{9} + \ldots$

Find the sum of the infinite geometric series. Remember the sum can only be found if $|r| < 1$. If the sum cannot be found, write NP.

13. $-625 + -125 + -25 + -5 + \ldots$
14. $2 + 4 + 8 + 16 + 32 + \ldots$
15. $32 + 8 + 2 + \frac{1}{2} + \frac{1}{8} + \ldots$
16. $1 + \frac{1}{2} + \frac{1}{4} + \frac{1}{8} + \frac{1}{16} + \ldots$
17. $-9 + -3 + -1 + -\frac{1}{3} + -\frac{1}{9} + \ldots$
18. $3 + 9 + 27 + 81 + 243 + \ldots$

16.5 Algorithms

An **algorithm** is a sequence of actions to accomplish a task. It is the process of performing specific steps to solve a problem. Long division is an example of an algorithm. A **flowchart** is a visual way of presenting an algorithm.

Example 11: Use the images below to carry out the algorithm.

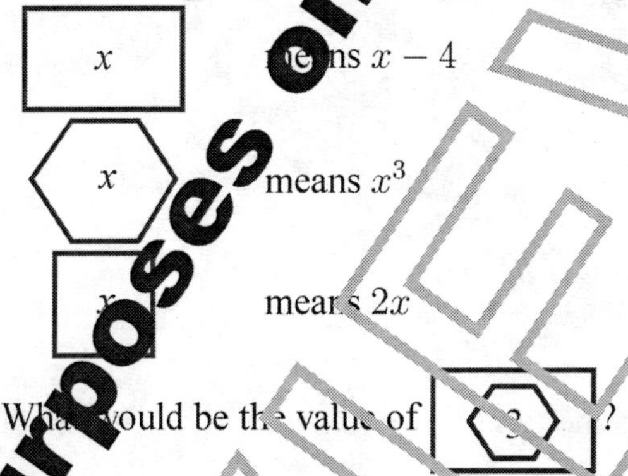

What would be the value of [3 in a hexagon in a rectangle]?

Step 1: First, locate the 3. The three is directly in a hexagon, which is in the rectangle. The algorithm must be solved beginning with the shape that the number, 3, is directly in. Since it is in the hexagon, you must look at what the hexagon symbolizes. The hexagon symbolizes x^3, so substitute 3 in for x.
$(3)^3 = 27$

Step 2: The hexagon is in a rectangle, so now you must perform the operation of the rectangle, $x - 4$, on the result from the hexagon's operation.
$27 - 4 = 23$
So the value of the algorithm is 23.

Use the images below to solve the algorithms.

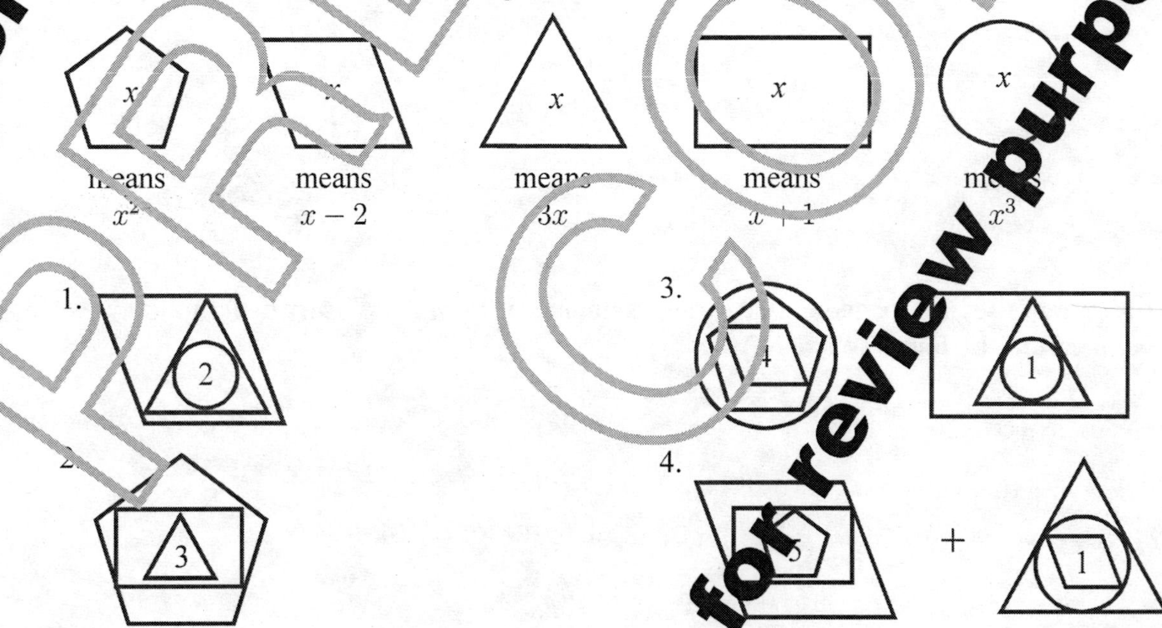

16.6 More Algorithms

Some algorithms use variables instead of flowcharts or other visual representatives.

Example 12: If $*$ is defined for all x and y such that $x * y = x^2y + 4y$, then $2 * 3 = ?$

Step 1: Since $x * y = x^2y + 4y$, you already have an equation to work with. Now, substitute 2 in for x and 3 in for y. The equation will look like
$2 * 3 = 2^2\,(3) + 4\,(3)$

Step 2: Solve.
$2 * 3 = 2^2\,(3) + 4\,(3) = 4\,(3) + 12 = 12 + 12 = 24$
$2 * 3 = 24$

For questions 1 through 4, use the following algorithm:

$$x @ y = 2xy^3 - 3x.$$

1. $2 @ 1 =$
2. $3 @ -1 =$
3. $1 @ 4 =$
4. $-2 @ 2 =$

For questions 5 through 8, use the following algorithm:

$$a * b = a\sqrt{b} - 4a.$$

5. $2 * 9 =$
6. $4 * 25 =$
7. $3 * 10 =$
8. $7 * 4 =$

For questions 9 through 12, use the following algorithm:

$$u \# v = 2(u + 3v - 1).$$

9. $-1 \# 3 =$
10. $2 \# 2 =$
11. $5 \# -2 =$
12. $0 \# 4 =$

For questions 13 through 16, use the following algorithm:

$$xy \& y = 3xy - 2x.$$

13. $4 \& 2 =$
14. $15 \& 3 =$
15. $8 \& 4 =$
16. $3 \& 1 =$

16.7 Inductive Reasoning and Patterns

Humans have always observed what happened in the past and used these observations to predict what would happen in the future. This is called **inductive reasoning**. Although mathematics is referred to as the "deductive science," it benefits from inductive reasoning. We observe patterns in the mathematical behavior of a phenomenon, then find a rule or formula for describing and predicting its future mathematical behavior. There are lots of different kinds of predictions that may be of interest.

Example 13: Nancy is watching her nephew, Drew, arrange his marbles in rows on the kitchen floor. The figure below shows the progression of his arrangement.

Row 1 ○
Row 2 ○○
Row 3 ○○○○
Row 4 ○○○○○○○○

Assuming this pattern continues, how many marbles would Drew place in a fifth row?

Solution: It appears that Drew doubles the number of marbles in each successive row. In the 4th row he had 8 marbles, so in the 5th row we can predict 16 marbles.

Example 14: Manuel drops a golf ball from the roof of his high school while Carla video tapes the motion of the ball. Later, the video is analyzed and the results are recorded concerning the height of each bounce of the ball.

What height do you predict for the fifth bounce?

Initial height	1st bounce	2nd bounce	3rd bounce	4th bounce
30 ft	18 ft	10.8 ft	6.48 ft	3.888 ft

To answer this question, we need to be able to relate the height of each bounce to the bounce immediately preceding it. Perhaps the best way to do this is with **ratios** as follows:

$\frac{\text{Height of 1st bounce}}{\text{Initial bounce}} = 0.6$ $\frac{\text{Height of 2nd bounce}}{\text{Height of 1st bounce}} = 0.6$

$\frac{\text{Height of 4th bounce}}{\text{Height of 3rd bounce}} = 0.6$

Since the ratio of the height of each bounce to the bounce before it appears constant, we have some basis for making predictions.

Using this, we can reason that the fifth bounce will be equal to 0.6 of the fourth bounce.
Thus we predict the fifth bounce to have a height of $0.6 \times 3.888 = 2.3328$ ft.

Which bounce will be the last one with a height of one foot or greater?

For this question, keep looking at predicted bounce heights until a bounce of less than 1 foot is reached.

The sixth bounce is predicted to be $0.6 \times 2.3328 = 1.399\,768$ ft.
The seventh bounce is predicted to be $0.6 \times 1.399\,768 = 0.839\,808$ ft.

Thus, the last bounce with a height greater than 1 ft is predicted to be the sixth one.

Read the following questions carefully. Use inductive reasoning to answer each question. You may wish to make a table or a diagram to help you visualize the pattern in some of the problems.

1. Bob and Alice have designed and created a website for their high school. The first week they had 5 visitors to the site; the second week they had 10 visitors; and during the third week, they had 20 visitors.

   (A) If current trends continue, how many visitors can they expect in the fifth week?

   (B) How many in the nth week?

   (C) How many weeks will it be before they get more than 500 visitors in a single week?

2. In 1979 (the first year of classes), there were 500 students at Brookstone High. In 1989, there were 1000 students. In 1999, there were 2000 students. How many students would you predict at Brookstone in 2009 if this pattern continues (and no new schools are built)?

3. The average combined (math and verbal) SAT score for students at Brookstone High was 1000 in 2001, 1100 in 2002, 1210 in 2003, and 1331 in 2004. Predict the combined SAT score for Brookstone seniors in 2005.

4. Marie has a daylily in her mother's garden. Every Saturday morning in the spring, she measures and records its height in the table below. What height do you predict for Marie's daylily on April 29? (Hint: Look at the *change* in height each week when looking for the pattern).

April 1	April 8	April 15	April 22
12 in	18 in	21 in	22.5 in

5. Bob puts a glass of water in the freezer and records the temperature every 15 minutes. The results are displayed in the table below. If this pattern of cooling continues, what will be the temperature at 2:15 P.M.? (Hint: Again, look at the changes in temperature in order to see the pattern.)

1:00 P.M.	1:15 P.M.	1:30 P.M.	1:45 P.M.
92° F	60° F	44° F	3[illegible]° F

Example 15: Mr. Applegate wants to put desks together in his math class so that students can work in groups. The diagram below shows how he wishes to do it.

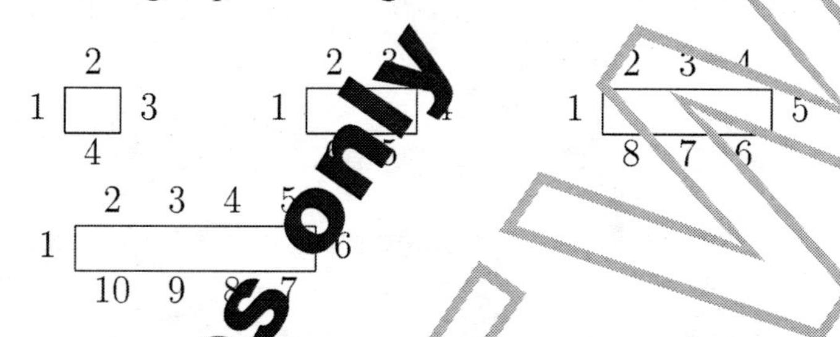

With one table he can seat 4 students, with two tables he can seat 6, with three tables 8, and with four tables 10.

How many students can he seat with 5 tables?

With 5 tables he could seat 5 students along the sides of the tables and 1 student on each end; thus, a total of 12 students could be seated.

Write a rule that Mr. Applegate could use to tell how many students could be seated at *n* tables. Explain how you got the rule.

For n tables, there would be n students along each of 2 sides and 2 students on the ends (1 on each end); thus, a total of $2n + 2$ students could be seated at n tables.

Example 16: When he isn't playing football for the Brookstone Bears, Tim designs web pages. A car dealership paid Tim $500 to start a site with photos of its cars. The dealer also agreed to pay Tim $50 for each customer who buys a car first viewed on the website.

Write and explain a rule that tells how much the dealership will pay Tim for the design of the website and the sale of *n* cars from the website.

Tim's payment will be the initial $500 plus $50 for each sale. Translated into mathematical language, if Tim sells n cars he will be paid a total of $500 + 50n$ dollars.

How many cars have to be sold from his site in order for Tim to get $1000 from the dealership?

He earned $500 just by establishing the site, so he only needs to earn an additional $500, which at $50 per car requires the sale of only 10 cars. (Note: Another way to solve this problem is to use the rule found in the first question. In that case, you simply solve the equation $500 + 50n = 1000$ for the variable n.)

Example 17: Eric is baking muffins to raise money for the Homecoming dance. He makes 18 muffins with each batch of batter, but he must give one muffin each to his brother, his sister, his dog, and himself each time a batch is finished baking.

Write a rule for the number of muffins Eric produces for the fund-raiser with n batches. He bakes 18 muffins with each batch, but only 14 are available for the fund-raiser. Thus with n batches he will produce $14n$ muffins for Homecoming. **The rule** $= 14n$.

Use your rule to determine how many muffins he will contribute if he makes 7 batches. The number of batches, n, equals 7. Therefore, he will produce $14 \times 7 = 98$ muffins with 7 batches.

Determine how many batches he must make in order to contribute at least 150 muffins. Ten batches will produce $10 \times 14 = 140$ muffins. Eleven batches will produce $11 \times 14 = 154$ muffins. To produce at least 150 muffins, he must bake at least 11 batches.

Determine how many muffins he would actually bake in order to contribute 150 muffins. Since Eric actually bakes 18 muffins per batch, 11 batches would result in Eric baking $11 \times 18 = 198$ muffins.

Carefully read and solve the problems below. Show your work.

1. Tito is building a picket fence along both sides of the driveway leading up to his house. He will have to place posts at both ends and at every 10 feet along the way because the fencing comes in prefabricated ten-foot sections.
   (A) How many posts will he need for a 180-foot driveway?
   (B) Write and explain a rule for determining the number of posts needed for n ten-foot sections.
   (C) How long of a driveway can he fence with 32 posts?

2. Dakota's beginning pay at his new job is \$300 per week. For every three months he continues to work there, he will get a \$10 per week raise.
   (A) Write a formula for Dakota's weekly pay after n three-month periods.
   (B) After n years?
   (C) How long will he have to work before his pay gets to \$400 per week?

3. Amanda is selling shoes this summer. In addition to her hourly wages, Amanda got a \$100 bonus just for accepting the position, and she gets a \$2 bonus for each pair of shoes she sells.
   (A) Write and explain a rule that tells how much she will make in bonuses if she sells n pairs of shoes.
   (B) How many pairs of shoes must she sell in order to make \$200 in bonuses?

4. The table below displays data relating temperature in degrees Fahrenheit to the number of chirps per minute for a cricket.

Temp (°F)	50	52	55	58	60	64	68
Chirps/min	40	48	60	72	80	96	112

Write a formula or rule that predicts the number of chirps per minute when the temperature is n degrees.

16.8 Mathematical Reasoning/Logic

The SAT math test calls for skill development in mathematical **reasoning** or **logic**. The ability to use logic is an important skill for solving math problems, but it can also be helpful in real-life situations. For example, if you need to get to Park Street, and the Park Street bus always comes to the bus stop at 3 PM, then you know that you need to get to the bus stop by at least 3 PM. This is a real-life example of using logic, which many people would call "common sense."

There are many different types of statements which are commonly used to describe mathematical principles. However, using the rules of logic, the truth of any mathematical statement must be evaluated. Below is a list of tools used in logic to evaluate mathematical statements.

Logic is the discipline that studies valid reasoning. There are many forms of valid arguments, but we will just review a few here.

A **proposition** is usually a declarative sentence which may be true or false.

An **argument** is a set of two or more related propositions, called **premises**, that provide support for another proposition, called the **conclusion**.

Deductive reasoning is an argument which begins with general premises and proceeds to a more specific conclusion. Most elementary mathematical problems use deductive reasoning.

Inductive reasoning is an argument in which the truth of its premises make it likely or probable that its conclusion is true.

ARGUMENTS

Most of logic deals with the evaluation of the validity of arguments. An argument is a group of statements that includes a conclusion and at least one premise. A premise is a statement that you know is true or at least you assume to be true. Then, you draw a conclusion based on what you know or believe is true in the premise(s). Consider the following example:

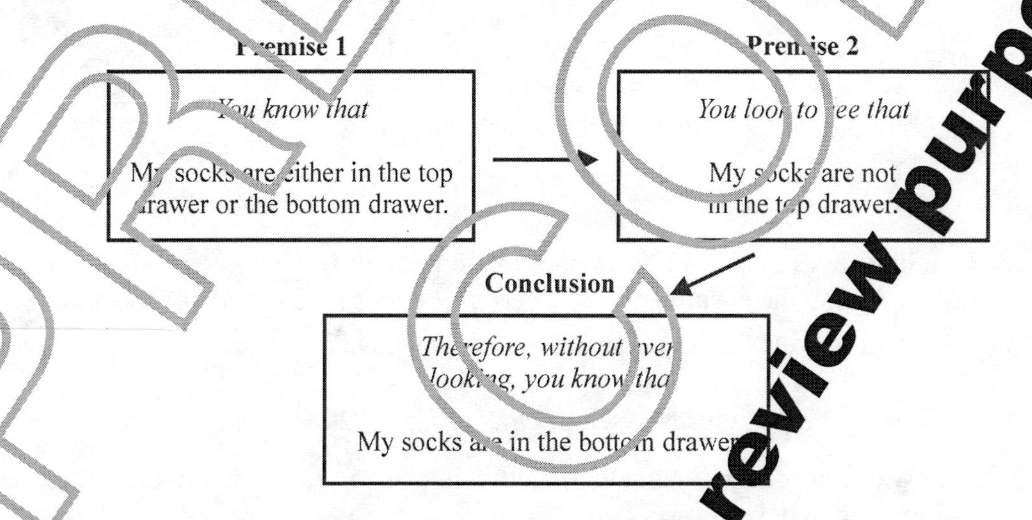

This argument is an example of deductive reasoning, where the conclusion is "deduced" from the premises and nothing else. In other words, if Premise 1 and Premise 2 are true, you don't even need to look in the bottom drawer to know that the conclusion is true.

16.9 Deductive and Inductive Arguments

In general, there are two types of logical arguments: **deductive** and **inductive**. Deductive arguments tend to move from general statements or theories to more specific conclusions. Inductive arguments tend to move from specific observations to general theories.

Deductive Reasoning

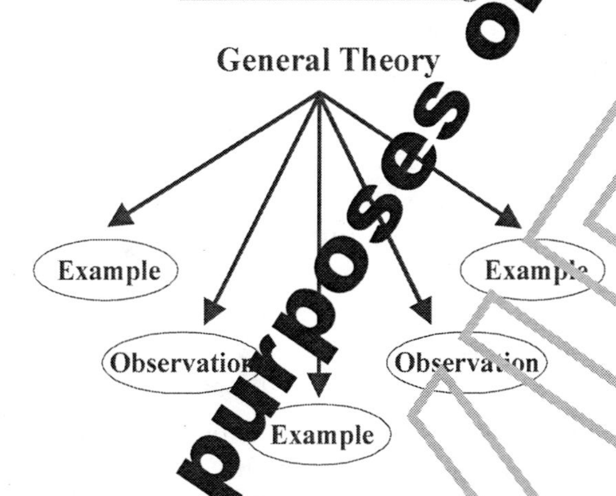

Inductive Reasoning

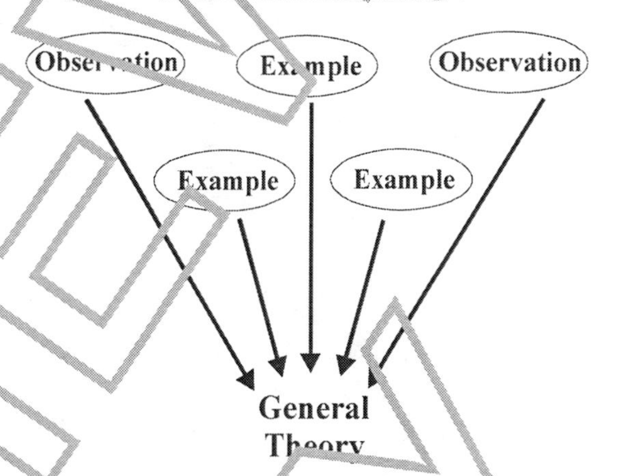

Compare the two examples below:

Deductive Argument

Premise 1	All men are mortal.
Premise 2	Socrates is a man.
Conclusion	Socrates is mortal.

Inductive Argument

Premise 1	The sun rose this morning.
Premise 2	The sun rose yesterday morning.
Premise 3	The sun rose two days ago.
Premise 4	The sun rose three days ago.
Conclusion	The sun will rise tomorrow.

An inductive argument cannot be proved beyond a shadow of a doubt. For example, it's a pretty good bet that the sun will come up tomorrow, but the sun not coming up presents no logical contradiction.

On the other hand, a deductive argument can have logical certainty, but it must be properly constructed. Consider the examples below.

True Conclusion from an Invalid Argument

All men are mortal.
Socrates is mortal.
Therefore Socrates is a man.

Even though the above conclusion is true, the argument is based on invalid logic. Both men and women are mortal. Therefore, Socrates could be a woman.

False Conclusion from a Valid Argument

All astronauts are men.
Julia Roberts is an astronaut.
Therefore Julia Roberts is a man.

In this case, the conclusion is false because the premises are false. However, the logic of the argument is valid because *if* the premises were true, then the conclusion would be true.

A **counterexample** is an example given in which the statement is true but the conclusion is false when we have assumed it to be true. If we said "All cocker spaniels have blonde hair," then a counterexample would be a red-haired cocker spaniel. If we made the statement, "If a number is greater than 10, it is less than 20," we can easily think of a counterexample, like 35.

Example 18: Which argument is valid?

If you speed on Hill Street, you will get a ticket.
If you get get a ticket, you will pay a fine.

(A) I paid a fine, so I was speeding on Hill Street.
(B) I got a ticket, so I was speeding on Hill Street.
(C) I exceeded the speed limit on Hill Street, so I paid a fine.
(D) I did not speed in Hill Street, so I did not pay a fine.

Solution: C is valid.
A is incorrect. I could have paid a fine for another violation.
B is incorrect. I could have gotten a ticket for some other violation.
D is incorrect. I could have paid a fine for speeding somewhere else.

Example 19: Assume the given proposition is true. Then, determine if each statement is true or false.

Given: If a dog is thirsty, he will drink.

(A) If a dog drinks, then he is thirsty. T or F
(B) If a dog is not thirsty, he will not drink. T or F
(C) If a dog will not drink, he is not thirsty. T or F

Solution: A is false. He is not necessarily thirsty; he could just drink because other dogs are drinking or drink to show others his control of the water. This statement is the **converse** of the original. The converse of the statement "If A, then B" is "If B, then A."

B is false. The reasoning from A applies. This statement is the **inverse** of the original. The inverse of the statement "If A, then B" is "If not A, then not B."

C is true. It is the **contrapositive**, or the complete opposite of the original. The contrapositive says "If not B, then not A."

For numbers 1–5, what conclusion can be drawn from each proposition?

1. All squirrels are rodents. All rodents are mammals. Therefore,
2. All fractions are rational numbers. All rational numbers are real numbers. Therefore,
3. All squares are rectangles. All rectangles are parallelograms. All parallelograms are quadrilaterals. Therefore,
4. All Chevrolets are made by General Motors. All Luminas are Chevrolets. Therefore,
5. If a number is even and divisible by three, then it is divisible by six. Eighteen is divisible by six. Therefore,

For numbers 6–9, assume the given proposition is true. Then, determine if the statements following it are true or false.

All squares are rectangles.

6. All rectangles are squares. T or F
7. All non-squares are non-rectangles. T or F
8. No squares are non-rectangles. T or F
9. All non-rectangles are non-squares. T or F

Chapter 16 Review

Find the pattern for the following number sequences, and then find the nth number requested.

1. 0, 1, 2, 3, 4 pattern______
2. 0, 1, 2, 3, 4 20th number_______
3. 1, 3, 5, 7, 9 pattern______
4. 1, 3, 5, 7, 9 25th number______
5. 3, 6, 9, 12, 15 pattern_______
6. 3, 6, 9, 12, 15 30th number_______

Find the limit of each sequence.

7. 0, 1, 2, 3, 4
8. $-32, -16, -8, -4, -2$
9. 3, 6, 9, 12, 15
10. $27, 9, 3, \frac{1}{3}, \frac{1}{9}$
11. $0, -1, -2, -3, -4$
12. $-1, -\frac{1}{2}, -\frac{1}{3}, -\frac{1}{4}, -\frac{1}{5}$

Identify the series as arithmetic or geometric, and find the sum of the finite series when $n = 10$.

13. $0 + 1 + 2 + 3 + 4 + \ldots$
14. $-32 + -16 + -8 + -4 + -2 + \ldots$
15. $3 + 6 + 9 + 12 + 15 + \ldots$
16. $27 + 9 + 3 + \frac{1}{3} + \frac{1}{9} + \ldots$
17. $0 + -1 + -2 + -3 + -4 + \ldots$
18. $2 + 5 + 8 + 11 + 14 + \ldots$

Find the sum of the infinite geometric series.

19. $-32 + -16 + -8 + -4 + -2 + \ldots$
20. $27 + 9 + 3 + \frac{1}{3} + \frac{1}{9} + \ldots$
21. $64 + 48 + 36 + 27 + 20.25 + \ldots$

Use the images below to solve the algorithms.

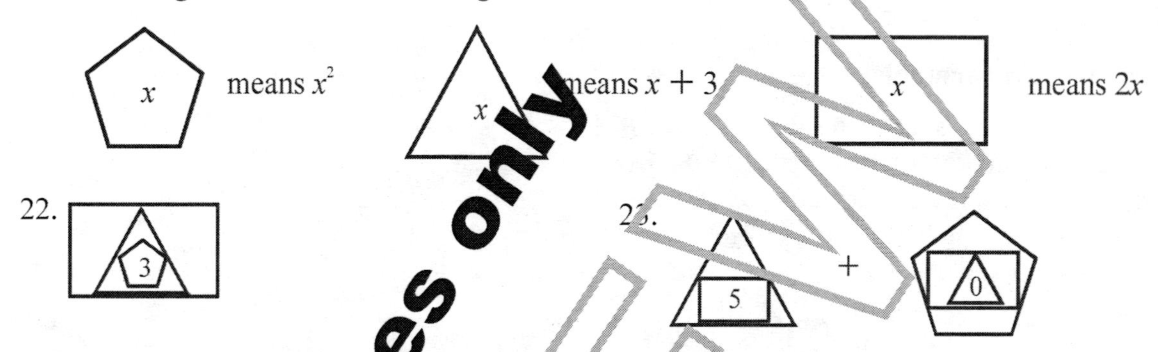

Justin receives a bill from his internet service provider. The first four months of service are charged according to the table below:

	January	February	March	April
Hours	0	10	5	25
Charge	\$4.95	\$14.45	\$9.70	\$28.70

24. Write a formula for the cost of n hours of internet service.

25. What is the greatest number of hours he can get on the internet and still keep his bill under \$20.00?

Lisa is baking cookies for the Fall Festival. She bakes 27 cookies with each batch of batter. However, she has a defective oven, which results in 5 cookies in each batch being burnt.

26. Write a formula for the number of cookies available for the festival as a result of Lisa baking n batches of cookies.

27. How many batches does she need in order to produce 300 cookies for the festival?

28. How many cookies (counting burnt ones) will she actually bake?

For questions 29 and 30, use the algorithm, # is defined for all a and b such that $a \# b = 4ab - b^2$.

29. 3 # 1 = ________ 30. 2 # −4 = ________

31. Find the object that will come sixth in the pattern. Circle your answer.

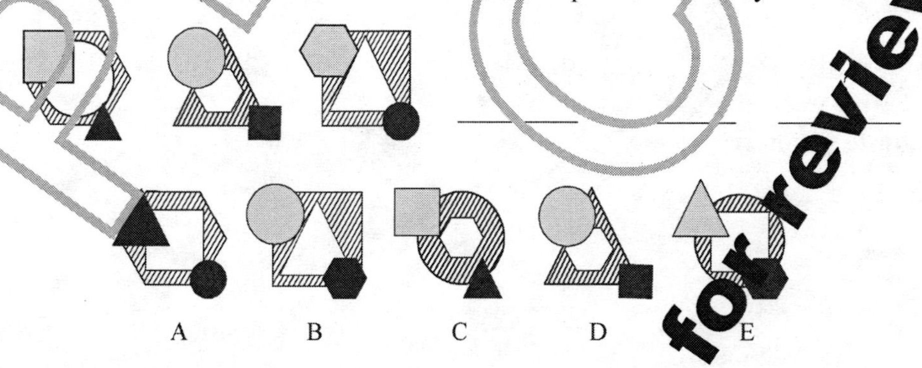

Chapter 17
Statistics

17.1 Range

In **statistics**, the difference between the largest number and the smallest number in a list is called the **range**.

Example 1: Find the range of the following list of numbers: 16, 73, 26, 15, and 35.

The largest number is 73, and the smallest number is 15. $73 - 15 = 58$
The range is 58.

Find the range for each list of numbers below.

1.	2.	3.	4.	5.	6.	7.
21	[illegible]	89	41	23	2	77
51	[illegible]7	22	3	20	38	94
48	31	65	56	64	29	27
42	55	36	41	38	[illegible]	46
12	8	20	19	21	[illegible]	63

8.	9.	10.	11.	12.	13.	14.
[illegible]1	6[illegible]	84	84	21	45	62
62	54	[illegible]9	65	78	[illegible]	39
32	5[illegible]	48	32	6	57	96
16	5	21	50	[illegible]	14	45
59	63	8[illegible]	71	45	61	14

15. 2, 1[illegible], 3, 25, and 17

16. 15, 48, [illegible]2, [illegible]1, and 8

17. [illegible], 74, 2, 86, and 75

18. [illegible]5, [illegible]1, 11, 22, and 65

19. [illegible]3, 18, 65, 12, and 74

20. 47, 1[illegible], 33, 25, and 19

21. 56, 10, 33, 7, 16, and 5

22. 4[illegible], 25, 78, 49, and 6

23. 4[illegible], 75, 63, and 21

24. 97, [illegible]3, 56, 12, and 6[illegible]

25. 87, 44, 63, and 1[illegible]

26. [illegible]4, [illegible]5, 66, 3[illegible], and 31

27. 35, 44, 8[illegible], [illegible], and 78

28. 95, 5[illegible], 62, 14, 8, and 3

17.2 Mean

In statistics, the arithmetic mean is the same as the average. To find the arithmetic mean of a list of numbers, first add together all of the numbers in the list, and then divide by the number of items in the list.

Example 2: Find the mean of 38, 72, 110, 548.

Step 1: First add: $38 + 72 + 110 + 548 = 768$

Step 2: There are 4 numbers in the list so divide the total by 4. $768 \div 4 = 192$
The mean is 192.

Practice finding the mean (average). Round to the nearest tenth if necessary.

1. Dinners served:
489 561 52[illegible] 450

2. Prices paid for shirts:
\$4.89 \$[illegible].97 \$5.90 \$8.64

3. Piglets born:
2[illegible] 19 15 21 22

4. Student absences:
6 5 13 8 9 12 7

5. Paychecks:
\$89.56 \$99.99 \$56.54

6. Choir attendance:
56 45 97 66 70

7. Long distance calls:
33 14 24 21 19

8. Train boxcars:
56 55 48 61 51

9. Cookies eaten:
[illegible] 6 8 9 2 4 3

Find the mean (average) of the following word problems.

10. Val's science grades are 95, 87, 65, 94, 78, and 97. What is her average?

11. Ann runs a business from her home. The number of orders for the last 7 business days are 17, 24, 13, 8, 11, 15, and 9. What is the average number of orders per day?

12. Melissa tracks the number of phone calls she has per day: 8, 2, 5, 4, 7, 3, [illegible]. What is the average number of calls she receives?

13. The Cheese Shop tracks the number of lunches they serve this week: 42, 55, 36, 41, 38, 33, and 46. What is the average number of lunches served?

14. Leah drives 364 miles in 7 hours. What is her average miles per hour?

15. Tim saves \$680 in 8 months. How much does his savings average each month?

16. Ken makes 117 passes in 13 games. How many passes does he average per game?

17.3 Finding Data Missing From the Mean

Example 3: Mara knew she had an 88 average in her biology class, but she lost one of her papers. The three papers she could find had scores of 98%, 84%, and 90%. What was the score on her fourth paper?

Step 1: Figure the total score on four papers with an 88% average. $.88 \times 4 = 3.52$

Step 2: Add together the scores from the three papers you have. $.98 + .84 + .9 = 2.72$

Step 3: Subtract the scores you know from the total score. $3.52 - 2.72 = .80$. She had 80% on her fourth paper.

Find the data missing from the following problems.

1. Gabriel earns 87% on his first geography test. He wants to keep a 92% average. What does he need to get on his next test to bring his average up?

2. Rian earned $68.00 on Monday. How much money must she earn on Tuesday to have an average of $80 earned for the two days?

3. Haley, Chuck, Dana, and Chris enter a contest to see who could bake the most chocolate chip cookies in an hour. They bake an average of 75 cookies. Haley bakes 55, Chuck bakes 70, and Dana bakes 90. How many does Chris bake?

4. Four wrestlers make a pact to lose some weight before the competition. They lose an average of 7 pounds each over the course of 3 weeks. Carlos loses 6 pounds, Steve loses 5 pounds, and Greg loses 9 pounds. How many pounds does Wes lose?

5. Three boxes are ready for shipment. The boxes average 26 pounds each. The first box weighs 30 pounds; the second box weighs 25 pounds. How much does the third box weigh?

6. The five jockeys running in the next race average 92 pounds each. Nicole weighs 89 pounds. Jon weighs 95 pounds. Jenny and Kasey weigh 90 pounds each. How much does Jordan weigh?

7. Jessica makes three loaves of bread that weigh a total of 45 ounces. What is the average weight of each loaf?

8. Celeste makes scented candles to give away to friends. She has 2 pounds of candle wax which she melted, scented, and poured into 8 molds. What is the average weight of each candle?

9. Each basketball player has to average a minimum of 5 points a game for the next three games to stay on the team. Ben is feeling the pressure. He scored 3 points the first game and 2 points the second game. How many points does he need to score in the third game to stay on the team?

17.4 Median

In a list of numbers ordered from lowest to highest, the **median** is the middle number. To find the **median**, first arrange the numbers in numerical order. If there is an odd number of items in the list, the **median** is the middle number. If there is an even number of items in the list, the **median** is the **average of the two middle numbers.**

Example 4: Find the median of 42, 35, 45, 37, and 41.

Step 1: Arrange the numbers in numerical order: 35 37 [41] 42 45

Step 2: Find the middle number. The median is 41.

Example 5: Find the median of 14, 53, 42, 6, 14, and 46.

Step 1: Arrange the numbers in numerical order: 6 14 [14 42] 46 53

Step 2: Find the average of the two middle numbers.
$(14 + 42) \div 2 = 28$. The median is 28.

Circle the median in each list of numbers.

1. [illegible]5, [illegible]5, 40, 30, and 45
2. 7, 2, 3, 6, 5, 1, and 8
3. 65, 42, 60, 46, and 90
4. 15, 16, 19, 25, 20
5. 75, 98, 87, 65, 82, 88, 100
6. 33, 42, 50, 22, and 19
7. 401, 758, and 254
8. 41, 23, 14, 21, and 19
9. 5, 8, 10, 13, 1, and 8

10.	11.	12.	13.	14.	15.	16.
19	9	45	52	20	8	15
14	3	32	54	21	17	40
12	10	66	19	25	13	[illegible]
15	17	55	63	18	14	32
18	6	61	20	16	22	28

Find the median in each list of numbers.

17. 10, 8, 21, 14, 9, and 12
18. 43, 36, 20, and 40
19. 5, 24, 9, 18, 12, and 3
20. 48, 13, 54, 82, 90, and 7
21. [illegible], 21, 36, and 27
22. 9, 4, 3, 1, 6, 2, 10, and 12

23.	24.	25.	26.	27.	28.
2	11	13	75	48	22
10	22	15	62	45	19
6	25	9	60	5[illegible]	15
18	28	35	52	[illegible]	43
20	10	29	80	[illegible]	34
23	23	33	50	58	28

17.5 Mode

In statistics, the mode is the number that occurs most frequently in a list of numbers.

Example 6: Exam grades for a math class were as follows:
70 88 92 85 99 85 70 85 99 100 88 70 99 88 88 99 88 92 85 88

Step 1: Count the number of times each number occurs in the list.

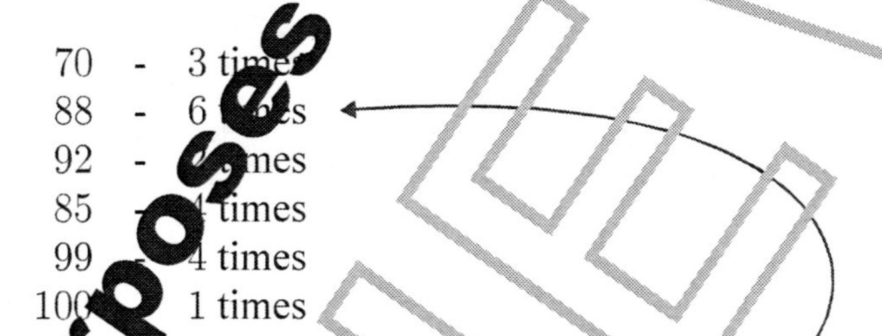

70 - 3 times
88 - 6 times
92 - 2 times
85 - 4 times
99 - 4 times
100 - 1 times

Step 2: Find the number that occurs most often.
The mode is 88 because it is listed 6 times. No other number is listed as often.

Find the mode in each of the following lists of numbers.

1.	2.	3.	4.	5.	6.	7.
88	54	21	56	64	5	12
15	42	16	67	22	4	41
88	44	15	67	22	9	45
17	56	78	19	15	8	32
18	44	21	56	14	4	16
88	44	16	67	14	7	12
17	56	21	20	22	4	12

8. 48, 32, 56, 32, 56, 48, 56
9. 12, 16, 54, 78, 16, 25, 20
10. 5, 4, 8, 3, 4, 2, 7, 8, 4, 2
11. 11, 9, 7, 11, 7, 5, 7, 7, 5
12. 84, 22, 79, 22, 87, 22, 22
13. 95, 87, 65, 94, 78, 95
14. 8, 2, 3, 4, 7, 2, 3, 6, 1
15. 89, 7, 11, 89, 17, 56
16. 15, 48, 52, 41, 8, 48
17. 22, 45, 48, 12, 22, 41, 22
18. 62, 44, 78, 62, 54, 44, 62
19. 54, 22, 54, 78, 22, 78, 22
20. 14, 17, 33, 21, 33, 17, 33
21. 65, 51, 8, 21, 8, 65, 70, 8
22. 17, 24, 13, 8, 11, 8, 15, 9
23. 51, 45, [illegible], 51, 65, 74, 51
24. 8, 74, 65, 15, 9, 10, 74
25. [illegible], 54, 2, 7, 89, 2, 7, 54, 2

17.6 Applying Measures of Central Tendency

On the SAT, you may be asked to solve real-world problems involving measures of central tendency.

Example 7: Aida is shopping around for the best price on a 17" computer monitor. She travels to seven stores and finds the following prices: \$199, \$159, \$249, \$329, \$199, \$209, and \$189. When Aida goes to the eighth and final store, she finds the price for the 17" monitor is \$549. Which of the measures of central tendency, mean, median, or mode, changes the most as a result of the last price Aida finds?

Step 1: **Solve for the three measures of the seven values.**

Mean: $\dfrac{\$199 + \$159 + \$249 + \$329 + \$199 + \$209 + \$189}{7} = \$219$

Median: From least to greatest: \$159, \$189, \$199, \$199, \$209, \$249, \$329. The 4th value = \$199

Mode: The number repeated the most is \$199.

Step 2: **Find the mean, median, and mode with the eighth value included.**

Mean: $\dfrac{\$199 + \$159 + \$249 + \$329 + \$199 + \$209 + \$189 + \$549}{8} = \$260.25$

Median: \$159, \$189, \$199, \$199, \$209, \$249, \$329, \$549. The avg. of 4th and 5th number = \$204

Mode: The number still repeated most is \$199.

Answer: The measure which changed the most by adding the 8th value is the **mean**.

1. The Realty Company has the selling prices for 10 houses sold during the month of July. The following prices are given in thousands of dollars:
   176 89 525 125 107 100 525 61 75 114
   Find the mean, median, and mode of the selling prices. Which measure is most representative for the selling price of such homes? Explain.

2. A soap manufacturing company wants to know if the weight of its product is on target, meaning 4.75 oz. With that purpose in mind, a quality control technician selects 15 bars of soap from production, 5 from each shift, and finds the following weights in oz.

   1st shift: 4.76, 4.75, 4.77, 4.77, 4.74
   2nd shift: 4.72, 4.72, 4.75, 4.76, 4.73
   3rd shift: 4.76, 4.76, 4.77, 4.76, 4.76

   (A) What are the values for the measures of central tendency for the sample from each shift?
   (B) Find the mean, median, and mode for the 24 hour production sample.
   (C) Which measure is the most accurate measure of central tendency for the 24 hour production?
   (D) Find the range of values for each shift. Is the range an effective tool for drawing a conclusion in this case? Why or why not?

17.7 Stem-and-Leaf Plots

A **stem-and-leaf plot** is a way to organize and analyze statistical data. To make a stem-and-leaf plot, first draw a vertical line.

Final Math Averages

85	92	87	62	75	84	[illegible]	52
45	77	98	75	71	79	85	82
87	74	76	68	93	77	[illegible]	84
79	65	77	82	86	[illegible]	92	60
99	75	88	74	79	[illegible]	63	84
87	90	75	81	73	[illegible]	73	75
31	86	89	65	[illegible]	75	79	76

Stem	Leaves
3	1
4	5
5	2
6	0,2,3,5,5,5,8,9,9
7	1,3,3,3,4,4,5,5,5,5,5,5,6,6,7,7,7,9,9,9
8	0,1,2,2,4,4,4,4,5,5,6,6,7,7,7,8,9
9	0,2,2,3,6,8,9

On the left side of the line, list all the numbers that are in the tens place from the set of data. Next, list each number in the ones place on the right side of the line in ascending order. It is easy to see at a glance that most of the students scored in the 70's or 80's with a majority having averages in the 70's. It is also easy to see that the maximum average is 99, and the lowest average is 31. Stem-and-leaf plots are a way to organize data making it easy to read.

1. Make a stem-and-leaf plot from the data below, and then answer the questions that follow.

Speeds on Turner Road

CAR SPEED, mph

45	52	47	35	[illegible]	50	51	43
40	51	32	24	55	41	32	33
36	59	49	52	34	28	69	47
[illegible]	15	63	42	35	42	58	59
39	41	25	34	22	16	40	31
55	10	46	38	50	52	48	36
21	32	36	41	52	49	45	32
52	45	50	35	55	65	20	41

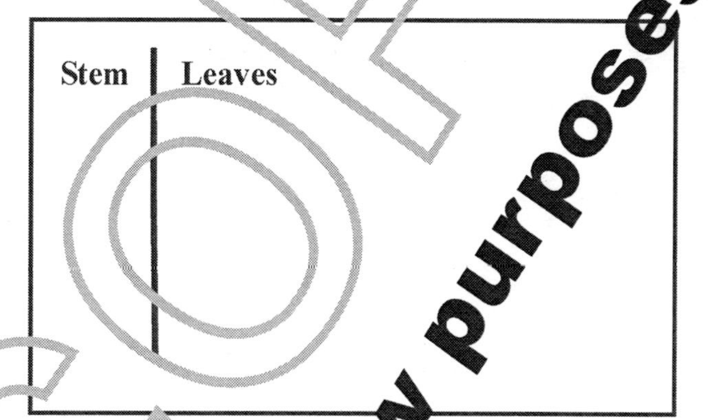

2. What was the fastest speed recorded?

3. What was the slowest speed recorded?

4. Which speed was most often recorded?

5. If the speed limit is 45 miles per hour, how many were speeding?

6. If the speed limit is 45 miles per hour, how many were at 20 mph over or under the speed limit?

17.8 More Stem-and-Leaf Plots

Two sets of data can be displayed on the same stem-and-leaf plot.

Example 8: The following is an example of a back-to-back stem-and-leaf plot.

Bryan's Math scores {60, 65, 72, 78, 85, 90}
Bryan's English scores {78, 88, 89, 89, 92, 95, 100}

Math		English
5, 0	6	
2, 8	7	8
5	8	8, 9, 9
	9	2, 5
	10	0

2|7 means 72 8|9 means 89

Read the stem-and-leaf plot below and answer the questions that follow.

3rd grade Boys' Weights		3rd grade Girls' Weights
8, 7, 5, 3, 2	4	0, 2, 4, 7
6, 4, 1, 0	5	1, 8, 8, 8, 9
5	6	0, 6, 6, 8, 8
0	[illegible]	8

4|5 means 45 6|8 means 68

1. What is the median for girls' weights?
2. What is the median for the boys' weights?
3. What is the mode for the girls' weights?
4. What is the weight of the lightest boy?
5. What is the weight of the heaviest boy?
6. What is the weight of the heaviest girl?

7. Create a stem-and-leaf plot for the data given below.

Automobile speeds on I-85

60	65	80	75	92	81	63
65	67	75	78	79	77	69
62	57	64	65	68	71	69
71	73	56	69	69	70	74

Automobile speeds on I-75

72	56	62	65	63	60	58
55	57	70	69	59	53	61
58	61	63	67	[illegible]	63	67
56	58	59	[illegible]	64	63	69

8. What is the median speed for I-75?
9. What is the median speed for I-85?
10. What is the mode speed for I-75?
11. What is the mode speed for I-85?
12. What was the fastest speed on either interstate?

17.9 Quartiles and Extremes

In statistics, large sets of data are separated into four equal parts. These parts are called **quartiles**. The **median** separates the date into two halves. Then, the median of the upper half is the **upper quartile**, and the median of the lower half is the **lower quartile**.

The **extremes** are the highest and lowest values in a set of data. The lowest value is called the **lower extreme**, and the highest value is called the **upper extreme**.

Example 9: The following set of data shows the high temperatures (in degrees Fahrenheit) in cities across the United States on a particular autumn day. Find the median, the upper quartile, the lower quartile, the upper extreme, and the lower extreme of the data.

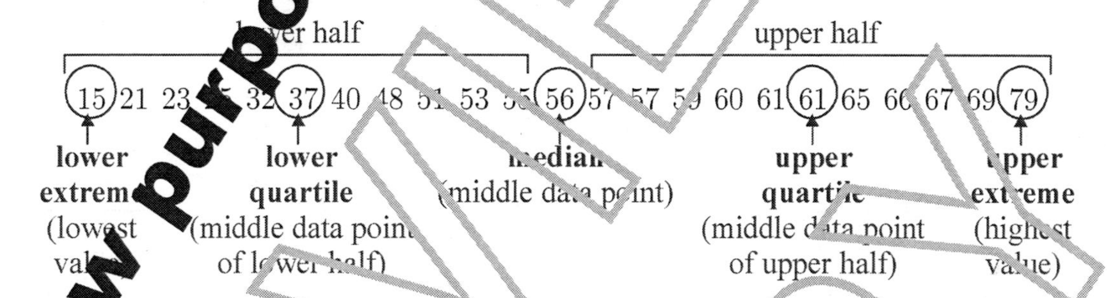

Example 10: The following set of data shows the fastest race car qualifying speeds in miles per hour. Find the medium, the upper quartile, the lower quartile, the upper extreme, and the lower extreme of the data.

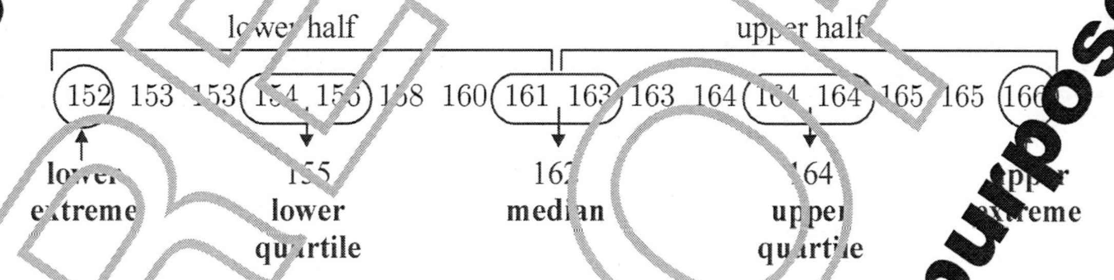

Note: When you have an even number of data points, the median is the average of the two middle points. The lower middle number is then included in the lower half of the data, and the upper middle number is included in the upper half.

Find the median, the upper quartile, the lower quartile, the upper extreme, and the lower extreme of each set of data given below.

1. 0 0 1 1 1 2 2 3 3 4 5
2. 15 16 18 20 22 22 23
3. 62 75 77 80 81 85 87 91 94
4. 74 74 76 76 77 78
5. 3 3 3 5 5 6 6 7 7 7 8 8
6. 190 191 192 192 194 195 196
7. 6 7 8 9 10 10 11 13 15
8. [illegible] 22 24 25 27 28 32 35

17.10 Box-and-Whisker Plots

Box-and-whisker plots are used to summarize data as well as to display data. A box-and-whisker plot summarizes data using the median, upper and lower quartiles, and the lower and upper extreme values. Consider the data below: a list of employees' ages at the Acme Lumber Company:

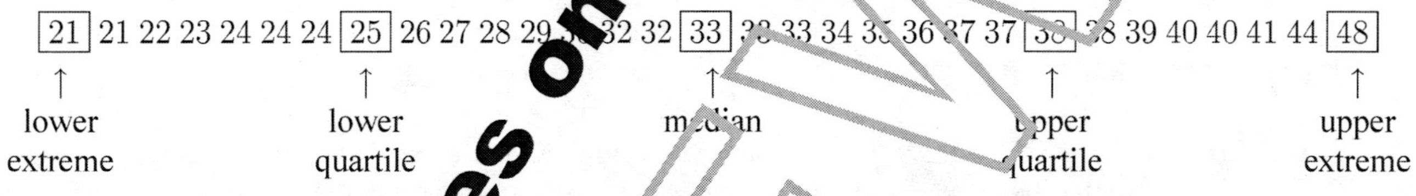

Step 1: Find the median, upper quartile, lower quartile, upper extreme, and lower extreme just like you did in the previous section.

Step 2: Plot the 5 data points found in step 1 above on a number line as shown below.

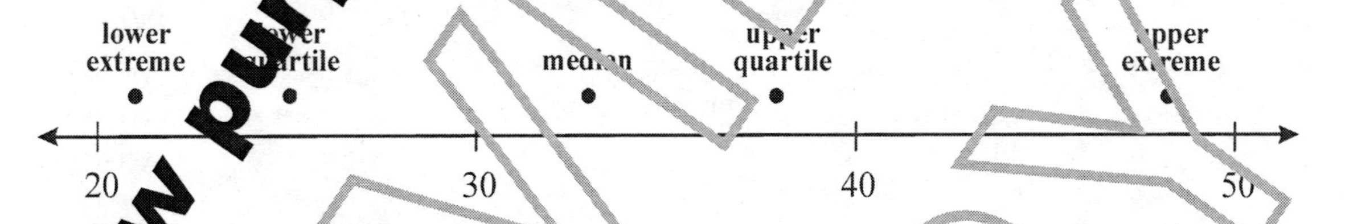

Step 3: Draw a box around the quartile values, and draw a vertical line through the median value. Draw whiskers from each quartile to the extreme value data points.

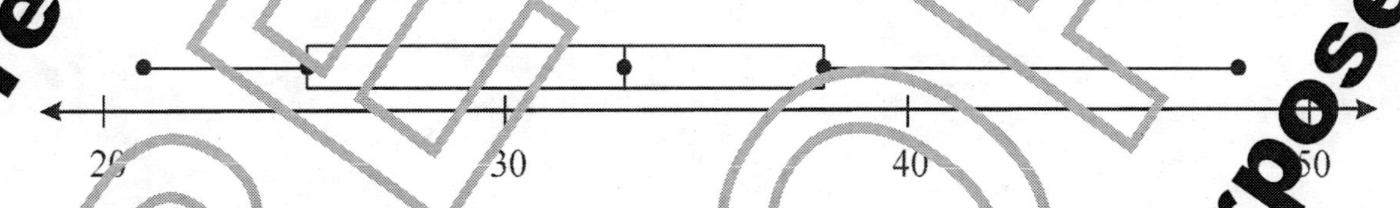

This box-and-whisker displays five types of information: lower extreme, lower quartile, median, upper quartile, and upper extreme.

Draw a box-and-whisker plot for the following sets of data.

1. 10 12 12 15 16 17 19 21 22 22 25 27 31 35 36 37 38 38 41 43 45 50 55 56 57 58 59

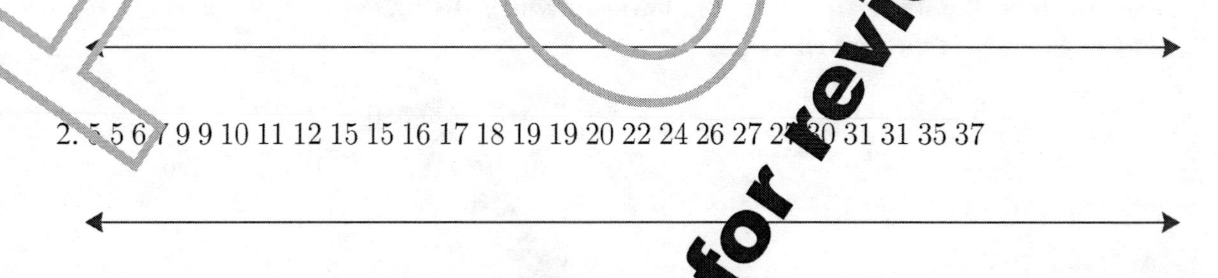

2. 5 5 6 7 9 9 10 11 12 15 15 16 17 18 19 19 20 22 24 26 27 27 30 31 31 35 37

17.11 Scatter Plots

A **scatter plot** is a graph of ordered pairs involving two sets of data. These plots are used to detect whether two sets of data, or variables, are truly related.

In the example to the right, two variables, income and education, are being compared to see if they are related or not. Twenty people were interviewed, age 25 and older, and the results were recorded on the chart.

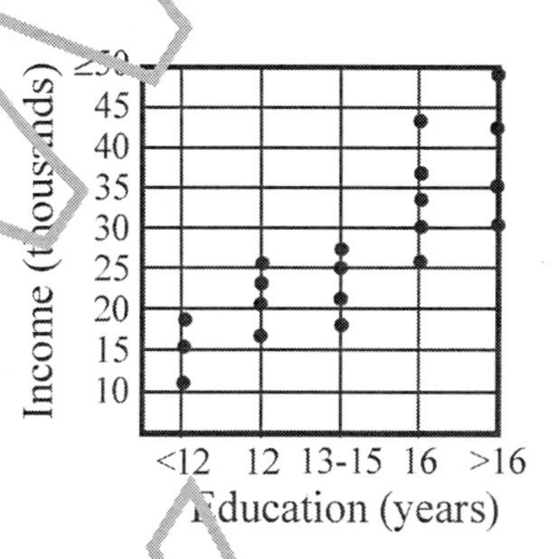

Imagine drawing a line on the scatter plot where half of the points are above the line and half the points are below it. In the plot on the right, you will notice that this line slants upward and to the right. This line direction means there is a **positive** relationship between education and income. In general, for every increase in education, there is a corresponding increase in income.

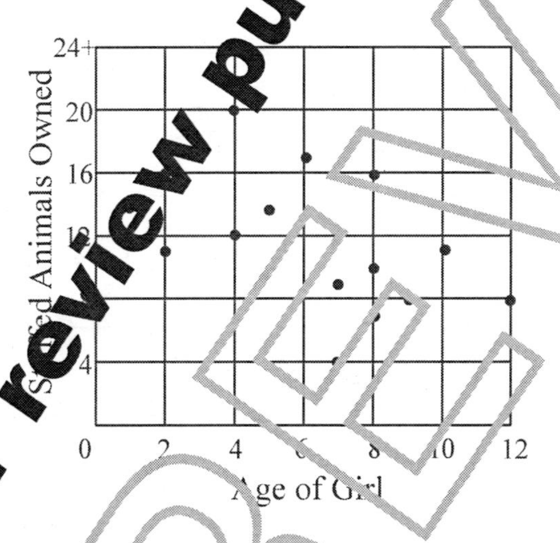

Now, examine the scatter plot on the left. In this case, 15 girls ages 2-12 were interviewed and asked, "How many stuffed animals do you currently have?" If you draw an imaginary line through the middle points, you will notice that the line slants downward and to the right. This plot demonstrates a **negative** relationship between the ages of girls and their stuffed animal ownership. In general, as the girls' ages increase, the number of stuffed animals owned decreases.

Finally, look at the scatter plot shown on the right. In this plot, Rita wanted to see the relationship between the temperature in the classroom and the grades she received on tests she took at that temperature. As you look to your right, you will notice that the points are distributed all over the graph. Because this plot is not in a pattern, there is no way to draw a line through the middle of the points. This type of point pattern indicates there is **no relationship** between Rita's grades on tests and the classroom temperature.

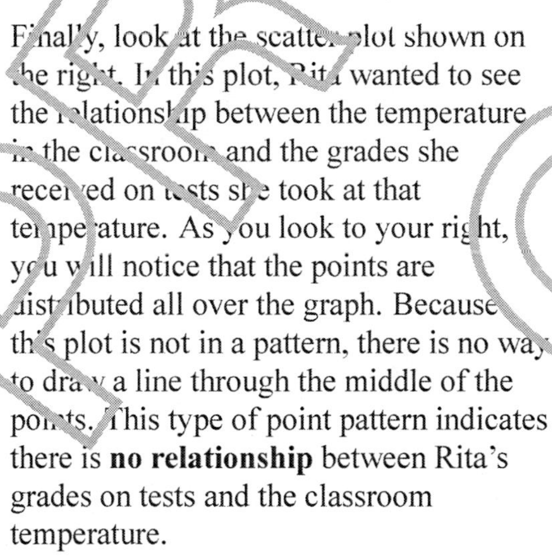

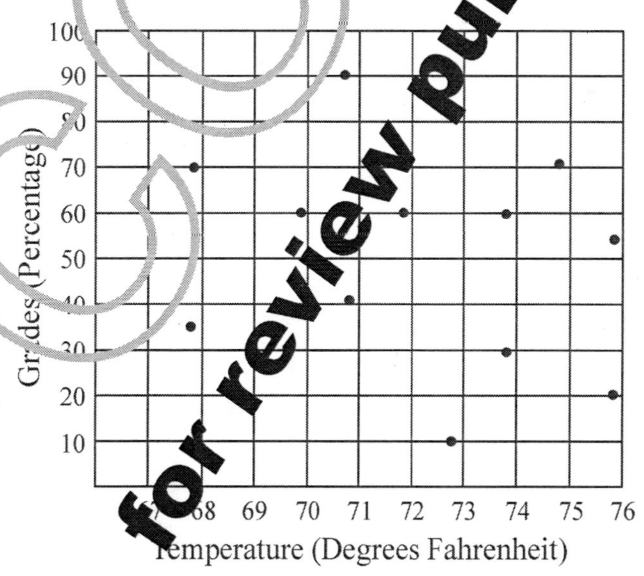

Examine each of the scatter plots below. Write whether the relationship shown between the two variables is "positive", "negative", or "no relationship".

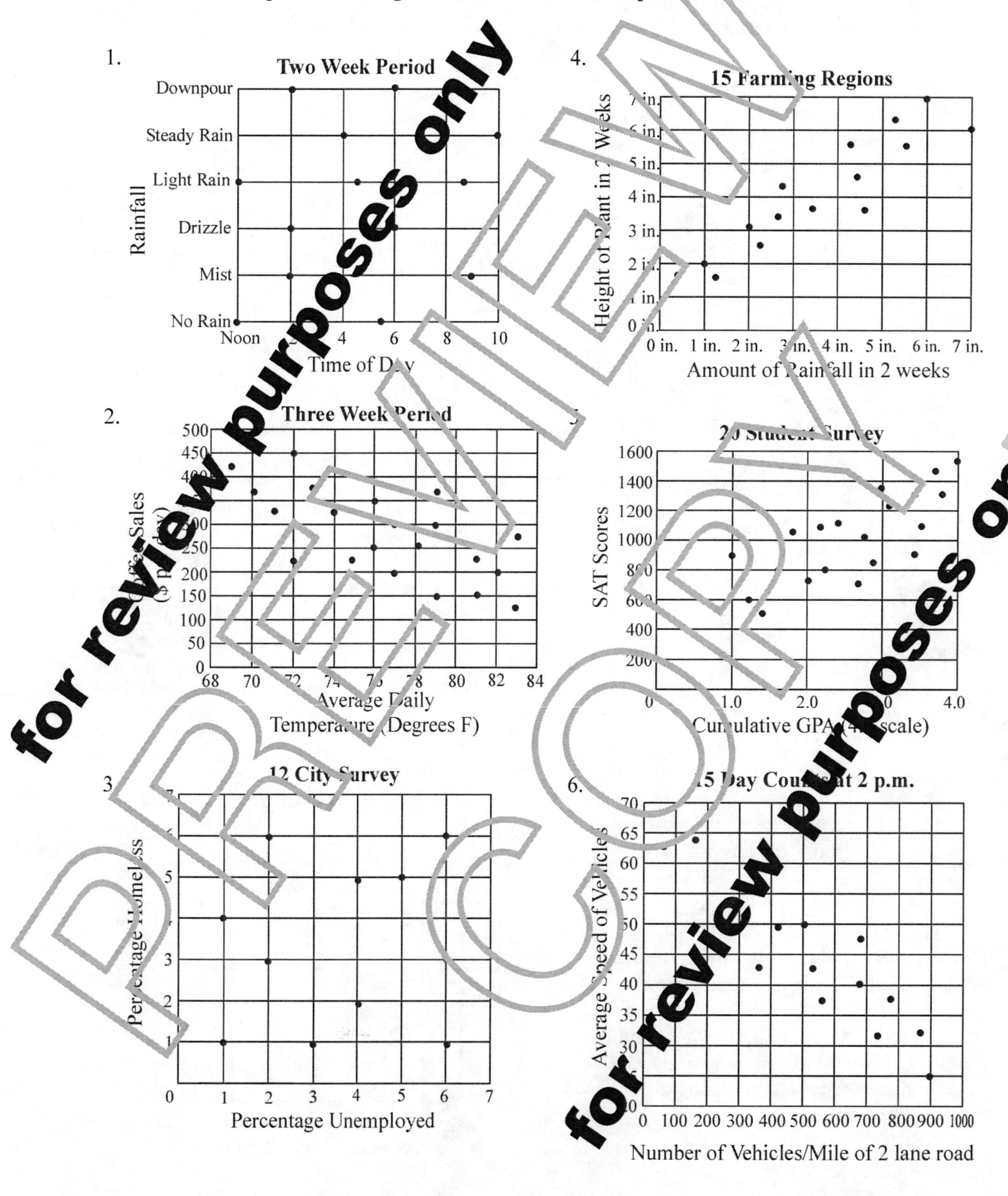

17.12 The Line of Best Fit

At this point, you now understand how to plot points on a Cartesian plane. You also understand how to find the data trend on a Cartesian plane. These skills are necessary to accomplish the next task, determining the line of best fit.

In order to find the line of best fit, you must first draw a scatter plot of all data points. Once this is accomplished, draw an oval around all of the points plotted. Draw a line through the points in such a way that the line separates half the points from one another. You may now use this line to answer questions.

Example 11: The following data set contains the heights of children between 5 and 13 years old. Make a scatter plot and draw the line of best fit to represent the trend. Using the graph, determine the height for a 14-year old child.

Age 5: 4'6", 4'4", [illegible]
Age 6: 4'7", 4'[illegible], 4'6"
Age 7: 4'9", [illegible], 4'6", 4'8"
Age 8: 4'8", 4'[illegible]", 4'7"
Age 9: 4'[illegible]", 4'7", 4'10"
Age 10: 4'9", 4'8", 4'10"
Age 11: 5'0", 4'10"
Age 12: 5'1", 4'11", 5'0", 5'3"
Age 13: 5'3", 5'2", 5'0", 5'1"

In this example, the data points lay in a positive sloping direction. To determine the line of best fit, all data points were circled, then a line of best fit was drawn. Half of the points lay below, half above the line of best fit drawn bisecting the narrow length of the oval.

To find the height of a 14-year old, simply continue the line of best fit forward. In this case, the height is 62 inches.

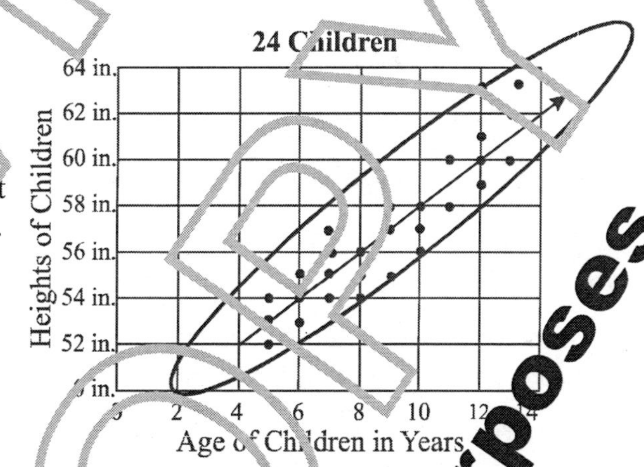

Plot the data sets below, then draw the line of best fit. Next, use the line to estimate the value of the next measurement.

1. Selected values of the Sleekster Brand Light Compact Vehicles: New Vehicle: $13,000.
   1 year old: $12,000, $11,000, $12,500
   2 year old: $9,[illegible]00, $10,500, $9,[illegible]00
   3 year old: $8,500, $8,000, $9,000
   4 year old: $7,500, $6,[illegible], $6,000
   5 year old: ?
2. The relationship between string length and kite height for the following kites:
   (L = 500 ft, H = 400 ft) (L = 250 ft, H = 150ft) (L = 100 ft, H = [illegible] ft) (L = 500 ft, H = 350 ft) (L = 250 ft, H = 200 ft) (L = 100 ft, H = 50 ft) (L = 600 ft, H = ?)
3. Relationship between Household Incomes(HI) and Household Property Values (HPV):
   (HI = $30,000, HPV = $100,000) (HI = $45,000, HPV = $120,000) (HI = $60,000, HPV = $135,000) (HI = $50,000, HPV = 115,000) (HI = $35,000, HPV = 105,000) (HI = 65,000, HPV = 155,000) (HI = $90,000, HPV = ?)

Chapter 17 Review

Find the mean, median, mode, and range for each of the following sets of data. Fill in the table below.

❶ Miles Run by Track Team Members

Name	Miles
Jeff	24
Eric	20
Craig	19
Simon	20
Elijah	25
Rich	19
Marcus	2[illegible]

❷ 1992 SUMMER OLYMPIC GAMES
Gold Medals Won

Country	Medals	Country	Medals
Unified Team	45	Hungary	11
United States	37	South Korea	12
Germany	33	France	8
China	16	Australia	7
Cuba	14	Japan	3
Spain	13		

❸ Hardware Store Payroll June Week 2

Name	Pay
Erica	$280
Dane	$206
Sam	$240
Nancy	$404
Elsie	$210
Gail	$305
David	$280

Data Set Number	Mean	Median	Mode	Range
❶				
❷				
❸				

4. [illegible]nica bowls three games and scores an average of 116 points per game. She scores 105 on her first game and 128 on her second game. What does she score on her third game?

5. Concession stand sales for each game in season are $320, $540, $230, $450, $280, and $[illegible]30. What is the mean sales per game?

6. Cendrick D'Amitrano works Friday and Saturday delivering pizza. He delivers 8 pizzas on Friday. How many pizzas must he deliver on Saturday to average 11 pizzas per day?

7. Long cooks three Vietnamese dinners that weigh a total of 40 ounces. What is the average weight for each dinner?

8. The Swamp Foxes scored an average of 7 points per soccer game. They scored 9 points in the first game, 4 points in the second game, and 5 points in the third game. What was their score for their fourth game?

9. Shondra is 66 inches tall, and DeWayne is 72 inches tall. How tall is Michael if the average height of these three students is 77 inches?

Nine cooks are asked, "If you use a thermometer, what is the actual temperature inside your oven when it is set at 350°F?" The responses are in the chart below.

Temperature (°F)	104	347	348	349	350	351	352
Number of Cooks	1	1	1	2		2	1

10. Find the mean of the data above.

11. Find the median of the data above.

Chapter 18
Data Interpretation

18.1 Tally Charts and Frequency Tables

Large lists can be tallied in a chart. To make a **tally chart**, record a tally mark in a chart for each time a number is repeated. To make a **frequency table**, count the times each number occurs in the list, and record the frequency.

Example 1: The age of each student in grades 6–8 in a local middle school are listed below. Make a tally chart and a frequency table for each age.

Student Ages grades 6–8							
10	11	11	12	14	12	11	12
13	13	13	[illegible]	14	11	12	11
12	14	12	[illegible]	15	11	13	14
12	10	12	[illegible]	12	13	12	12
13	12	[illegible]	12	11	10	13	11
14	14	[illegible]	15	12	13	14	13

TALLY CHART	
Age	Tally
10	IIII
11	~~IIII~~ ~~IIII~~
12	~~IIII~~ ~~IIII~~ ~~IIII~~
13	~~IIII~~ ~~IIII~~
14	~~IIII~~ II
15	II

FREQUENCY TABLE	
Age	Frequency
10	4
11	10
12	15
13	10
14	7
15	2

Make a chart to record tallies and frequencies for the following problems.

1. The sheriff's office monitors the speed of cars traveling on Turner Road for one week. The following data is the speed of each car that travels Turner Road during the week. Tally the data in 10 mph increments starting with 10–19 mph, and record the frequency in a chart.

Car Speed, mph									
45	52	47	35	48	50	51	43	52	41
40	51	32	24	55	41	32	33	45	
36	39	[illegible]	[illegible]	34	[illegible]	39	47	56	
[illegible]	15	63	42	[illegible]	42	58	59	35	
39	41	[illegible]	[illegible]	22	[illegible]	40	31	55	
[illegible]	[illegible]	46	38	50	52	48	36	65	
21	[illegible]	[illegible]	41	52	49	45	32	20	

Speed	Tally	Frequency
10-19		
20-29		
30-39		
40-49		
50-59		
60-69		

2. The following data gives final math averages for Ms. Kirby's class. In her class, an average of 90–100 is an A, 80–89 is B, 70–79 is a C, 60–69 is a D, and an average below 60 is an F. Tally and record the frequency of A's, B's, C's D's, and F's.

Final Math Averages									
[illegible]	92	87	62	75	84	96	52	31	79
[illegible]	[illegible]	98	75	71	79	85	82	86	76
87	[illegible]	76	68	93	77	65	84	89	
79	65	77	82	86	84	92	60	65	
99	75	88	74	79	80	63	84	69	
87	90	75	81	73	69	73	75	75	

Grade	Tally	Frequency
A		
B		
C		
D		
F		

18.2 Histograms

A **histogram** is a bar graph of the data in a frequency table.

Example 2: Draw a histogram for the customer sales data presented in the frequency table.

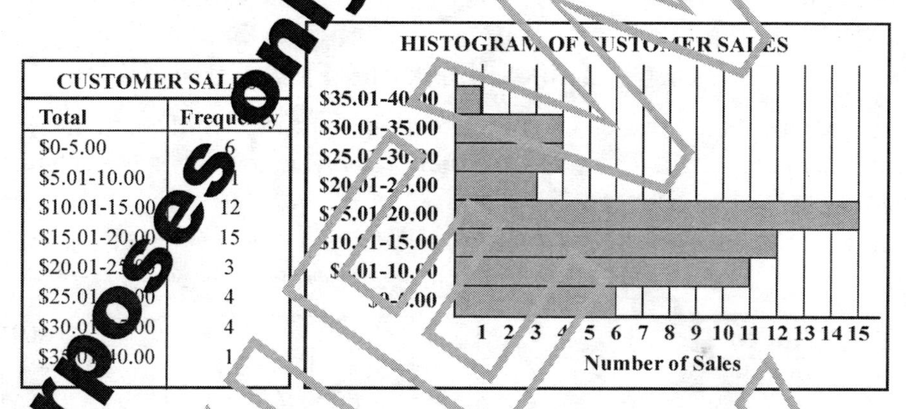

CUSTOMER SALES	
Total	Frequency
$0-5.00	6
$5.01-10.00	11
$10.01-15.00	12
$15.01-20.00	15
$20.01-25.00	3
$25.01-30.00	4
$30.01-35.00	4
$35.01-40.00	1

Use the frequency charts that you filled in the previous section to draw histograms for the same data.

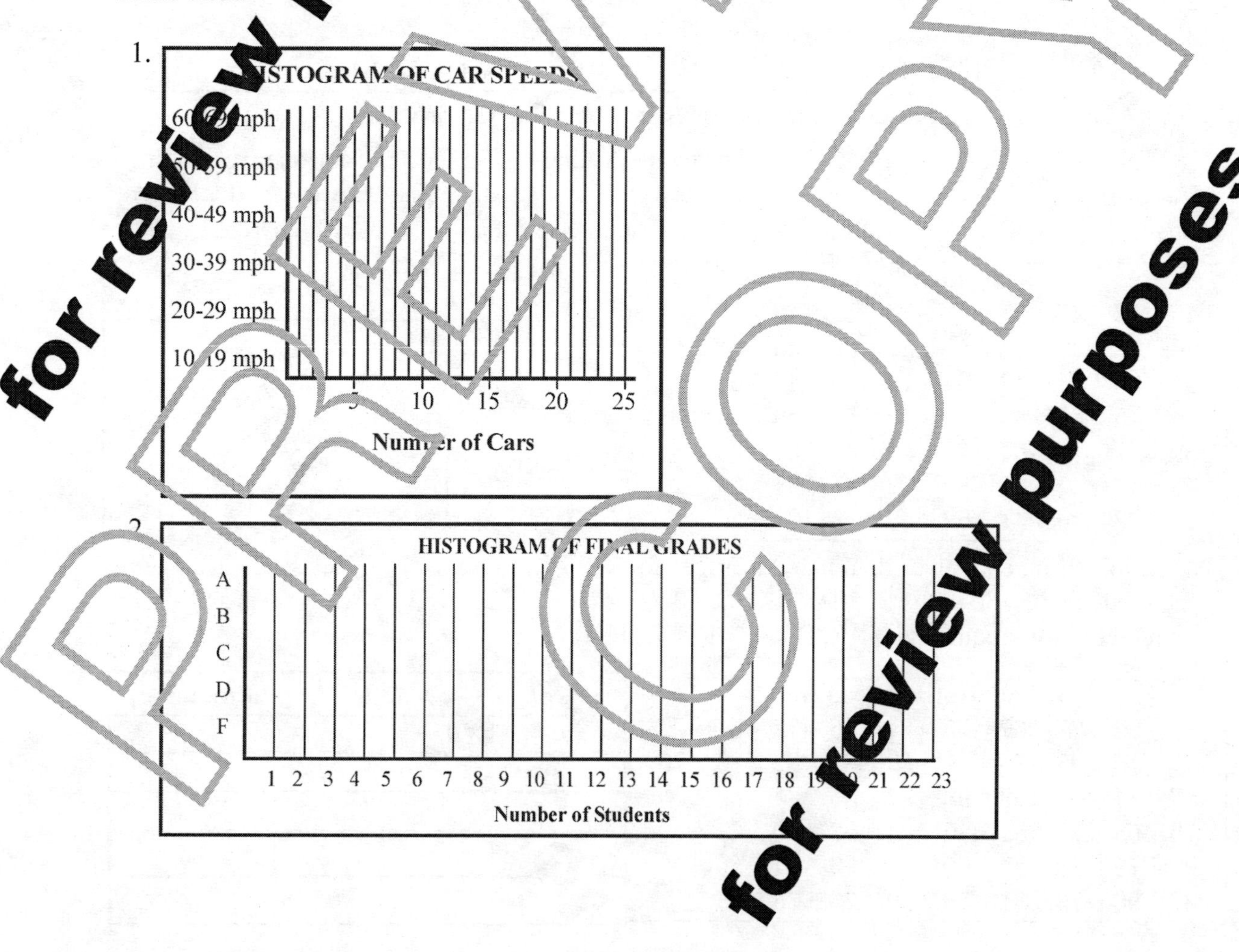

18.3 Reading Tables

A **table** is a concise way to organize large quantities of information using rows and columns. **Read each table carefully, and then answer the questions that follow.**

Some employers use a tax table like the one below to figure how much Federal Income Tax should be withheld from a single person paid weekly. The number of withholding allowances claimed is also commonly referred to as the number of deductions claimed.

Federal Income Tax Withholding Table **SINGLE** Persons - **WEEKLY** Payroll Period					
If the wages are –		And the number of withholding allowances claimed is –			
		0	1	2	3
At least	But less than	The amount of income tax to be withheld is –			
$250	**260**	31	23	16	9
$260	**270**	32	25	17	10
$270	**280**	34	26	19	12
$280	**290**	35	28	20	13
$290	**300**	37	29	22	15

1. David is single, claims 2 withholding allowances, and earned $275 last week. How much Federal Income Tax was withheld?

2. Cecily is single, claims 0 deductions, and earned $297 last week. How much Federal Income Tax was withheld?

3. Sherri is single, claims 3 deductions, and earned $268 last week. How much Federal Income Tax was withheld from her check?

4. Mitch is single and claims 1 allowance. Last week, he earned $291. How much was withheld from his check for Federal Income Tax?

5. Ginger is single, earns $275 this week, and claims 0 deductions. How much Federal Income Tax is withheld from her check?

6. Bill is single and earns $263 per week. He claims 1 withholding allowance. How much Federal Income Tax is withheld each week?

18.4 Bar Graphs

Bar graphs can be either vertical or horizontal. There may be just one bar or more than one bar for each interval. Sometimes each bar is divided into two or more parts. In this section, you will work with a variety of bar graphs. Be sure to read all titles, keys, and labels to completely understand all the data that is presented. **Answer the questions about each graph.**

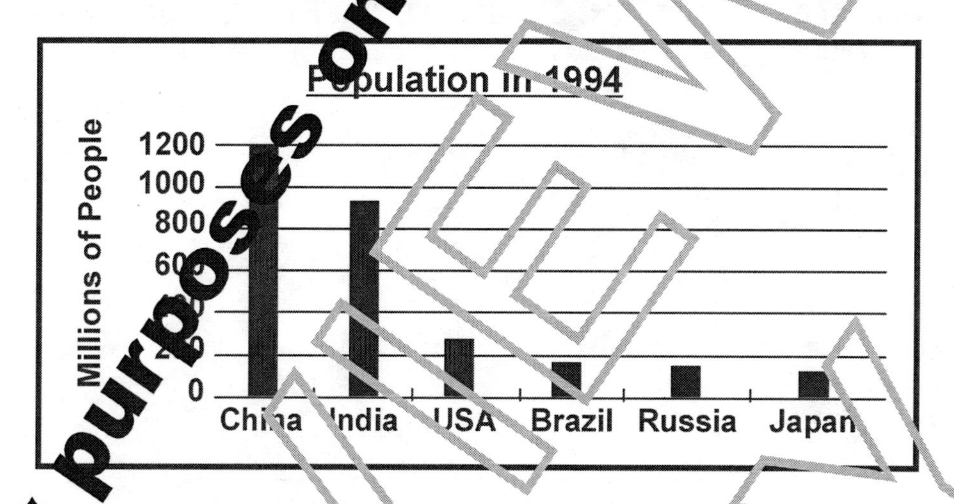

1. Which country has over 1 billion people?
2. How many countries have fewer than 200, 000, 000 people?
3. How many more people does India have than Japan?
4. If you added together the populations of the USA, Brazil, Russia, and Japan, would would the sum be closer to the population of India or China?

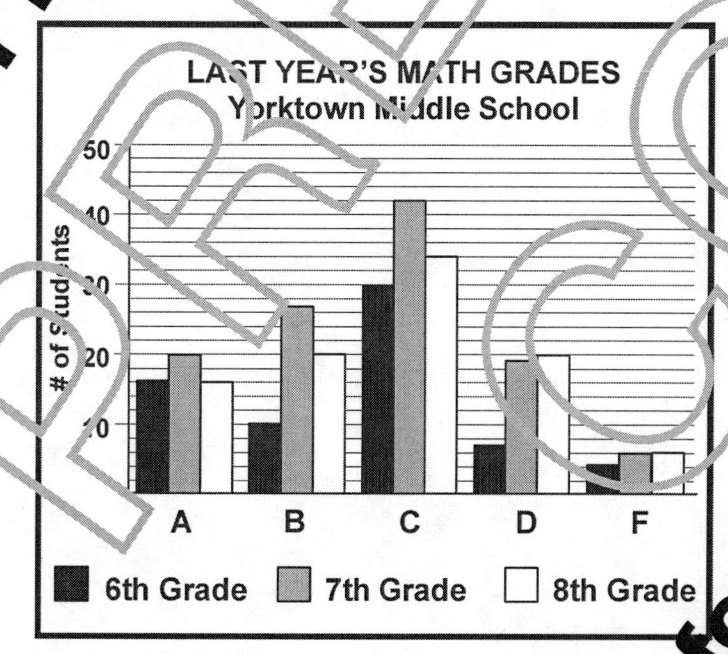

5. How many of last year's [illegible]th graders made C's in math?
6. How many more math students made B's in the [illegible]th grade than in the 8th grade?
7. Which letter grade occurs the most number of times in the data?
8. How many 8th graders took math last year?
9. How many students made A's in math last year?

18.5 Line Graphs

The first line graph below is shown with a globe marking the lines of latitude to make the line graphs more understandable. Study the line graph below, and then answer questions 1–5.

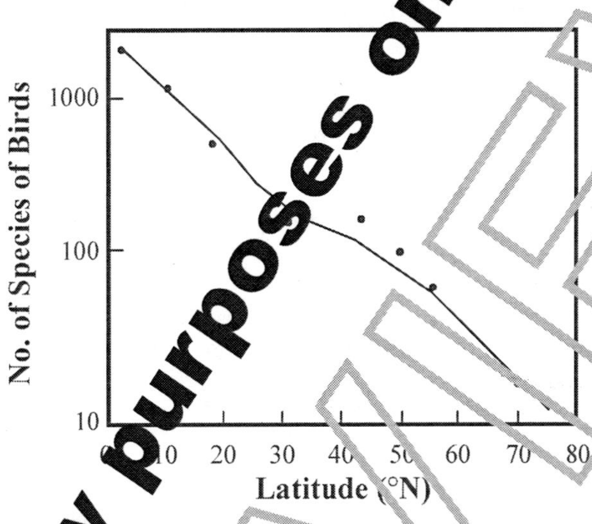

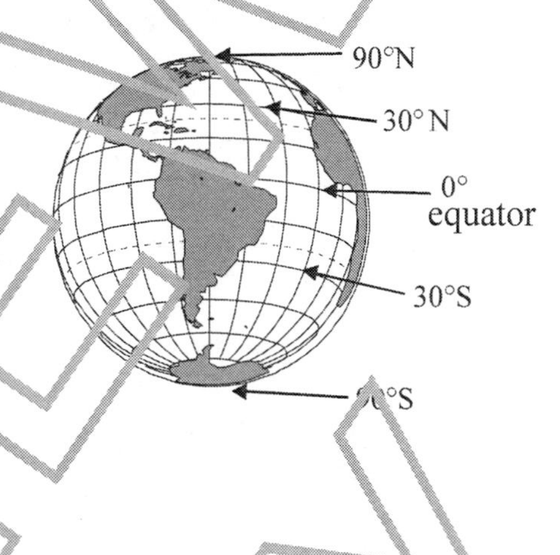

After reading the graph above, label each of the following statements as true or false.

1. There are more species of birds at the North Pole than at the equator.
2. There are more species of birds in Mexico than in Canada.
3. As the latitude increases, the number of species of birds decreases.
4. At 30°N there are over 100 species of birds.
5. The warmer the climate, the fewer kinds of birds there are.

These true or false statements, 6–10, refer to the graph on the left.

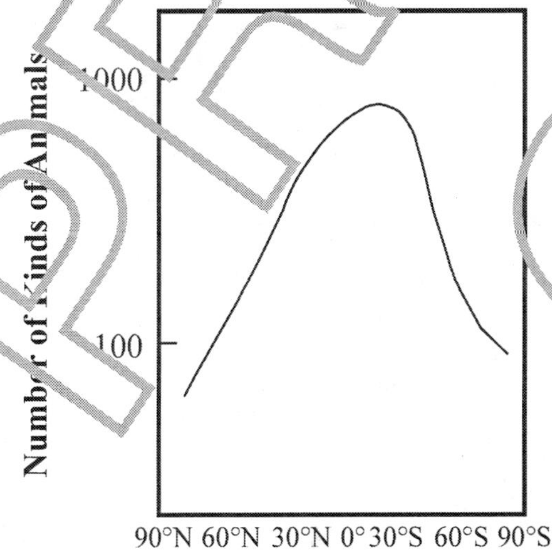

6. The farther north and south you go from the equator, the greater the variety of animals there are.
7. The closer you get to the equator, the greater the variety of animals there are.
8. There are fewer kinds of animals at 30°S than at 60°S latitude.
9. The number of kinds of animals increases as the latitude increases.
10. The number of kinds of animals increases at the poles.

18.6 Circle Graphs

Circle graphs represent data expressed in percentages of a total. The parts in a circle graph should always add up to 100%. Circle graphs are sometimes called **pie graphs** or **pie charts**.

To figure the value of a percent in a circle graph, multiply the percent by the total. Use the circle graphs below to answer questions. The first question is worked for you as an example.

Tina's Monthly Spending Habits

Monthly Spending Allowance = $80

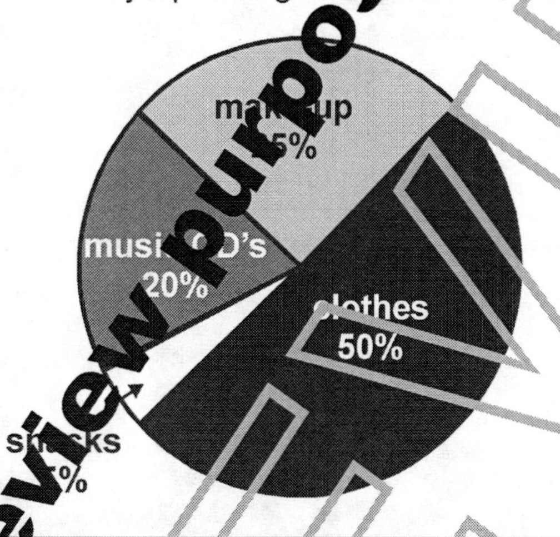

1. How much does Tina spend each month on music CDs?

   $80 × 0.20 = $16.00 **$16.00**

2. How much does Tina spend each month on make-up? ________

3. How much does Tina spend each month on clothes? ________

4. How much does Tina spend each month on snacks? ________

Fill in the following chart.

Favorite Activity	Number of Students
5. watching TV	
6. talking on the phone	
7. playing video games	
8. surfing the internet	
9. playing sports	
10. reading	

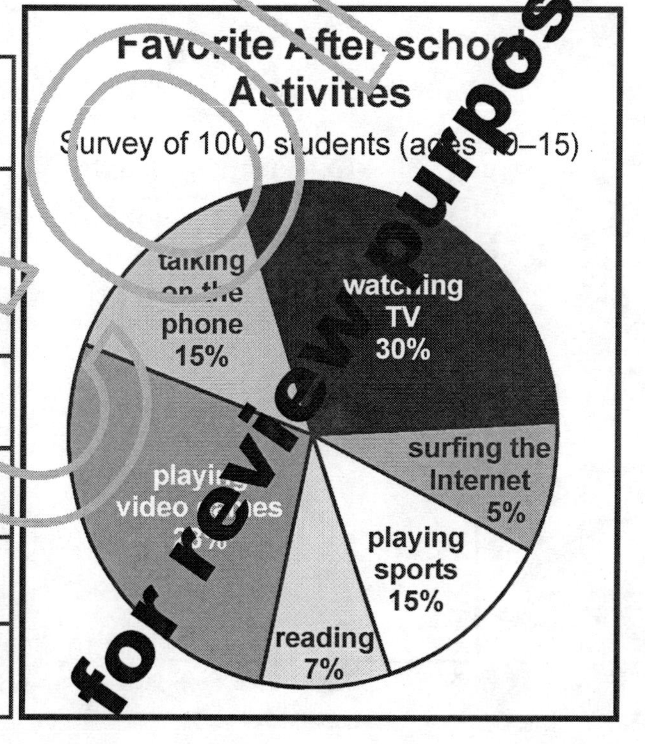

Chapter 18 Review

Use the data given to answer the questions that follow.

The 6th grade did a survey on the number of pets each student had at home. The following give the data produced by the survey.

NUMBER OF PETS PER STUDENT																								
0	2	6	2	1	0	4	2	3	3	0	2	3	5	1	4	2	0	5	2	3	3	4	3	6	2
5	1	2	3	5	6	3	2	2	5	2	3	4	3	0	1	4	1	2	4	5	7	6	1	4	7

1. Fill in the frequency table.

Number of Pets	Frequency

2. Fill in the histogram.

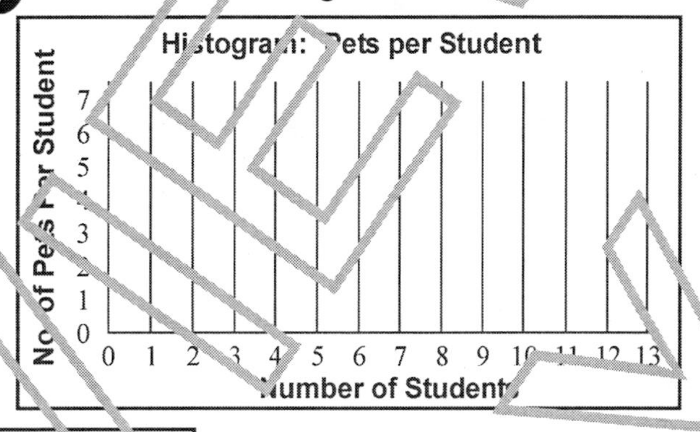

KNIGHTS BASKETBALL Points Scored				
Player	game 1	game 2	game 3	game 4
[illegible]	5	2	4	8
Jason	10	[illegible]	10	12
Brandon	2	6	5	6
Ned	1	[illegible]	6	2
Austin	0	4	7	8
David	7	2	[illegible]	4
Zac	8	6	7	4

3. How many points did the Knights basketball team score in game 1?

4. How many more points does David score in game 3 than in game 1?

5. How many points does Jason score in the first 4 games?

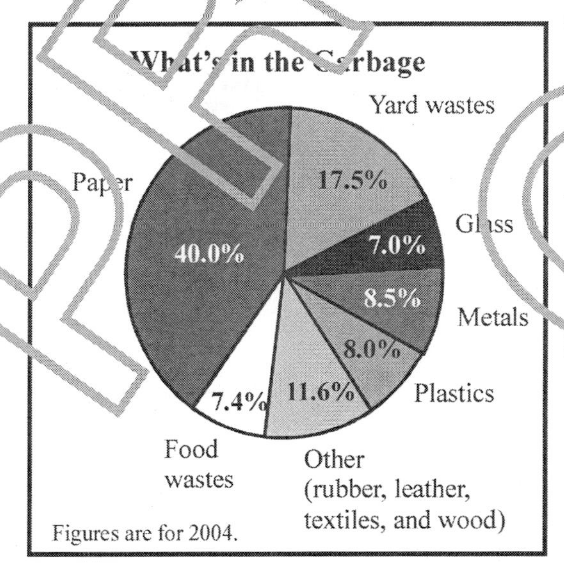

6. In 2004, the United States produced 160 million metric tons of garbage. According to the pie chart, how much glass was in the garbage?

7. Out of the 160 million metric tons of garbage, how much was glass, plastic, and metal?

8. If in 2006, the garbage reaches 200 million metric tons, and the percentage of wastes remains the same as in 2004, how much food in metric tons will be in the 2006 garbage?

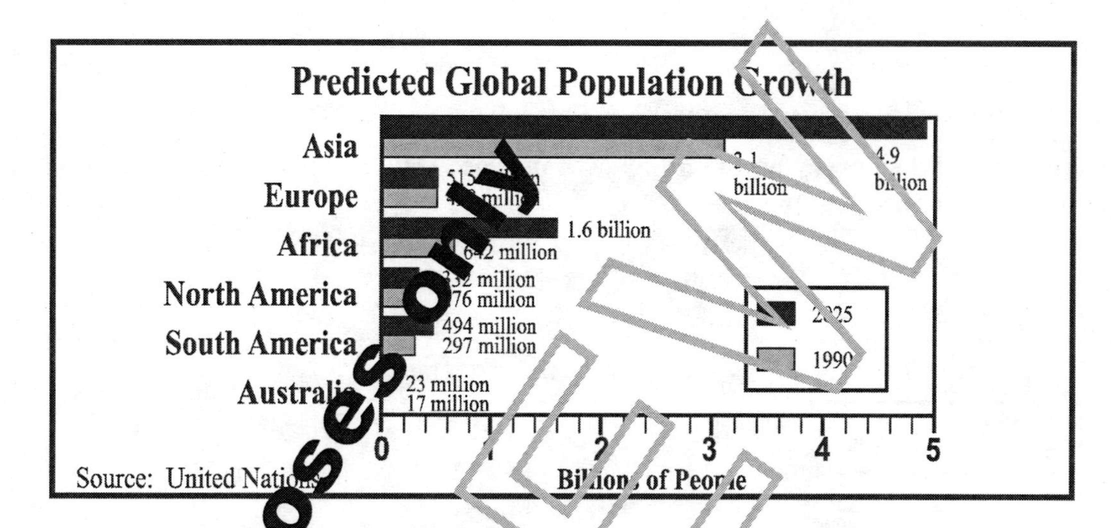

9. By how many is Asia's population predicted to increase between 1990 and 2025?

10. In 1990, how much larger was Africa's population than Europe's?

11. Where is the population expected to more than double between 1990 and 2025?

12. In the space below, draw a line graph showing a population increasing over time.

13. In the space below, draw a line graph of a population headed for extinction.

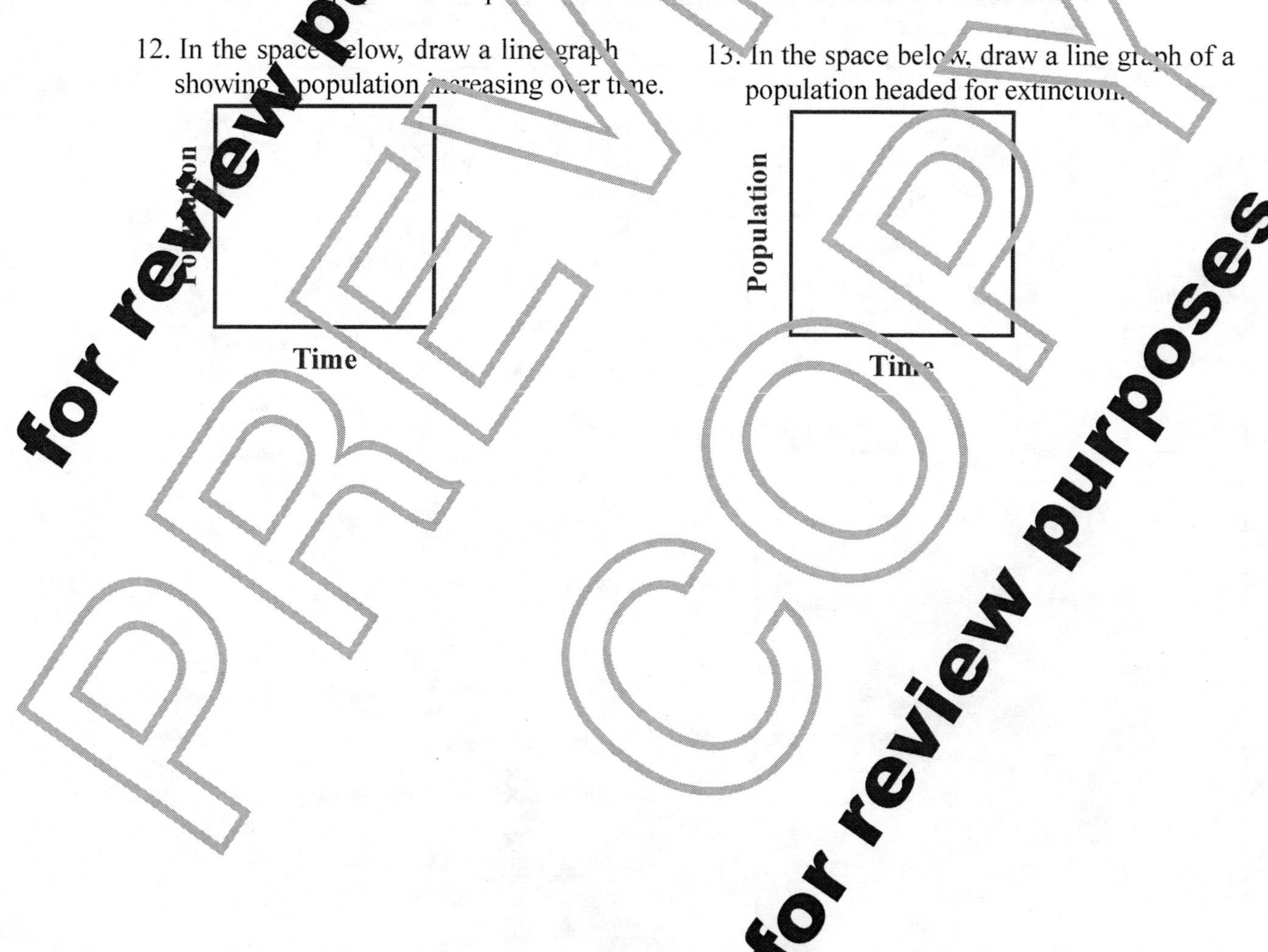

Chapter 19
Probability and Counting

19.1 Probability

Probability is the chance something will happen. Probability is most often expressed as a fraction, a decimal, a percent, or can also be written out in words.

Example 1: Billy has 3 red marbles, 5 white marbles, and 4 blue marbles on the floor. His cat comes along and bats one marble under the chair. What is the **probability** it is a red marble?

Step 1: The number of red marbles will be the top number of the fraction. → $\mathbf{\frac{3}{12}}$

Step 2: The total number of marbles is the bottom number of the fraction.

The answer may be expressed in lowest terms. $\frac{3}{12} = \frac{1}{4}$.

Expressed as a decimal, $\frac{1}{4} = 0.25$, as a percent, $\frac{1}{4} = 25\%$, and written out in words $\frac{1}{4}$ is one out of four.

Example 2: Determine the probability that the pointer will stop on a shaded wedge or the number 1.

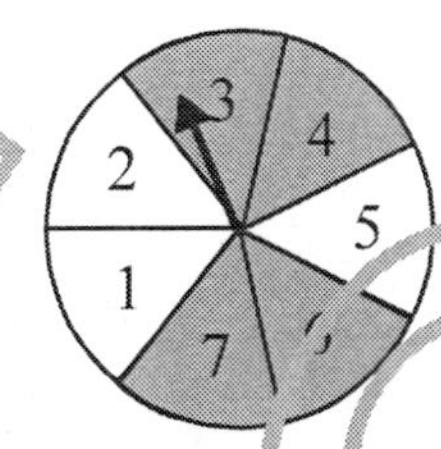

Step 1: Count the number of possible wedges that the spinner can stop on to satisfy the above problem. There are 5 wedges that satisfy it (4 shaded wedges and one number 1). The top number of the fraction is 5.

Step 2: Count the total number of wedges, 7. The bottom number of the fraction is 7. The answer is $\frac{5}{7}$ or **five out of seven.**

Example 3: Refer to the spinner above. If the pointer stops on the number 7, what is the probability that it will **not** stop on 7 the next time?

Step 1: Ignore the information that the pointer stopped on 7 the previous spin. The probability of the next spin does not depend on the outcome of the previous spin. Simply find the probability that the spinner will **not** stop on 7. Remember, if P is the probability of an event occurring, $1 - P$ is the probability of an event **not** occurring. In this example, the probability of the spinner landing on 7 is $\frac{1}{7}$.

Step 2: The probability that the spinner will not stop on 7 is $1 - 7$ which equals $\frac{6}{7}$. The answer is $\frac{6}{7}$ or **six out of seven.**

Find the probability of the following problems. Express the answer as a percent.

1. A computer chooses a random number between 1 and 50. What is the probability that you will guess the same number that the computer chose in 1 try?

2. There are 24 candy-coated chocolate pieces in a bag. Eight have defects in the coating that can be seen only with close inspection. What is the probability of pulling out a defective piece without looking?

3. Seven sisters have to choose which day each will wash the dishes. They put equal-sized pieces of paper in a hat, each labeled with a day of the week. What is the probability that the first sister who draws will choose a weekend day?

4. For his garden, Clay has a mixture of 12 white corn seeds, 24 yellow corn seeds, and 16 bicolor corn seeds. If he reaches for a seed without looking, what is the probability that Clay will plant a bicolor corn seed first?

5. Mom just got a new department store credit card in the mail. What is the probability that the last digit is an odd number?

6. Alex has a paper bag of cookies that holds 8 chocolate chip, 4 peanut butter, 6 butterscotch chip, and 12 ginger. Without looking, his friend John reaches in the bag for a cookie. What is the probability that the cookie is peanut butter?

7. An umpire at a little league baseball game has 14 balls in his pockets. Five of the balls are brand A, 6 are brand B, and 3 are brand C. What is the probability that the next ball he throws to the pitcher is a brand C ball?

8. What is the probability that the spinner's arrow will land on an even number?

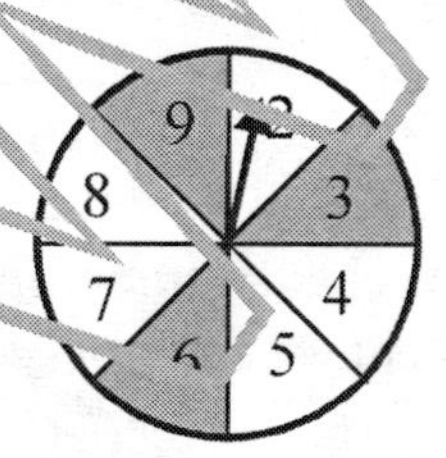

9. The spinner in the problem above stopped on a shaded wedge on the first spin and stopped on the number 2 on the second spin. What is the probability that it will not stop on a shaded wedge or on the 2 on the third spin?

10. A company is offering 1 grand prize, 3 second place prizes, and 25 third place prizes based on a random drawing of contest entries. If your entry is one of the 500 total entries, what is the probability you will win a third place prize?

11. In the contest problem above, what is the probability that you will win the grand prize or a second place prize?

12. A box of a dozen doughnuts has 3 lemon cream-filled, 5 chocolate cream-filled, and 4 vanilla cream-filled. If the doughnuts look identical, what is the probability of picking a lemon cream-filled?

19.2 Independent and Dependent Events

In mathematics, the outcome of an event may or may not influence the outcome of a second event. If the outcome of one event does not influence the outcome of the second event, these events are **independent.** However, if one event has an influence on the second event, the events are **dependent**. When someone needs to determine the probability of two events occurring, he or she will need to use an equation. These equations will change depending on whether the events are independent or dependent in relation to each other. When finding the probability of two **independent** events, multiply the probability of each favorable outcome together.

Example 4: One bag of marbles contains 1 white, 1 yellow, 2 blue, and 3 orange marbles. A second bag of marbles contains 2 white, 3 yellow, 1 blue, and 2 orange marbles. What is the probability of drawing a blue marble from each bag?

Solution: Probability of favorable outcomes

Bag 1: $\frac{2}{7}$

Bag 2: $\frac{1}{8}$

Probability of a blue marble from each bag: $\frac{2}{7} \times \frac{1}{8} = \frac{2}{56} = \frac{1}{28}$

In order to find the probability of two **dependent** events, you will need to use a different set of rules. For the first event, you must divide the number of favorable outcomes by the number of possible outcomes. For the second event, you must subtract one from the number of favorable outcomes **only** if the favorable outcome is the **same**. However, you must subtract one from the number of total possible outcomes. Finally, you must multiply the probability for event one by the probability for event two.

Example 5: One bag of marbles contains 3 red, 4 green, 7 black, and 2 yellow marbles. What is the probability of drawing a green marble, removing it from the bag, and then drawing another green marble?

	Favorable Outcomes	Total Possible Outcomes
Draw 1	4	16
Draw 2	3	15
Draw 1 × Draw 2	12	240

Answer: $\frac{12}{240}$ or $\frac{1}{20}$

Example 6: Using the same bag of marbles, what is the probability of drawing a red marble and then drawing a black marble?

	Favorable Outcomes	Total Possible Outcomes
Draw 1	3	16
Draw 2	7	15
Draw 1 × Draw 2	21	240

Answer $\frac{21}{240}$ or $\frac{7}{80}$

Find the probability of the following problems. Express the answer as a fraction.

1. Prithi has two boxes. Box 1 contains 3 red, 2 silver, 4 gold, and 2 blue combs. She also has a second box containing 1 black and 1 clear brush. What is the probability that Prithi selects a red brush from box 1 and a black brush from box 2?

2. Steve Marduke has two spinners in front of him. The first one is numbered 1 – 6, and the second is numbered 1 – 3. If Steve spins each spinner once, what is the probability that the first spinner will show an odd number and the second spinner will show a "1"?

3. Carrie McCallister flips a coin twice and gets heads both times. What is the probability that Carrie will get tails the third time she flips the coin?

4. Artie Drake turns a spinner which is evenly divided into 11 sections numbered 1 – 11. On the first spin, Artie's pointer lands on "8". What is the probability that the spinner lands on an even number the second time he turns the spinner?

5. Leanne Davis plays a game with a street entertainer. In this game, a ball is placed under one of three coconut halves. The vendor shifts the coconut halves so quickly that Leanne can no longer tell which coconut half contains the ball. She selects one and misses. The entertainer then shifts all three around once more and asks Leanne to pick again. What is the probability that Leanne will select the coconut half containing the ball?

6. What is the probability that Jane Robelot reaches into a bag containing 1 daffodil and 2 gladiola bulbs and pulls out a daffodil bulb, and then reaches into a second bag containing 6 tulip, 3 lily, and 2 gladiola bulbs and pulls out a lily bulb?

7. Terrell casts his line into a pond containing 7 catfish, 8 bream, 3 trout, and 6 northern pike. He immediately catches a bream. What are the chances that Terrell will catch a second bream the next time he casts his line?

8. Gloria Quintero enters a contest in which the person who draws his or her initials out of a box containing all 26 letters of the alphabet wins the grand prize. Gloria reaches in, draws a "G", keeps it, then draws another letter. What is the probability that Gloria will next draw a "Q"?

9. Vince Macaluso is pulling two socks out of a washing machine in the dark. The washing machine contains three tan, one white, and two black socks. If Vince reaches in and pulls out the socks one at a time, what is the probability that he will pull out two tan socks on his first two tries?

10. John Salome has a bag containing 2 yellow plums, 2 red plums, and 3 purple plums. What is the probability that he reaches in without looking and pulls out a yellow plum and eats it, then reaches in again without looking and pulls out a red plum to eat?

19.3 Tree Diagrams

Drawing a tree diagram is another method of determining the probability of events occurring.

Example 7: If you toss two six-sided numbered cubes that have 1, 2, 3, 4, 5, or 6 on each side, what is the probability you will get two cubes that add up to 9? One way to determine the probability is to make a tree diagram.

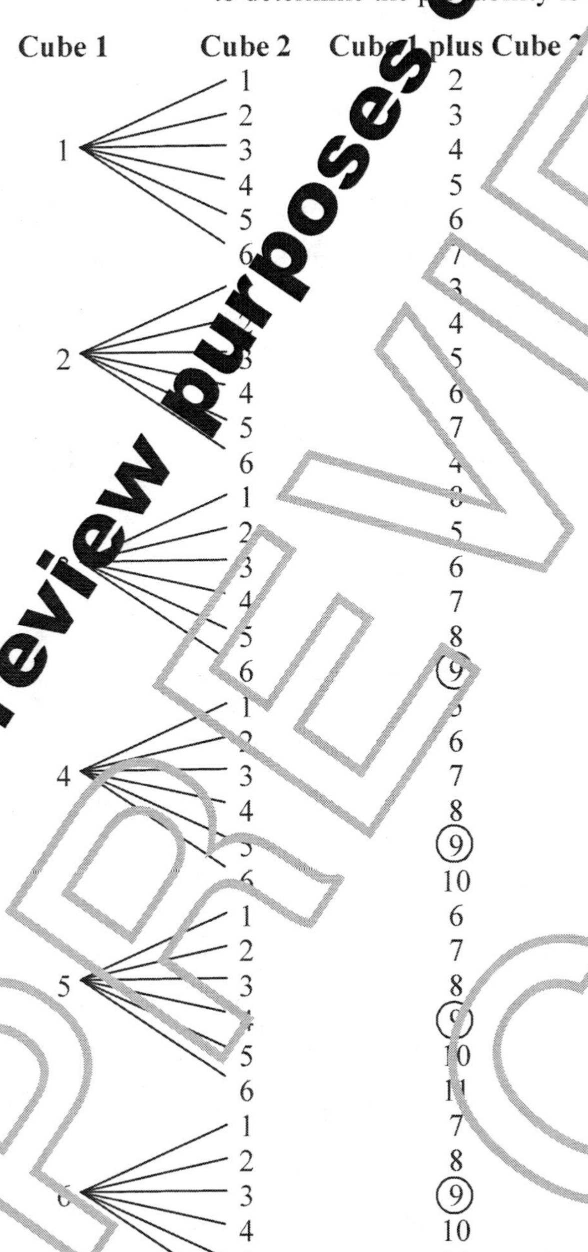

Alternative method

Write down all of the numbers on both cubes which would add up to 9.

Cube 1	Cube 2
4	5
5	4
6	3
3	6

Numerator = 4 combinations

For denominator: Multiply the number of sides on one cube times the number of sides on the other cube.

$6 \times 6 = 36$

Numerator: / Denominator: $\frac{4}{36} = \frac{1}{9}$

There are 36 possible ways the cubes could land. Out of those 36 ways, the two cubes add up to 9 only 4 times. The probability you will get two cubes that add up to 9 is $\frac{4}{36}$ or $\frac{1}{9}$.

Read each of the problems below. Then answer the questions.

1. Jake has a spinner. The spinner is divided into eight equal regions numbered 1–8. In two spins, what is the probability that the numbers added together will equal 12?

2. Charlie and Libby each spin one spinner one time. The spinner is divided into 5 equal regions numbered 1–5. What is the probability that these two spins added together would equal 7?

3. Gail spins a spinner twice. The spinner is divided into 9 equal regions numbered 1–9. In two spins, what is the probability that the difference between the two numbers will equal 4?

4. Diedra throws two 10-sided numbered polyhedrons. What is the probability that the difference between the two numbers will equal 7?

5. Cameron throws two six-sided numbered cubes. What is the probability that the difference between the two numbers will equal 3?

6. Tesla spins one spinner twice. The spinner is divided into 11 equal regions numbered 1–11. What is the probability that the two numbers added together will equal 11?

7. Samantha decides to roll two five-sided numbered cubes. What is the probability that the two numbers added together will equal 4?

8. Mary Ellen spins a spinner twice. The spinner is divided into 7 equal regions numbered 1–7. What is the probability that the product of the two numbers equals 10?

9. Conner decides to roll two six-sided numbered cubes. What is the probability that the product of the two numbers equals 4?

10. Tabitha spins one spinner twice. The spinner is divided into 9 equal regions numbered 1–9. What is the probability that the sum of the two numbers equals 10?

11. Darnell decides to roll two 15-sided numbered polyhedrons. What is the probability that the difference between the two numbers is 13?

12. Inez spins one spinner twice. The spinner is divided into 12 equal regions numbered 1–12. What is the probability that the sum of two numbers equals 10?

13. Gina spins one spinner twice. The spinner is divided into 8 equal regions numbered 1–8. What is the probability that the two numbers added together equals 9?

14. Celia rolls two six-sided numbered cubes. What is the probability that the difference between the two numbers is 2?

15. Brett spins one spinner twice. The spinner is divided into 4 equal regions numbered 1–4. What is the probability that the difference between the two numbers will be 3?

19.4 Probability Distributions

Many times in business or science situations, when data is collected, decisions are made by assigning probabilities to all possible outcomes of events taking place and evaluating the results. There are two types of outcomes that can be described: **discrete** and **continuous** probabilities. **Discrete probabilities** are those that must be counted, such as the toss of a coin or the roll of a die. **Continuous probabilities** are those that must be measured, such as time spent waiting in line at the bank or the height of a team of basketball players.

A **discrete probability distribution** is simply a table or graph which represents all possible outcomes of an experiment with the associated probability of each outcome. Consider rolling one six-sided die; there are six possible outcomes, each equally likely.

Outcome	1	2	3	4	5	6
$P(A)$	$\frac{1}{6}$	$\frac{1}{6}$	$\frac{1}{6}$	$\frac{1}{6}$	$\frac{1}{6}$	$\frac{1}{6}$

Using this data, you can see the probability of rolling a four is 5. What is the probability of rolling 3 or less? It is the probability of rolling a three plus the probability of rolling a two plus the probability of rolling a one, which equals $\frac{1}{6} + \frac{1}{6} + \frac{1}{6} = \frac{1}{2}$.

A **continuous probability distribution** is a function of measured values rather than a table of all outcomes. The most frequently used of these is the **normal curve distribution**, which is commonly known as the bell curve.

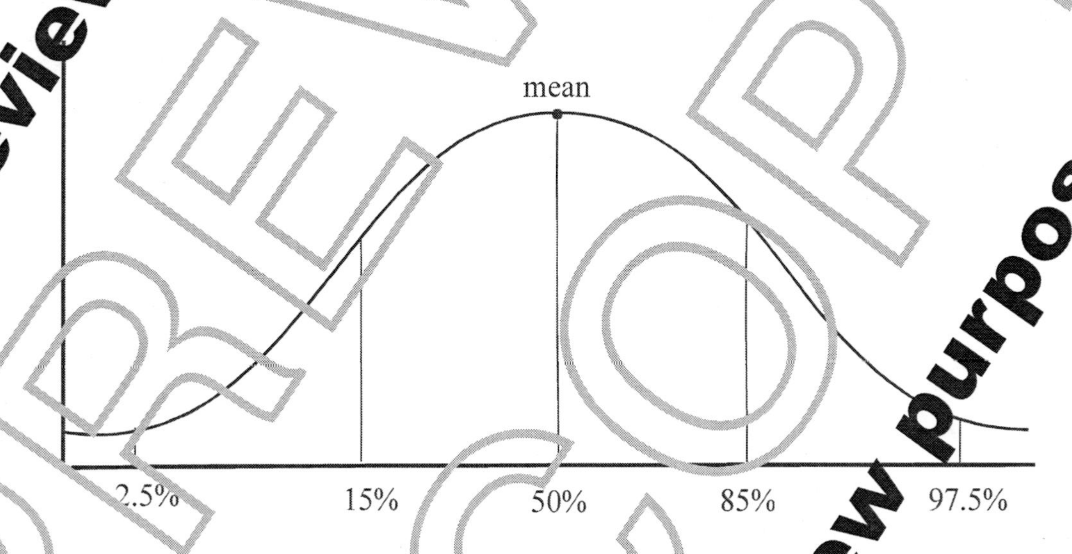

The **mean** of the data is the center of the curve, where the probability of the event is 50%. If this curve above measured the wait in minutes for a table at a restaurant instead of percent, then you can see that a few people waited 2.5 minutes or less, a few people waited 97.5 minutes or more, but most of the people waited about 50 minutes. These percentages are obtained using something known as the **empirical rule**, which determines how data is distributed in normal curves.

NOTE: You **CANNOT** find a probability of a specific value like the percent of people who waited exactly 15 minutes to get a table.

The regions of the normal curve can be used to solve continuous probability questions such as the one posed on the previous page. One of interest to students is the distribution of grades in a particular class. Statisticians and educators alike are interested in the IQ distribution of a given population sample. These can be modeled by continuous probability distributions because these are outcomes that must be measured, not counted.

Work the following problems.

1. What are the probabilities of the number of heads in three coin tosses? Construct the chart.

Outcome	HHH							
$P(A)$	$\frac{1}{8}$							

2. What is the probability that only one head is obtained in three coin tosses? Use the chart constructed in Problem 1.

Use the curve below to answer the following questions. Apply the empirical rule percentages from the previous page.

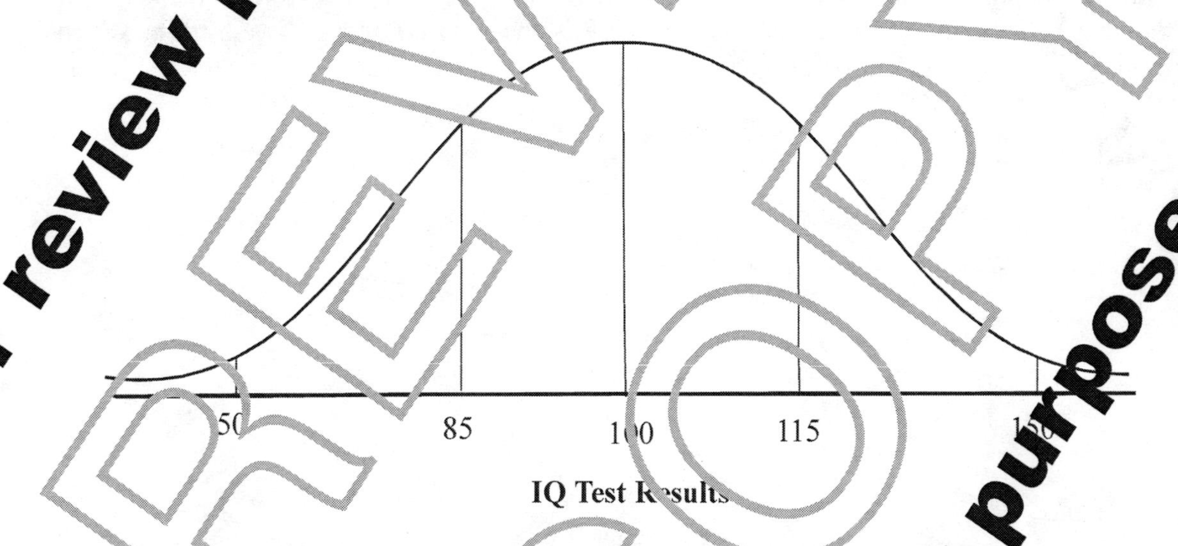

IQ Test Results

3. What percentage of adults taking the IQ test scored less than 115?

4. What percentage scored 85 or below?

5. What percentage scored 115 or above?

6. What percentage scored between 85 and 115?

7. How many scored exactly 95?

19.5 Geometric Probability

Two-step area problems involved finding the area of two geometric figures (either the same type of figure or different figures). We then added or subtracted the area to arrive at the correct answer. In this section, we will calculate two geometric areas and then use these calculated areas to set up and solve a probability problem. A geometric probability is simply the area of the event in question divided by the total area in which the event can take place. The following examples will show how these concepts can be combined.

Example 8: If a circular stack of hay was located in a field as shown in the figure below, what is the probability that lightning will strike the hay in a thunderstorm equally distributed over the field?

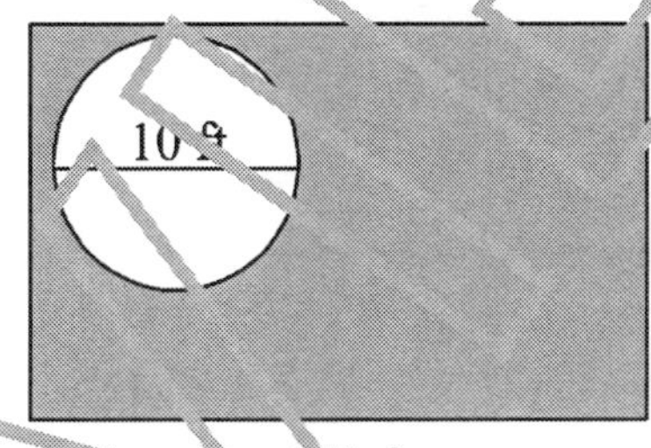

Step 1: The target area is the circle. The area of a circle is $A = \pi r^2$. The diameter of the circle is given as 10 feet, so the radius is one half of this, or 5 feet. The area is then

$A = \pi \times r^2 = 3.14 \times 5 \times 5 = 78.5 \text{ ft}^2$

This is the event that we want to occur, so this will be the top part of the fraction for the probability calculation.

Step 2: Find the total area in which lightning could strike. This is the sample space, or the bottom of the fraction for the probability calculation. The area of the rectangle, including the circle, will be the sample space.

$A = l \times w = 150 \times 100 = 1500 \text{ ft}^2$

Step 3: The probability calculation is then the area of the event taking place (lightning striking in the circle) divided by the entire possible area (the rectangular field).

$$P(A) = \frac{78.5 \text{ ft}^2}{1500 \text{ ft}^2} = 0.052$$

In other words, lightning has a 0.052 chance of striking in the circular area of the field. Converting this to a percent by moving the decimal two places to the right, lightning has a 5.2% chance of striking the hay.

Work the following geometric probability problems.

1. The rectangular park has a fountain 6 feet in diameter in the center. If a small meteor fell in the park last night, what is the probability that it hit the fountain?

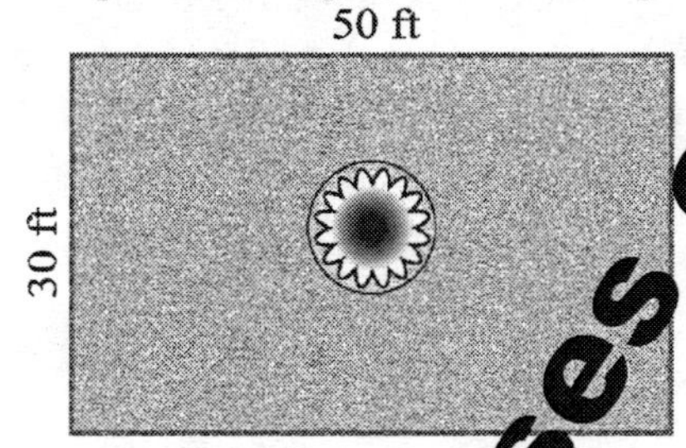

2. Two friends enjoy challenging each other by making up dart boards.
   (A) What is the probability of landing a dart in the black part of the board?
   (B) What is the probability of landing in the white part?

3. In a popular board game, a circular wheel is spun to determine the dollar values of certain letters. There are 18 divisions of the wheel broken down as follows:

   (A) What is the probability of spinning $500?
   (B) What is the probability of spinning $5,000?

4. These spotlights are illuminating circles on the dance floor with a diameter of 2.5 feet. The entire dance floor of 15 feet by 25 feet is shown. What is the probability that a spec of floating dust lands on the unlit part of the floor?

5. A satellite which has been orbiting the earth for many years has failed and is slowly falling. Engineers on the ground are no longer able to control the satellite, and are warning that it is going to fall to the earth sometime in the next week. What is the probability that the satellite will hit on land instead of in the water? (Hint: the earth is approximately 71% water.)

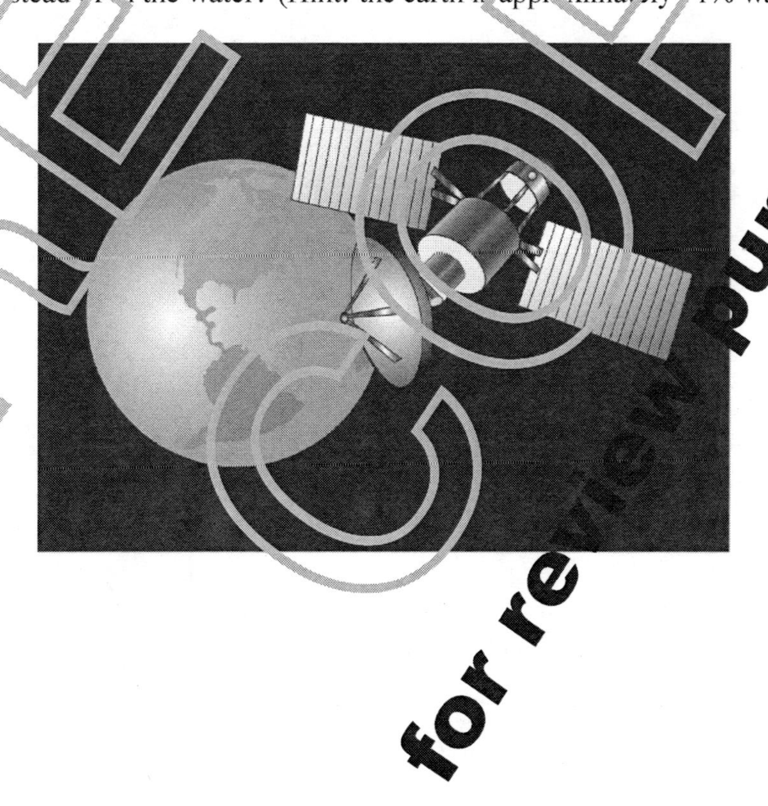

19.6 Permutations

A **permutation** is an arrangement of items in a specific order. If a problem asks how many ways you can arrange 6 books on a bookshelf, it is asking how many permutations there are for 6 items.

Example 1: Ron has 4 items: a model airplane, a trophy, and autographed football, and a toy sports car. How many ways can he arrange the 4 items on a shelf?

Solution: Ron has 4 choices for the first item on the shelf. He then has 3 choices left for the second item. After choosing the second item, he has 2 choices left for the third item and only one choice for the last item. The diagram below shows the permutations for arranging the 4 items on a shelf if he chose to put the trophy first.

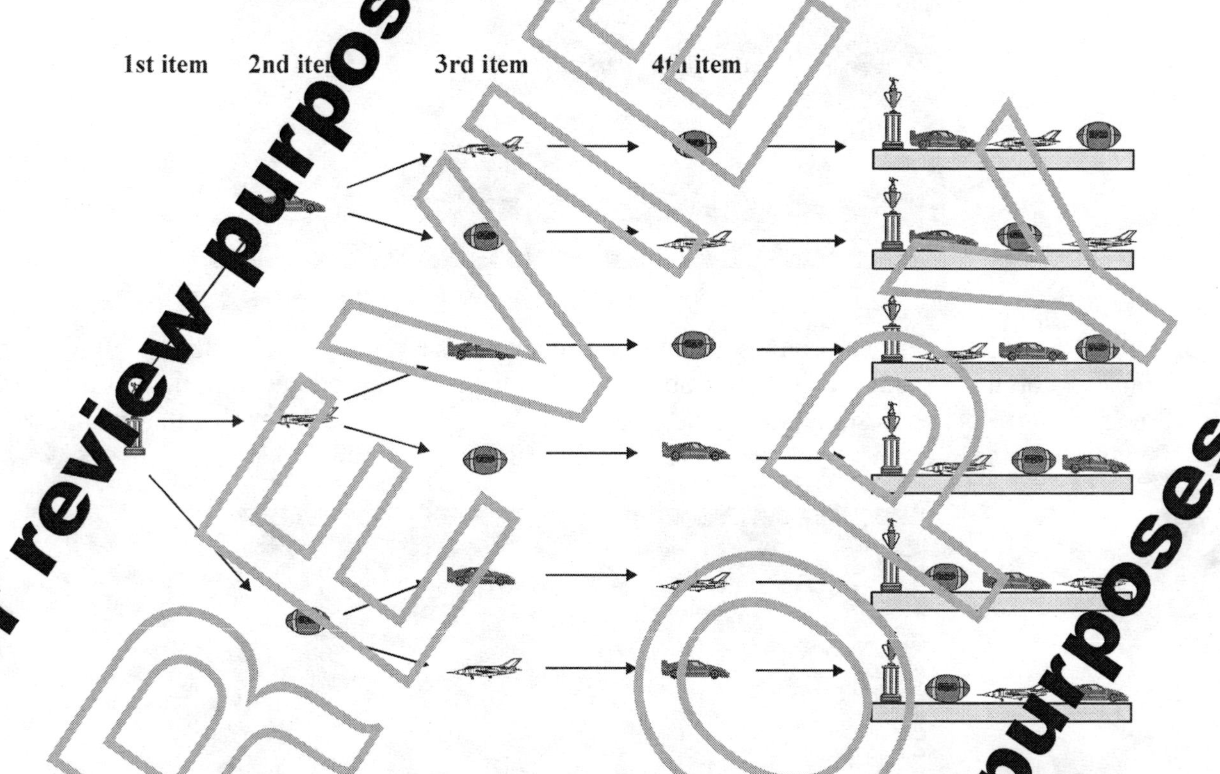

Count the number of permutations if Ron chooses the trophy as the first item. There are 6 permutations. Next, you could construct a pyramid of permutations choosing the model car first. That pyramid would also have 6 permutations. Then, you could construct a pyramid choosing the airplane first. Finally, you could construct a pyramid choosing the football first. You would then have a total of 4 pyramids each having 6 permutations. The total number of permutations is $6 \times 4 = 24$. There are 24 ways to arrange the 4 items on a bookshelf.

You probably don't want to draw pyramids for every permutation problem. What if you want to know the permutations for arranging 30 objects? Fortunately, mathematicians have come up with a formula for calculating permutations.

For the above problem, Ron has 4 items to arrange. Therefore, multiply $4 \times 3 \times 2 \times 1 = 24$. Another way of expressing this calculation is 4!, stated as 4 factorial. $4! = 4 \times 3 \times 2 \times 1$.

Example 2: How many ways can you line up 6 students?

Solution: The number of permutations for 6 students $= 6! = 6 \times 5 \times 4 \times 3 \times 2 \times 1 = 720$. There are 6 choices for the first position, 5 for the second position, 4 for the third, 3 for the fourth, 2 for the fifth, and 1 for the sixth.

Example 3: Shelley and her mom, dad, and brother are having cake for her birthday. Since it is Shelley's birthday she gets a piece first. How many ways are there to pass out the pieces of cake?

Solution: Since Shelley gets the first piece, the first spot is fixed. The second, third, and fourth spots are not fixed and anyone left can be in one of the three spots.

Spot	1	2	3	4
Choices of people	1	3	2	1

Now multiply the choices together, $1 \times 3 \times 2 \times 1 = 6$ ways to pass out cake.

Work the following permutation problems.

1. How many ways can you arrange 5 books on a bookshelf?
2. Myra has 6 novels to arrange on a book shelf. How many ways can she arrange the novels?
3. Seven sprinters signed up for the 100 meter dash. How many ways can the seven sprinters line up on the start line?
4. Keri wants an ice cream cone with one scoop of chocolate, one scoop of vanilla, and one scoop of strawberry. How many ways can the scoops be arranged on the cone?
5. How many ways can you arrange the letters A, B, C, and D?
6. At Sam's party, the DJ has four song requests. In how many different orders can he play the 4 songs?
7. Yvette has 5 comic books. How many different ways can she stack the comic books?
8. Sandra's couch can hold three people. How many ways can she and her two friends sit on the couch?
9. How many ways can you arrange the numbers 2, 3, 5?
10. At a busy family restaurant, four tables open up at the same time. How many different ways can the hostess seat the next four families waiting to be seated?
11. How many ways can you arrange the numbers 1, 2, 3, 4, 5, 6, 7, 8, 9, 10 and always have 3 at position 1 and 10 at position 5?

19.7 More Permutations

Example 4: If there are 6 students, how many ways can you line up any 4 of them?

Solution: Multiply $6 \times 5 \times 4 \times 3 = 360$. There are 6 choices for the first position in line, 5 for the second position, 4 for the third position, and 3 for the last position. There are 360 ways to line up 4 of the 6 students.

Find the number of permutations for each of the problems below.

1. How many ways can you arrange 4 out of 8 books on a shelf?

2. How many 3 digit numbers can be made using the numbers 2, 3, 5, 8, and 9?

3. How many ways can you line up 4 students out of a class of 20?

4. Kim worked in the linen department of a store. Eight new colors of towels came in. Her job was to line up the new towels on a long shelf. How many ways can she arrange the 8 colors?

5. Terry's CD player holds 5 CDs. Terry owns 12 CDs. How many different ways can he arrange his CDs in the CD player?

6. Erik has 11 shirts he wears to school. How many ways can he choose a different shirt to wear on Monday, Tuesday, Wednesday, Thursday, and Friday?

7. Deb has a box of 12 markers. The art teacher told her to choose three markers and line them up on her desk. How many ways can she line up 3 markers from the 12?

8. Jeff went into an ice cream store serving 32 flavors of ice cream. He wanted a cone with two different flavors. How many ways could he order 2 scoops of ice cream, one on top of the other?

9. In how many ways can you arrange any 3 letters from the 26 letters in the alphabet?

19.8 Combinations

In a **permutation**, objects are arranged in a particular order. In a **combination**, the order does not matter. In a **permutation**, if someone picked two letters of the alphabet, **k, m** and **m, k** would be considered 2 different permutations. In a **combination**, **k, m** and **m, k** would be the same combination. A different order does not make a new combination.

Example 5: How many combinations of 3 letters from the set {a, b, c, d, e} are there?

Step 1: Find the **permutation** of 3 out of 5 objects.

Step 2: Divide by the permutation of the **number of objects** to be chosen from the total (3). This step eliminates the duplicates in finding the permutations.

Step 3: Cancel common factors and simplify.

$$\frac{5 \times \overset{2}{\cancel{4}} \times \cancel{3}}{\cancel{3} \times \cancel{2} \times 1} = 10$$

There can be 10 combinations of three letters from the set {a, b, c, d, e}.

Find the number of combinations for each problem below.

1. How many combinations of 4 numbers can be made from the set of numbers {2, 4, 6, 7, 8, 9}?

2. Johnston Middle School wants to choose 3 students at random from the 7th grade to take an opinion poll. There are 124 seventh graders in the school. How many different groups of 3 students could be chosen? (Use a calculator for this one.)

3. How many combinations of 3 students can be made from a class of 20?

4. Fashion Ware catalog has a sweater that comes in 8 colors. How many combinations of 2 different colors does a shopper have to choose from?

5. Angelo's Pizza offers 10 different pizza toppings. How many different combinations can be made of pizzas with four toppings?

6. How many different combinations of 5 flavors of jelly beans can you make from a store that sells 25 different flavors of jelly beans?

7. The track team is running the relay race in a competition this Saturday. There are 14 members of the track team. The relay race requires 4 runners. How many combinations of 4 runners can be formed from the track team?

8. Kerri got to pick 2 prizes from a grab bag containing 12 prizes. How many combinations of 2 prizes are possible?

19.9 More Combinations

Another kind of combination involves selection from several categories.

Example 6: At Joe's Deli, you can choose from 4 kinds of bread, 5 meats, and 3 cheeses when you order a sandwich. How many different sandwiches can be made with Joe's choices for breads, meats, and cheeses if you choose 1 kind of bread, 1 meat, and 1 cheese for each sandwich?

JOE'S SANDWICHES

Breads	Meats	Cheeses
White	Roast beef	Swiss
Pumpernickel	Corned beef	American
Light rye	Pastrami	Mozzarella
Whole wheat	Roast chicken	
	Roast turkey	

Solution: Multiply the number of choices in each category. There are 4 breads, 5 meats, and 3 cheeses, so $4 \times 5 \times 3 = 60$. There are 60 combinations of sandwiches.

Find the number of combinations that can be made in each of the problems below.

1. Angie has 4 pairs of shorts, 6 shirts, and 2 pairs of tennis shoes. How many different outfit combinations can be made with Angie's clothes?

2. Raymond has 7 baseball caps, 2 jackets, 10 pairs of jeans, and 2 pairs of sneakers. How many combinations of the 4 items can he make?

3. Claire has 6 kinds of lipstick, 4 eye shadows, 2 kinds of lip liner, and 2 mascaras. How many combinations can she use to make up her face?

4. Clarence's dad is ordering a new truck. He has a choice of 5 exterior colors, 3 interior colors, 2 kinds of seats, and 3 sound systems. How many combinations does he have to pick from?

5. A fast food restaurant has 8 kinds of sandwiches, 3 kinds of French fries, and 5 kinds of soft drinks. How many combinations of meals could you order if you ordered a sandwich, fries, and a drink?

6. In summer camp, Tyrone can choose from 4 outdoor activities, 3 indoor activities, and 3 water sports. He has to choose one of each. How many combinations of activities can he choose?

7. Jackie won a contest at school and gets to choose one pencil and one pen from the school store and an ice cream from the lunch room. There are 5 colors of pencils, 3 colors of pens, and 4 kinds of ice cream. How many combinations of prize packages can she choose?

Chapter 19 Review

1. There are 50 students in the school orchestra in the following sections:

   25 string section
   15 woodwind
   5 percussion
   5 brass

   One student will be chosen at random to present the orchestra director with an award. What is the probability the student will be from the woodwind section?

2. Fluffy's cat treat box contains 6 chicken-flavored treats, 5 beef-flavored treats, and 7 fish-flavored treats. If Fluffy's owner reaches in the box without looking, and chooses one treat, what is the probability that Fluffy will get a chicken-flavored treat?

3. The spinner in figure A stopped on the number 5 on the first spin. What is the probability that it will not stop on 5 on the second spin?

Fig. A

Fig. B

4. Sherri turns the spinner in figure B above 3 times. What is the probability that the pointer always lands on a shaded number?

5. Three cakes are sliced into 20 pieces each. Each cake contains 1 gold ring. What is the probability that one person who eats one piece of cake from each of the 3 cakes will find 3 gold rings?

6. Brianna tosses a coin 4 times. What is the probability she gets all tails?

Read the following, and answer questions 7–11.

There are 9 slips of paper in a hat, each with a number from 1 to 9. The numbers correspond to a group of students who must answer a question when the number for their group is drawn. Each time a number is drawn, the number is put back in the hat.

7. What is the probability that the number 6 will be drawn twice in a row?

8. What is the probability that the first 5 numbers drawn will be odd numbers?

9. What is the probability that the second, third, and fourth numbers drawn will be even numbers?

10. What is the probability that the first five times a number is drawn it will be the number 5?

11. What is the probability that the first five numbers drawn will be 1, 2, 3, 4, 5 in that order?

Solve the following word problems. For questions 12–14, write whether the problem is "dependent" or "independent."

12. Felix Perez reaches into a 10-piece puzzle and pulls out one piece at random. This piece has two places where it could connect to other pieces. What is the probability that he will select another piece which fits the first one if he selects the next piece at random?

13. Barbara Stein is desperate for a piece of chocolate candy. She reaches into a bag which contains 8 peppermint, 5 butterscotch, 7 toffee, 3 mint, and 6 chocolate pieces and pulls out a toffee piece. Disappointed, she throws it back into the bag and then reaches back in and pulls out one piece of candy. What is the probability that Barbara pulls out a chocolate piece on the second try?

14. Christen Solis goes to a pet shop and immediately decides to purchase a guppy she saw swimming in an aquarium. She reaches into the tank containing 5 goldfish, 6 guppies, 4 miniature catfish, and 3 minnows and accidently pulls up a goldfish. Breathing a sigh, Christen places the goldfish back in the water. The fish are swimming so fast, it is impossible to tell what fish Christen would catch. What is the probability that Christen will catch a guppy on her second try?

Answer the following geometric probability problem.

15. Mrs. Stabler has a pool in her back yard. The pool's dimensions are 25 ft by 15 ft, and the yard's dimensions are 35 ft by 58 ft. Mrs. Stabler's son is in the backyard launching the rockets that he made with his father. Assuming the next rocket shot off lands in the backyard, what is the probability that it will land in the pool? Express your answer as a percent.

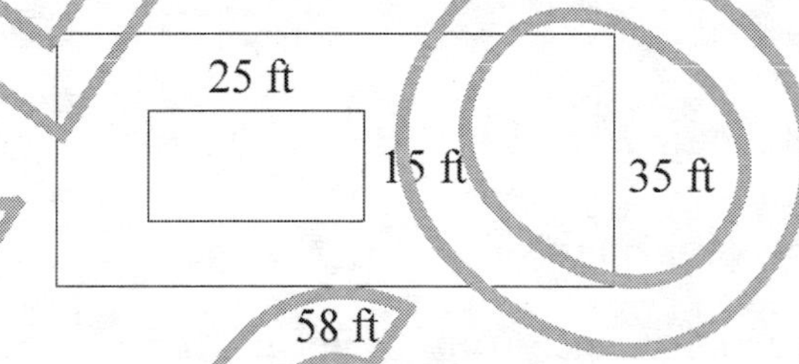

Answer the following permutation and combination problems.

16. Daniel has 7 trophies he has won playing soccer. How many different ways can he arrange them in a row on his bookshelf?

17. Missy has 12 colors of nail polish. She wears 1 color each day, 7 different colors a week. How many combinations of 7 colors can she make before she has to repeat the same 7 colors in a week?

18. Eileen has a collection of 12 antique hats. She plans to donate 5 of the hats to a museum. How many combinations of hat are possible for her donation?

19. Julia has 5 porcelain dolls. How many ways can she arrange 3 of the dolls on a display shelf?

20. Ms. Randal has 10 students. Every day she randomly draws the names of 2 students out of a bag to turn in their homework for a test grade. How many combinations of 2 students can she draw?

21. In the lunch line, students can choose 1 out of 3 meats, 1 out of 4 vegetables, 1 out of 3 desserts, and 1 out of 5 drinks. How many lunch combinations are there?

22. Andrea has 7 teddy bears in a row on a shelf in her room. How many ways can she arrange the bears in a row on her shelf?

23. Adrianna has 4 hats, 8 shirts, and 9 pairs of pants. Choosing one of each, how many different clothes combinations can she make?

24. The buffet line offers 5 kinds of meat, 3 different salads, a choice of 4 desserts, and 5 different drinks. If you choose one food from each category, from how many combinations would you have to choose?

25. How many pairs of students can Mrs. Smith choose to go to the library if she has 20 students in her class?

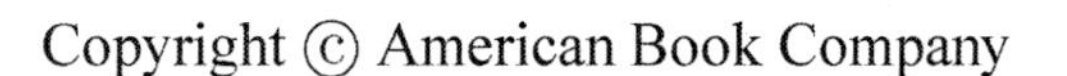

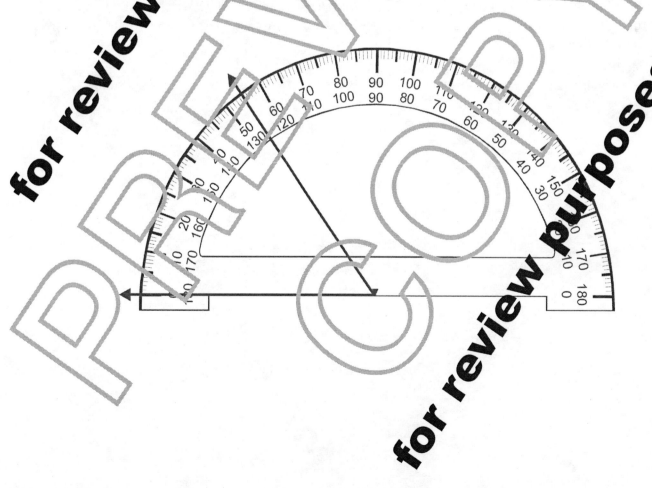

Chapter 20
Angles

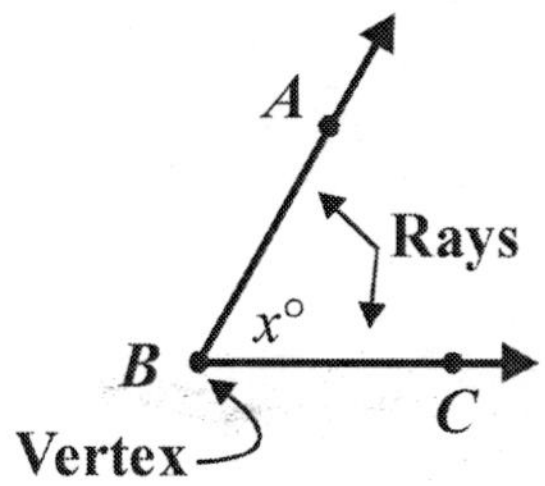

Angles are made up of two rays with a common endpoint. Rays are named by the endpoint B and another point on the ray. Ray $\overrightarrow{BA}$ and ray $\overrightarrow{BC}$ share a common endpoint.

Angles are usually named by three capital letters. The middle letter names the vertex. The angle to the left can be named $\angle ABC$ or $\angle CBA$. An angle can also be named by a lower case letter between the sides, $\angle x$, or by the vertex alone, $\angle B$.

A protractor, , is used to measure angles. The protractor is divided evenly into a half circle of 180 degrees (180°). When the middle of the bottom of the protractor is placed on the vertex, and one of the rays of the angle is lined up with 0°, the other ray of the angle crosses the protractor at the measure of the angle. The angle below has the ray pointing left lined up with 0° (the outside numbers), and the other ray of the angle crosses the protractor at 55°. The angle measures 55°.

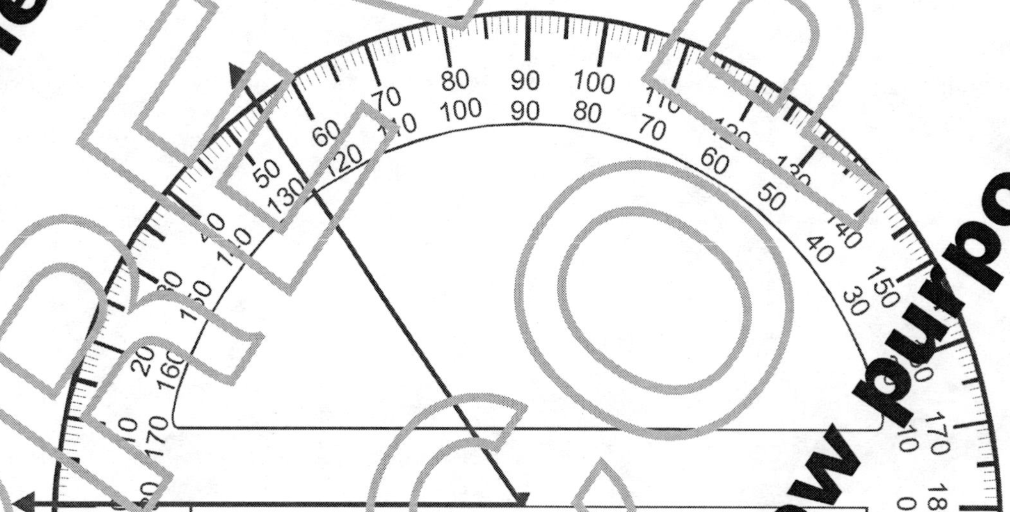

20.1 Types of Angles

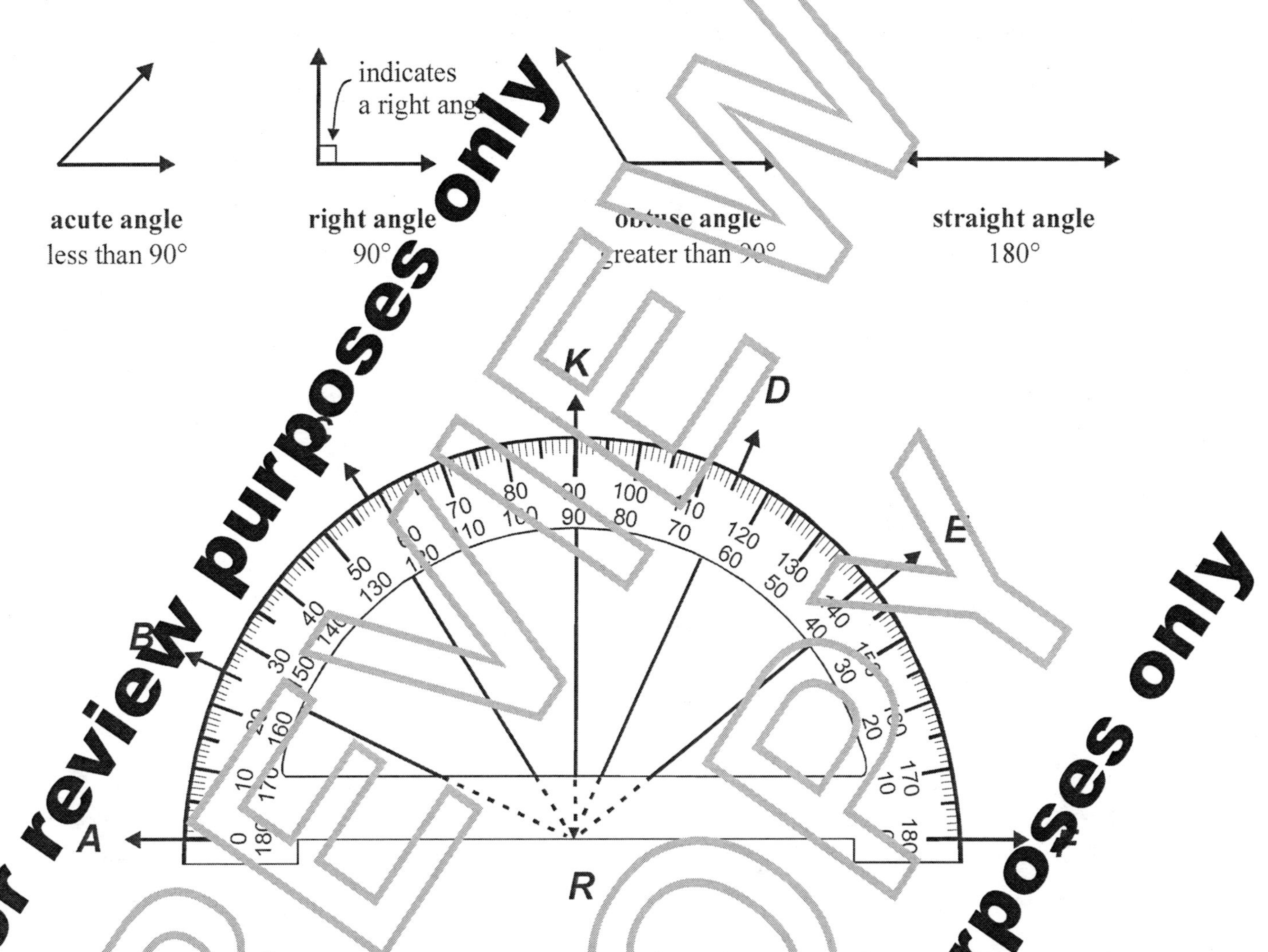

Using the protractor above, find the measure of the following angles. Then, tell what type of angle it is: acute, right, obtuse, or straight.

		Measure	Type of Angle
1.	What is the measure of angle ARF?	________	________
2.	What is the measure of angle CRF?	________	________
3.	What is the measure of angle BRF?	________	________
4.	What is the measure of angle ERF?	________	________
5.	What is the measure of angle ARB?	________	________
6.	What is the measure of angle KRA?	________	________
7.	What is the measure of angle CRA?	________	________
8.	What is the measure of angle DRF?	________	________
9.	What is the measure of angle ARD?	________	________
10.	What is the measure of angle FRK?	________	________

20.2 Measuring Angles

Estimate the measure of the following angles. Then, use your protractor to record the actual measure.

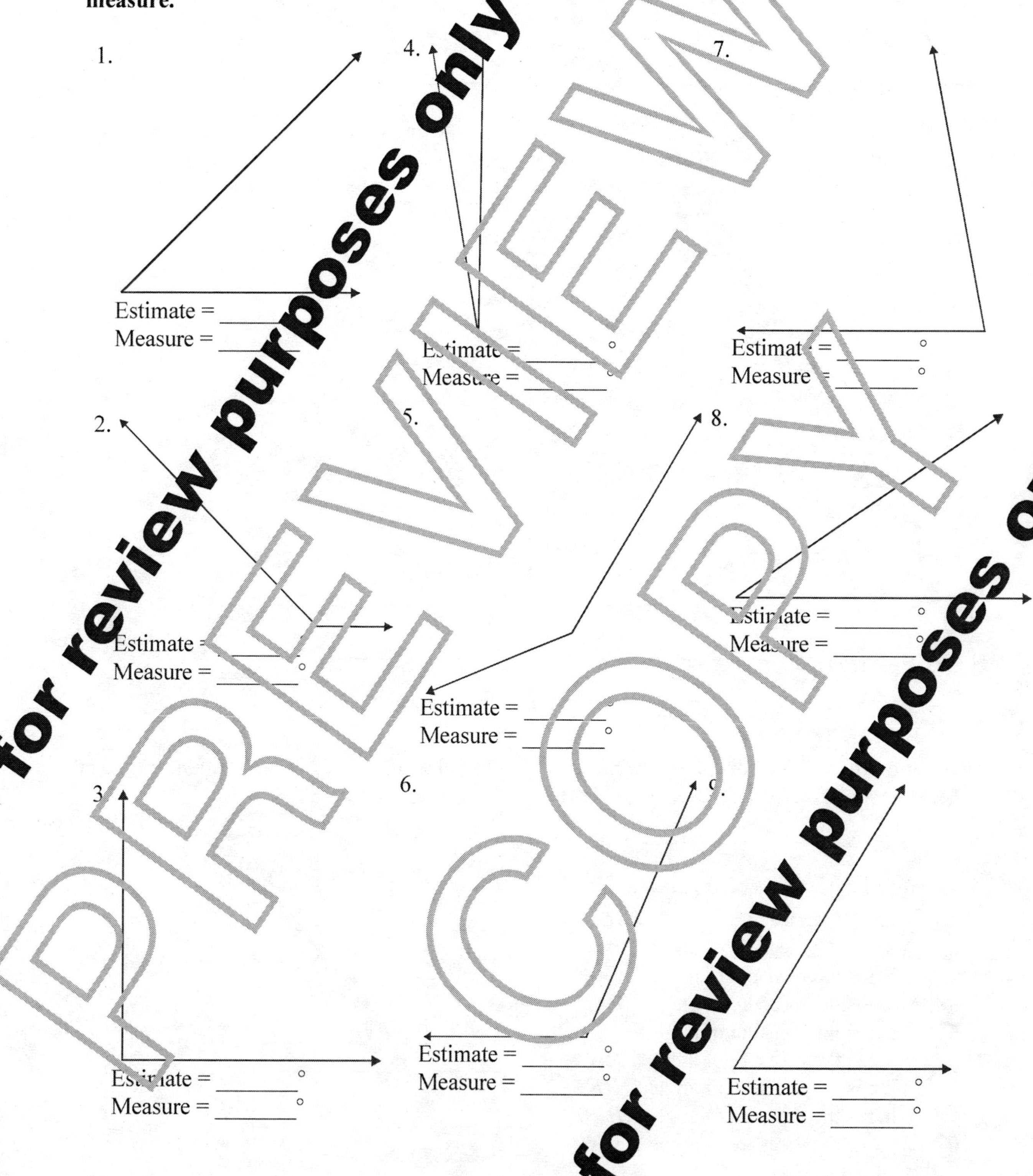

20.3 Central Angles

In this section, you will learn about central angles and why they are important when making circle graphs. A central angle is the angle formed by each "piece of the pie." The vertex of a central angle is in the center of the circle. Look at the diagram below.

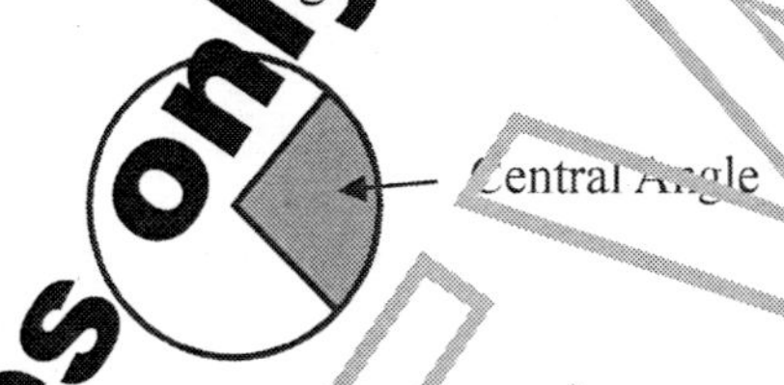

In circle graphs, the percentages have to add up to 100%, and the angles for each "piece of pie" must add up to 360°. Notice the pie graph to the right. Each percent of the pie is marked with a corresponding angle measure.

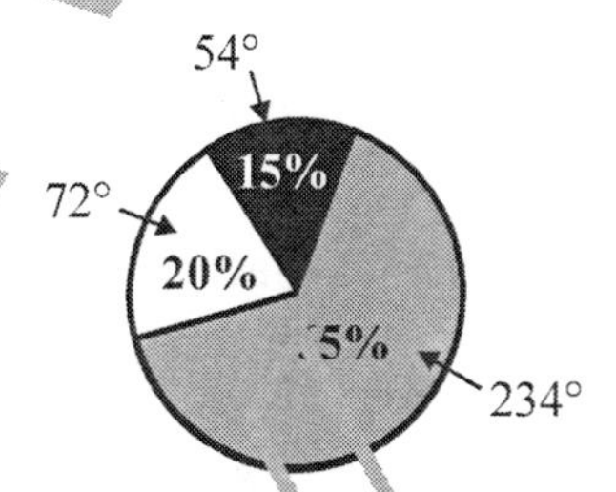

Example 1: Consider the following data:

Survey of fiction reading preferences among 300 high school students	
Sports stories	70
Science fiction	60
Fantasy	90
Historical novels	40
Poetry	10
Mysteries	30

Step 1: First, you need to find what percent of the total each type of fiction represents. Divide each item in the survey by the total number of students. For example, according to the chart, 70 out of 300 students prefer sports stories, so to find the percent, divide 70 by 300. Rounding to the nearest percent, you get 23%. Repeat for each type of fiction.

Step 2: Multiply the percent of each type of fiction by 360° to figure out how many degrees each "piece of pie" should be. For sports stories, 23% of 360° is 83° (rounded to the nearest degree).

	Number of students	% of total	Central angle of the circle
Sports stories	70	$23\frac{1}{3}\%$	84°
Science fiction	60	20%	72°
Fantasy	90	30%	108°
Historical novels	40	$13\frac{1}{3}\%$	48°
Poetry	10	$3\frac{1}{3}\%$	12°
Mysteries	30	10%	36°
	300	100%	360°

Complete the following exercises on central angles.

A dartboard has been divided into wedges according to the following color percentages.

blue	30%
red	25%
yellow	10%
green	35%

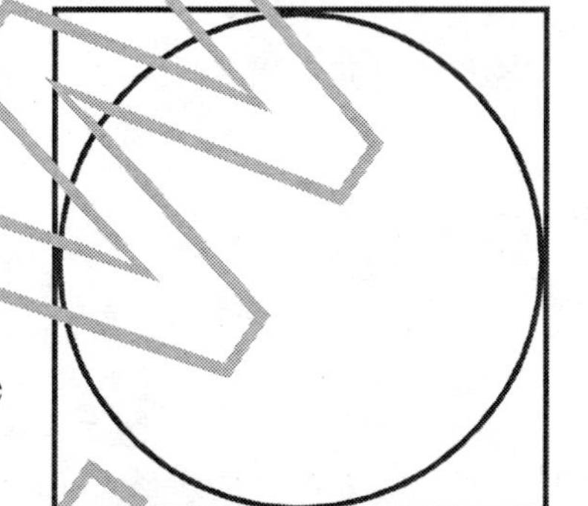

1. Find the central angle measurements of each color.

2. Complete the dartboard by drawing each color wedge to scale. Label each color on the dartboard and indicate the measure of each central angle.

The students at Maverick High School voted on Teacher of the Year. The Student Council tallied the 720 total votes and created a pie chart.

Mr. Perry	252
Mrs. Nance	180
Miss Murphy	144
Mr. Bard	87
Mr. Olson	36
All Others	21

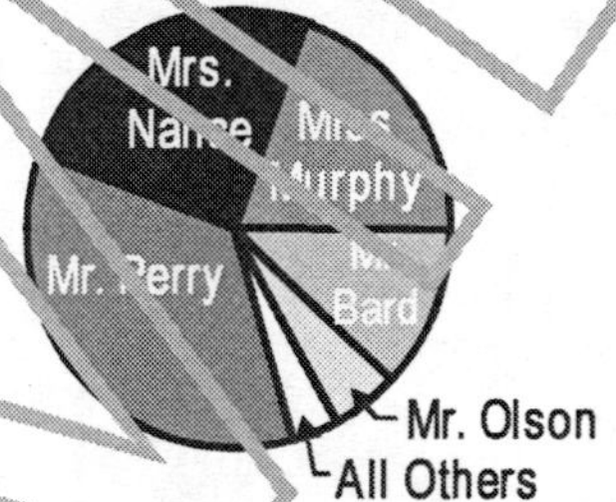

3. Calculate the percent of the votes that each teacher received.

4. Calculate the measures of each central angle on the pie chart.

Other central angle type questions involve a clock. These are very similar to the pie chart.

Example 2: The second hand on a clock goes from the 11 to the 2. How many degrees has the second hand traveled?

Step 1: You know that there are 60 seconds in a minute. The second hand traveled 15 seconds out of 60 as it went from the 11 to the 2. $15 \div 60 = 0.25$ or 25%

Step 2: Multiply the percent by 360° just like you did for the pie chart. $0.25 \times 360 = 90°$. So, the second hand traveled 90°.

Calculate how many degrees the second hand travels for the following.

5. 12 to 3
6. 9 to 11
7. 4 to 10
8. 6 to 3
9. 12 to 1
10. 3 to 5

20.4 Adjacent Angles

Adjacent angles are two angles that have the same vertex and share one ray. They do not share space inside the angles.

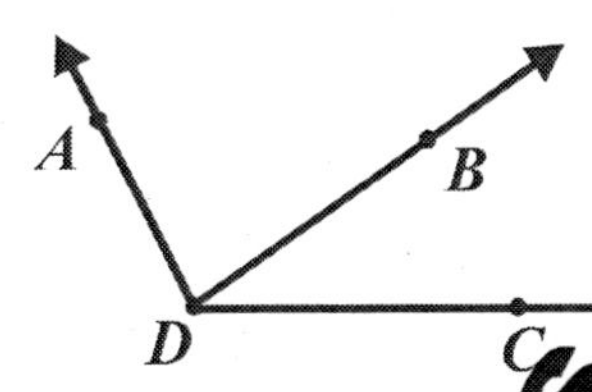

$\angle ADB$ is **adjacent** to $\angle BDC$. However, $\angle ADB$ is **not adjacent** to $\angle ADC$ because adjacent angles do not share any space inside the angles.

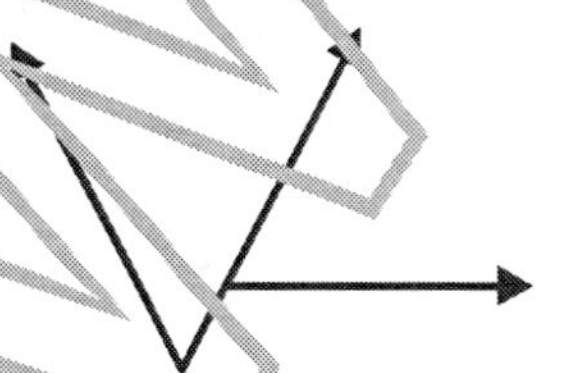

These two angles are **not adjacent**. They share a common ray but do not share the same vertex.

For each diagram below, name the angle that is adjacent to it.

1. $\angle CDB$ is adjacent to $\angle$_____

2. $\angle TUV$ is adjacent to $\angle$_____

3. $\angle SRP$ is adjacent to $\angle$_____

4. $\angle PQR$ is adjacent to $\angle$_____

5. $\angle YOP$ is adjacent to $\angle$_____

6. $\angle XVY$ is adjacent to $\angle$_____

7. $\angle DCF$ is adjacent to $\angle$_____

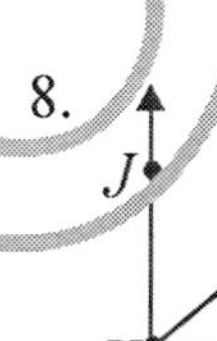

8. $\angle JKL$ is adjacent to $\angle$_____

20.5 Vertical Angles

When two lines intersect, two pairs of vertical angles are formed. Vertical angles are not adjacent. Vertical angles have the same measure.

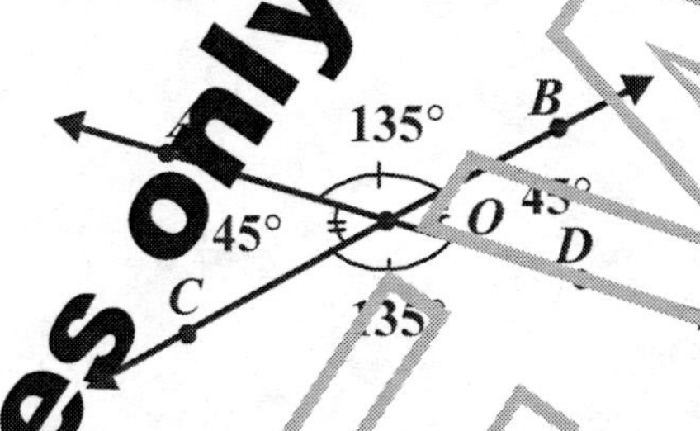

$\angle AOB$ and $\angle COD$ are vertical angles. $\angle AOC$ and $\angle BOD$ are vertical angles. **Vertical angles** are **congruent**. Congruent means they have the same measure.

In the diagram below, name the second angle in each pair of vertical angles.

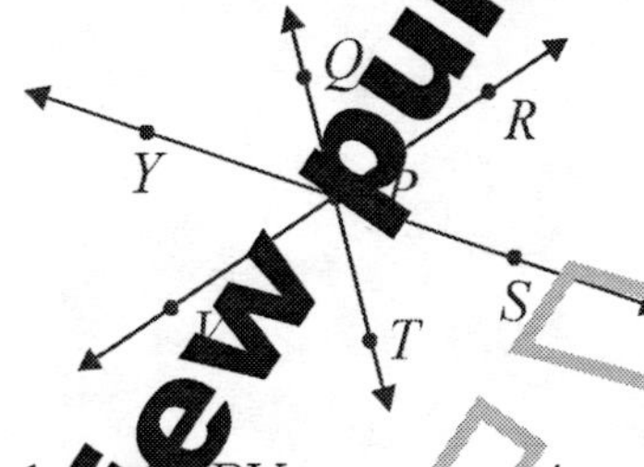

1.	$\angle YPV$	_____	4.	$\angle VPT$	_____	7.	$\angle MLN$	_____	10.	$\angle GLM$	_____
2.	$\angle QPR$	_____	5.	$\angle RPT$	_____	8.	$\angle KLH$	_____	11.	$\angle KLM$	_____
3.	$\angle SPT$	_____	6.	$\angle VPS$	_____	9.	$\angle GLN$	_____	12.	$\angle HLG$	_____

Use the information given to find the measure of each unknown vertical angle.

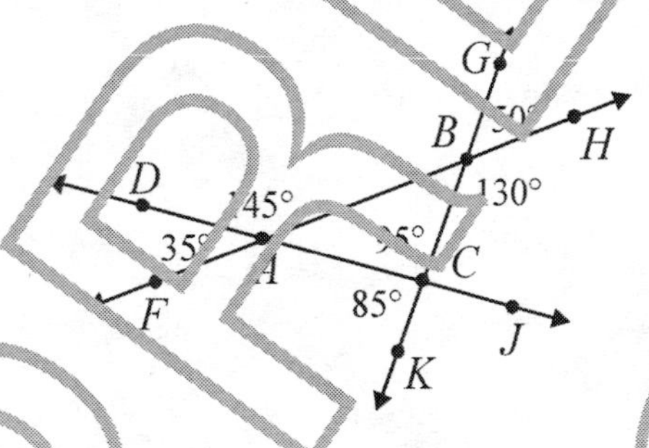

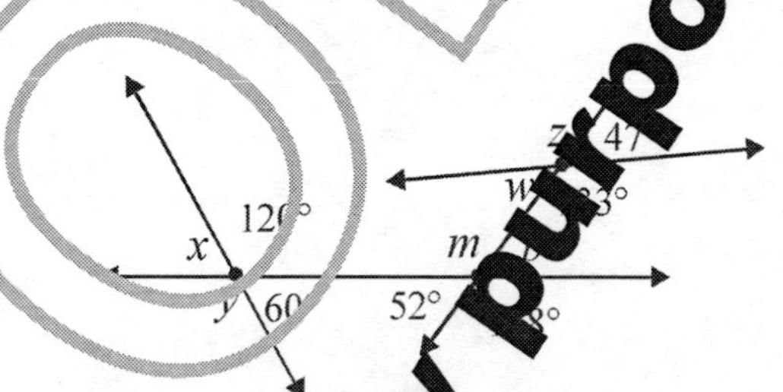

13. $\angle CAF =$ ______
14. $\angle ABC =$ ______
15. $\angle KCJ =$ ______
16. $\angle ABG =$ ______
17. $\angle BCJ =$ ______
18. $\angle CAB =$ ______

19. $\angle x =$ ______
20. $\angle y =$ ______
21. $\angle z =$ ______
22. $\angle w =$ ______
23. $\angle m =$ ______
24. [illegible] = ______

20.6 Complementary and Supplementary Angles

Two angles are **complementary** if the sum of the measures of the angles is 90°.
Two angles are **supplementary** if the sum of the measures of the angles is 180°.
The angles may be adjacent but do not need to be.

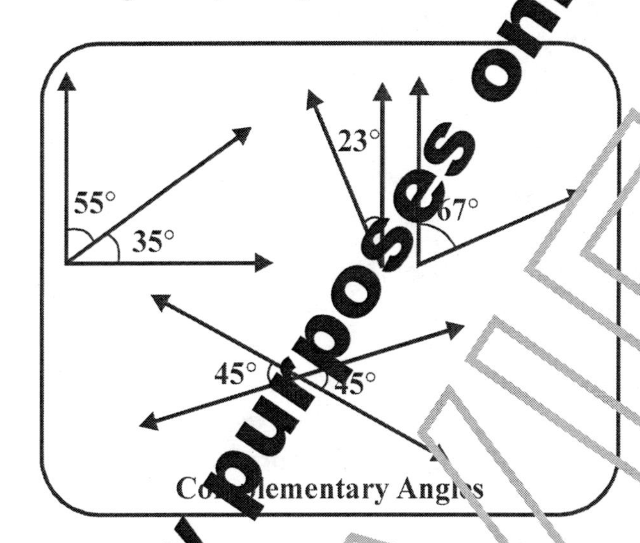

Complementary Angles

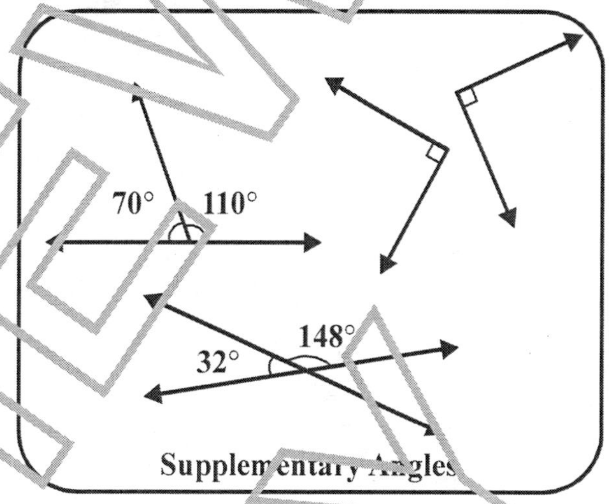

Supplementary Angles

Calculate the measure of each unknown angle.

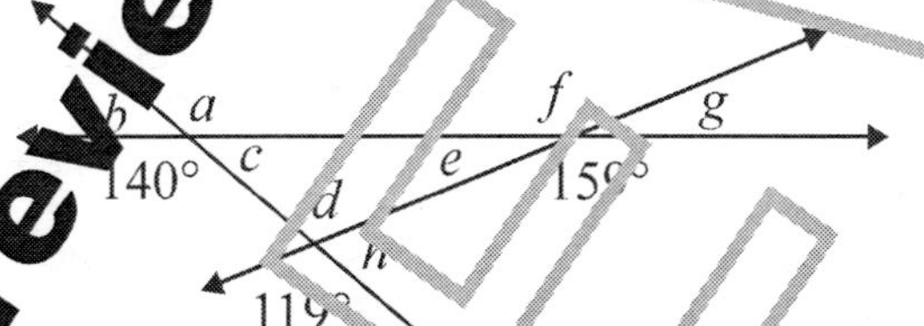

1. $\angle a =$ ______
2. $\angle b =$ ______
3. $\angle c =$ ______
4. $\angle d =$ ______
5. $\angle e =$ ______
6. $\angle f =$ ______
7. $\angle g =$ ______
8. $\angle h =$ ______

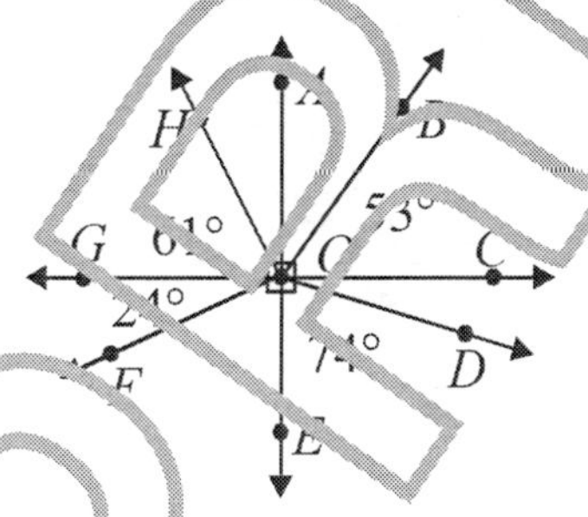

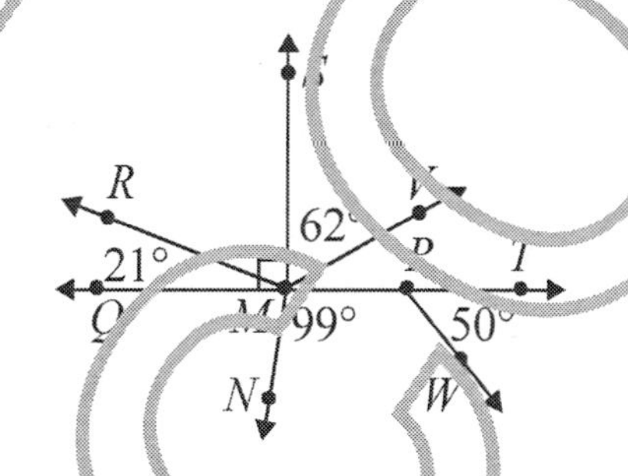

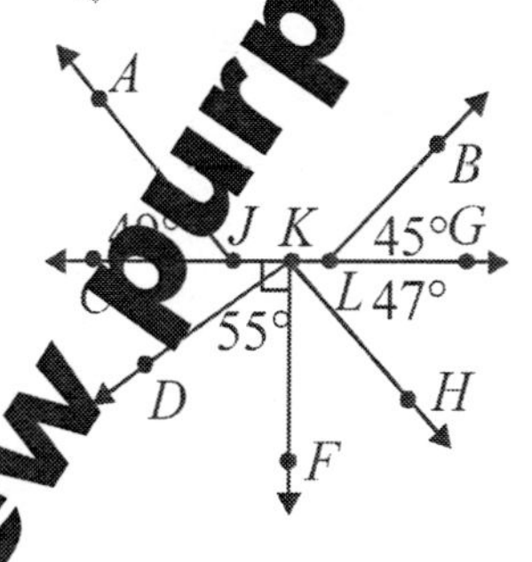

9. $\angle AOB =$ ______
10. $\angle COD =$ ______
11. $\angle EOF =$ ______
12. $\angle AOH =$ ______
13. $\angle RMS =$ ______
14. $\angle VMT =$ ______
15. $\angle QMN =$ ______
16. $\angle WPQ =$ ______
17. $\angle AJK =$ ______
18. $\angle CKD =$ ______
19. $\angle FKH =$ ______
20. $\angle BLC =$ ______

20.7 Corresponding, Alternate Interior, and Alternate Exterior Angles

If two parallel lines are intersected by a **transversal**, a line passing through both parallel lines, the **corresponding angles** are congruent.

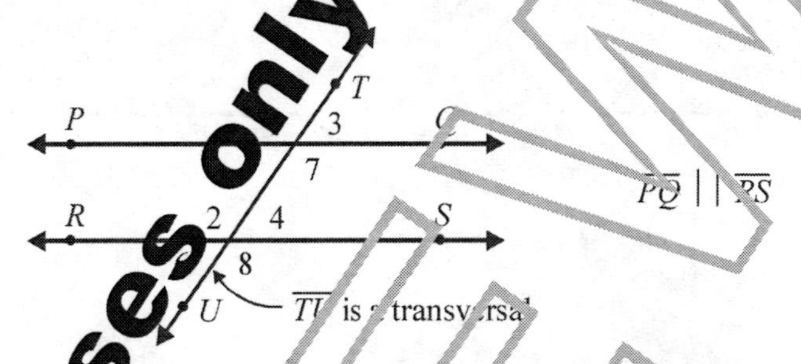

$\angle 1$ and $\angle 2$ are congruent.	They are corresponding angles.
$\angle 3$ and $\angle 4$ are congruent.	They are corresponding angles.
$\angle 5$ and $\angle 6$ are congruent.	They are corresponding angles.
$\angle 7$ and $\angle 8$ are congruent.	They are corresponding angles.

Alternate interior angles are also congruent. They are on the opposite sides of the transversal and inside the parallel lines.

$\angle 5$ and $\angle 4$ are congruent.	They are alternate interior angles.
$\angle 7$ and $\angle 2$ are congruent.	They are alternate interior angles.

Alternate exterior angles are also congruent. They are on the opposite sides of the transversal and above and below the parallel lines.

$\angle 1$ and $\angle 8$ are congruent.	They are alternate exterior angles.
$\angle 3$ and $\angle 6$ are congruent.	They are alternate exterior angles.

Look at the diagram below. For each pair of angles, state whether they are corresponding (C), alternate interior (I), alternate exterior (E), vertical (V), or supplementary angles (S).

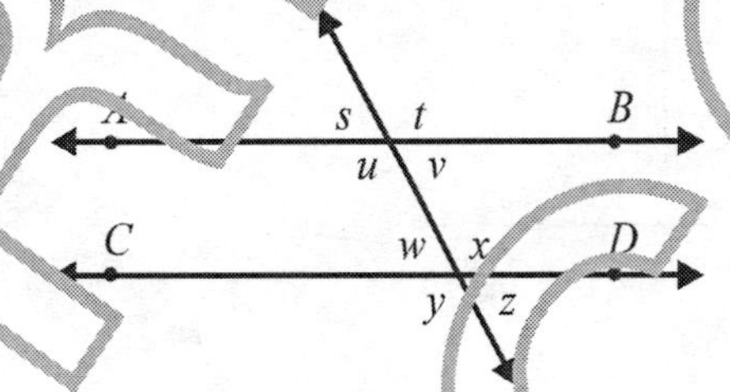

$\overline{AB}$ and $\overline{CD}$ are parallel.

1. $\angle u, \angle x$
2. $\angle w, \angle s$
3. $\angle t, \angle y$
4. $\angle s, \angle t$
5. $\angle w, \angle y$
6. $\angle t, \angle x$
7. $\angle w, \angle z$
8. $\angle v, \angle w$
9. $\angle v, \angle z$
10. $\angle s, \angle z$
11. $\angle t, \angle u$
12. $\angle w, \angle x$
13. $\angle u, \angle s$
14. $\angle s, \angle v$
15. $\angle x, \angle z$

20.8 Sum of Interior Angles of a Polygon

Given a polygon, you can find the sum of the measures of the interior angles using the following formula: Sum of the measures of the interior angles $= 180°(n-2)$, where n is the number of sides of the polygon.

Example 3: Find the sum of the measures of the interior angles of the following polygon:

Solution: The figure has 8 sides. Using the formula we have $180°\,(8-2) = 180°\,(6) = 1080°$

Using the formula, $180°(n-2)$, find the sum of the interior angles of the following figures.

1. 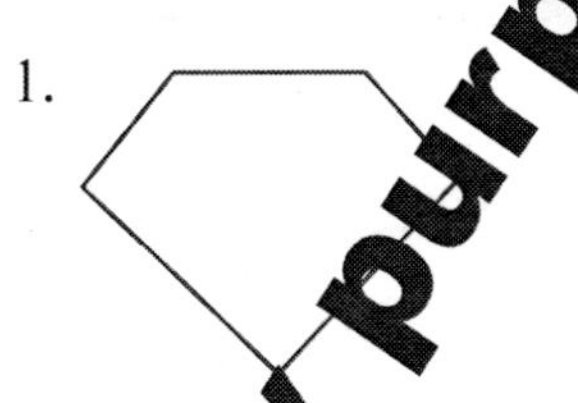

4. 7. 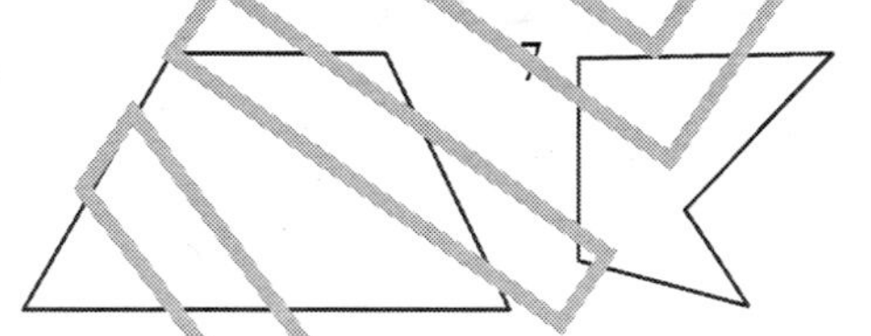

10. 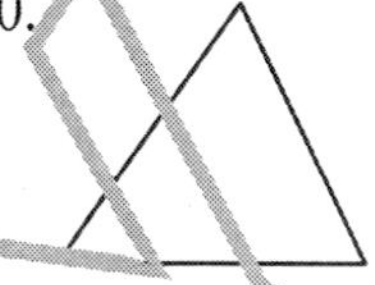

2. 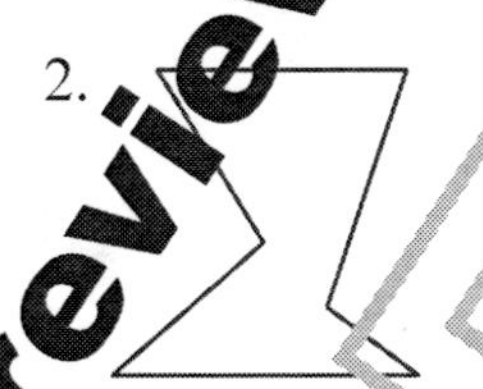

5. 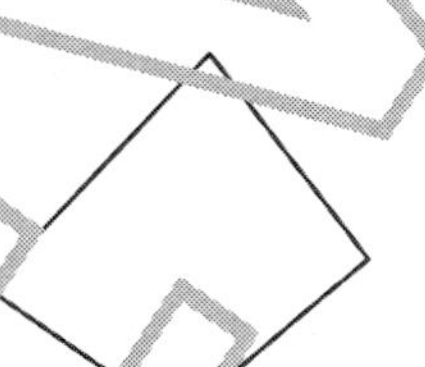

8. 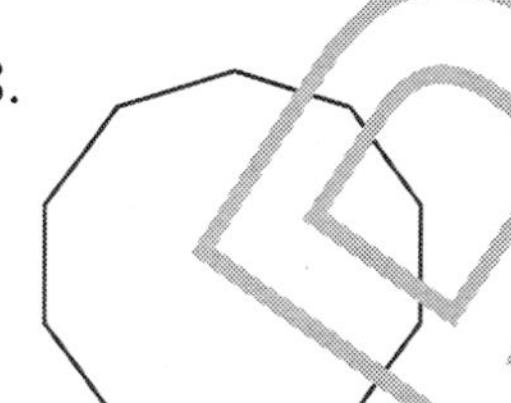

11. 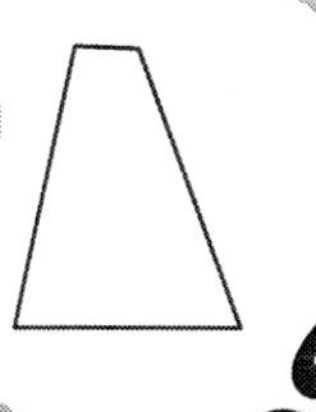

3. 

6.

9. 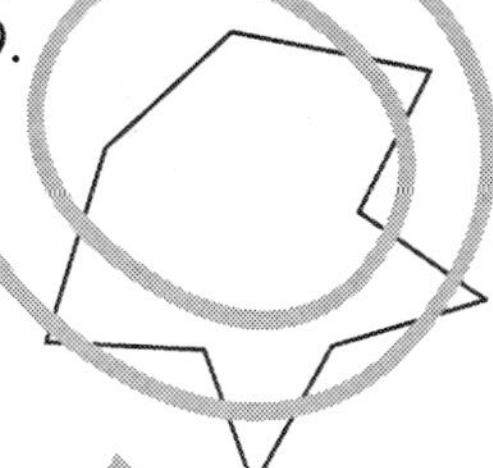

12. 

Find the measure of $\angle G$ in the regular polygons shown below. Remember that the sides of a regular polygon are equal.

13. 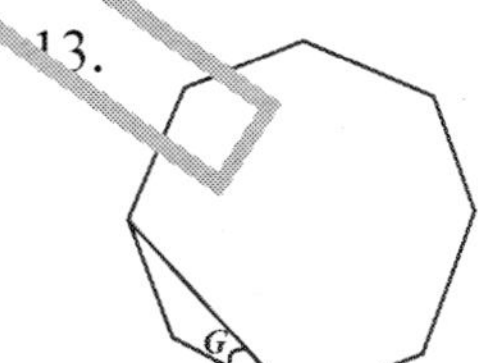

14. 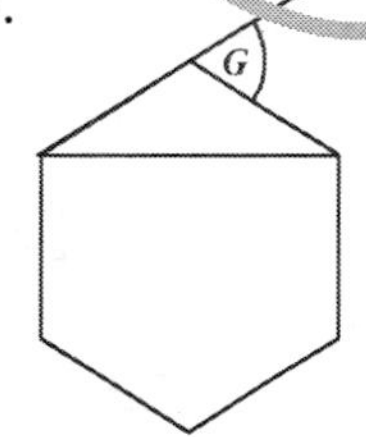

15. 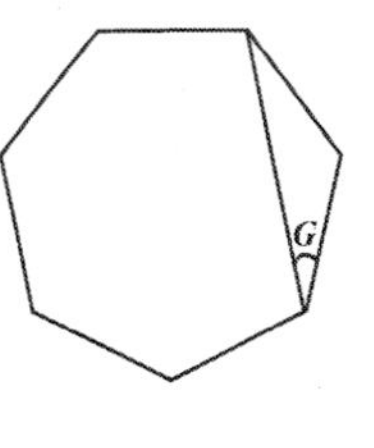

20.9 Congruent Figures

Two figures are **congruent** when they are exactly the same size and shape. If the corresponding sides and angles of two figures are congruent, then the figures themselves are congruent. For example, look at the two triangles below.

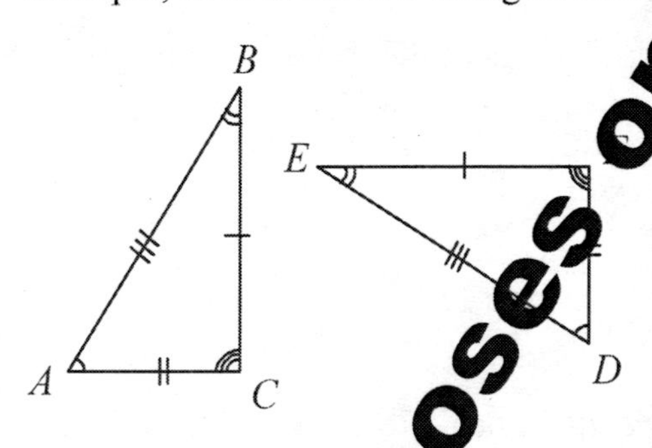

Compare the lengths of the sides of the triangles. The hash marks indicate that $\overline{AB}$ and $\overline{ED}$ have the same length. Therefore, they are congruent, which can be expressed as $\overline{AB} \cong \overline{ED}$. We can also see that $\overline{BC} \cong \overline{EF}$ and $\overline{AC} \cong \overline{FD}$. In other words, the corresponding sides are congruent. Now, compare the corresponding angles. The arc markings show that the corresponding angles have the same measure and are, therefore, congruent: $\angle A \cong \angle D$, $\angle B \cong \angle E$, and $\angle C \cong \angle F$. Because the corresponding sides and angles of the triangles are congruent, we say that the triangles are congruent: $\triangle ABC \cong \triangle DEF$.

Example 4: Decide whether the figures in each pair below are congruent or not.

PAIR 1

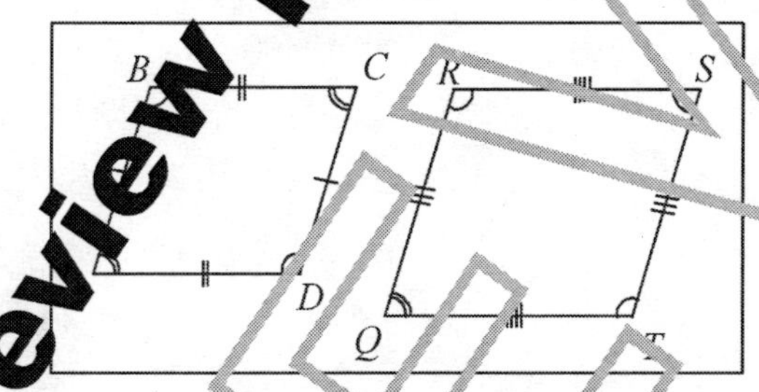

In Pair 1, the two parallelograms have congruent corresponding angles. However, because the corresponding sides of the parallelogram are not the same size, the figures are not congruent.

PAIR 2

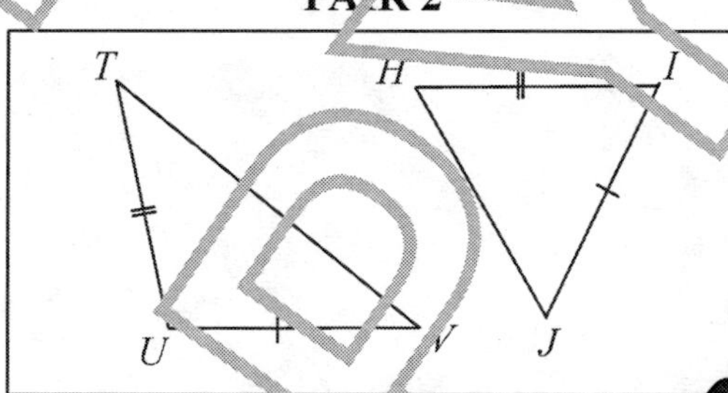

In Pair 2, the two triangles have two corresponding sides which are congruent. However, the hypotenuse of these triangles are not congruent (indicated by the lack of a triple hash mark).

PAIR 3

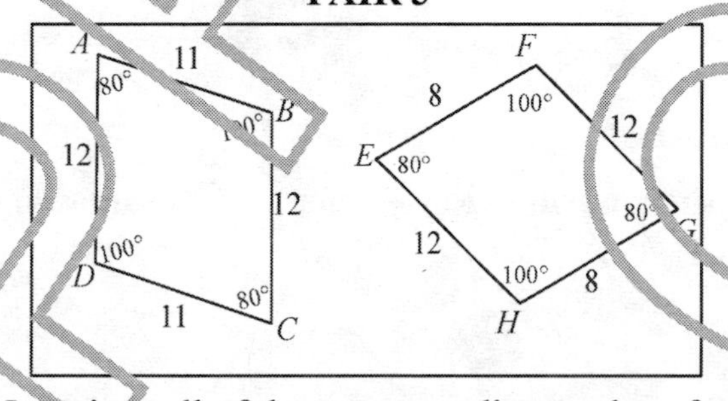

In Pair 3, all of the corresponding angles of these parallelograms are congruent; however, the corresponding sides are not congruent. Therefore, these figures are not congruent.

PAIR 4

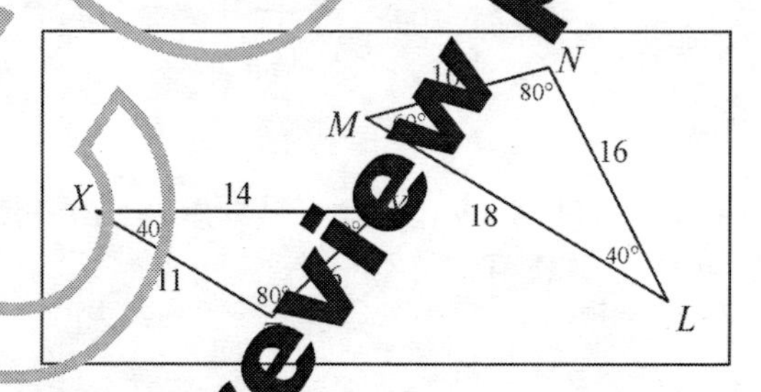

In Pair 4, the triangles share congruent corresponding angles, but the measures for all three corresponding sides of the triangles are not congruent. Therefore, the triangles are not congruent.

Examine the pairs of corresponding figures below. On the first line below the figures, write whether the figures are congruent or not congruent. On the second line, write a brief explanation of how you chose your answer.

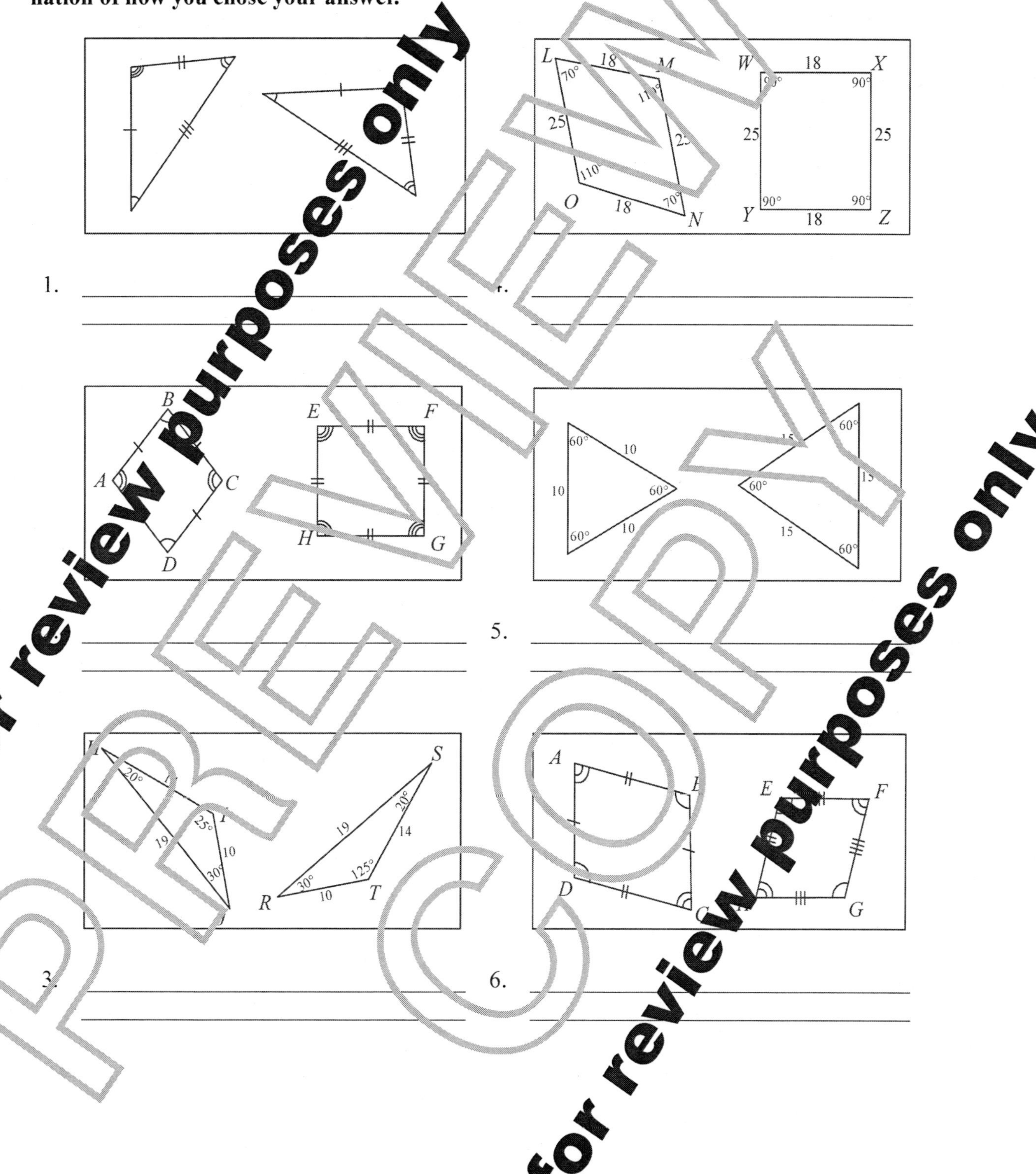

20.10 Similar and Congruent

Similar figures have the same shape but are two different sizes. Their corresponding sides are proportional. **Congruent figures** are exactly alike in size and shape and their corresponding sides are equal. See the examples below.

SIMILAR
CONGRUENT
Label each pair of figures below as either S if they are similar, C if they are congruent, or N if they are neither.
1.
2.
3.
4.
5.
6.
7.
8.
9.
10.
11.
12.
13.
14.

Chapter 20 Review

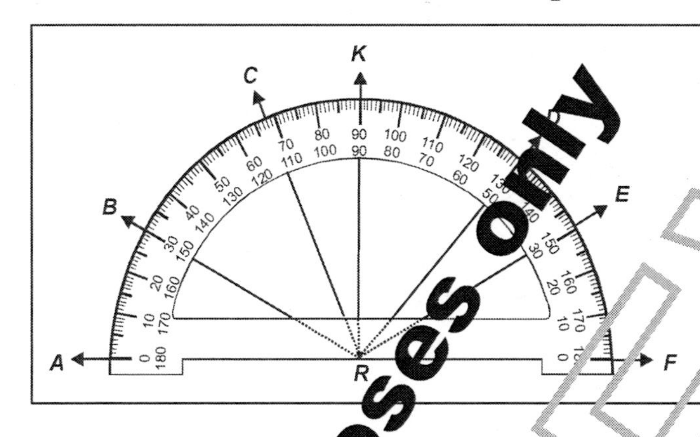

1. What is the measure of $\angle DRA$?

2. What is the measure of $\angle CRF$?

3. What is the measure of $\angle ARB$?

Use the following diagram for questions 4–14.

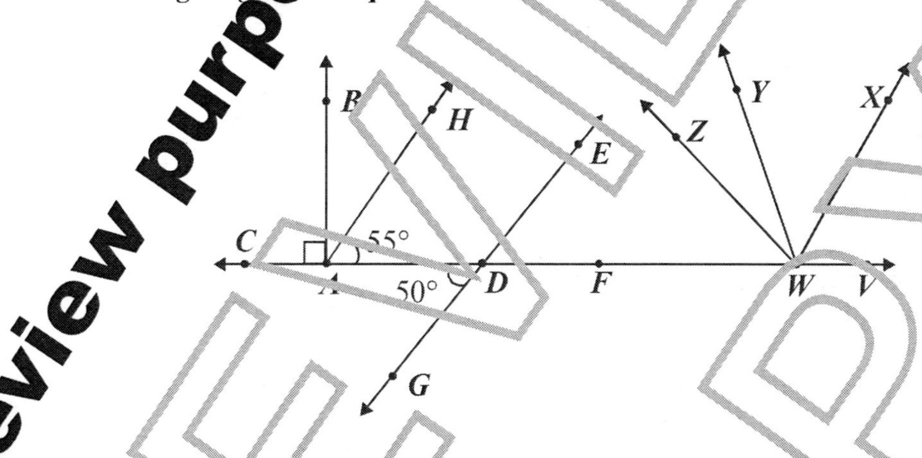

4. Which angle is a supplementary angle to $\angle EDF$?

5. What is the measure of $\angle CDF$?

6. Which two angles are right angles?

7. What is the measure of $\angle EDF$?

8. Which angle is adjacent to $\angle BAD$?

9. Which angle is a complementary angle to $\angle HAD$?

10. What is the measure of $\angle HAB$?

11. What is the measure of $\angle CAB$?

12. What kind of angle is $\angle FDA$?

13. What kind of angle is $\angle GDA$?

14. Which angles are adjacent to $\angle EDA$?

Look at the diagram below. For each pair of angles, state whether they are corresponding (C), alternate interior (I), alternate exterior (E), vertical (V), or supplementary (S) angles.

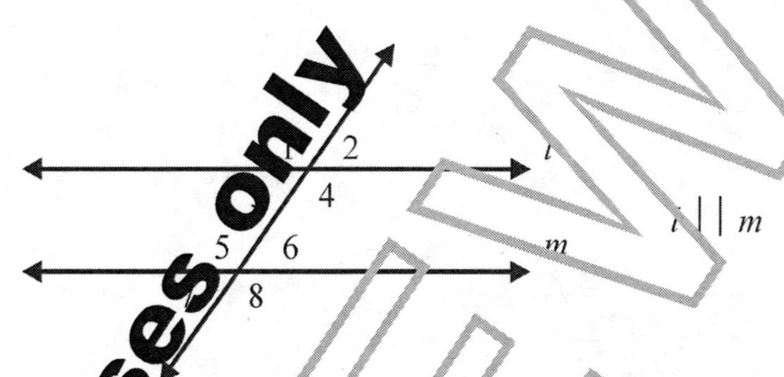

15. ∠1 and ∠4
16. ∠2 and ∠6
17. ∠1 and ∠3
18. ∠5 and ∠8
19. ∠5 and ∠7
20. ∠6 and ∠5
21. ∠2 and ∠7
22. ∠1 and ∠2
23. ∠4 and ∠5
24. ∠6 and ∠8
25. ∠3 and ∠6
26. ∠4 and ∠8
27. ∠1 and ∠5
28. ∠2 and ∠3

29. What is the sum of the measures of the interior angles in the figure below?

Chapter 21
Triangles

21.1 Types of Triangles

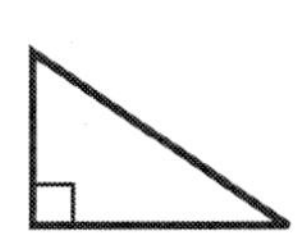

right triangle
contains 1 right ∠

acute triangle
all angles are acute
(less than 90°)

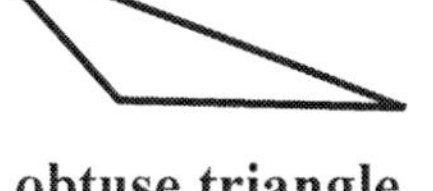

obtuse triangle
one angle is obtuse
(greater than 90°)

equilateral triangle
all three sides equal
all angles are 60°

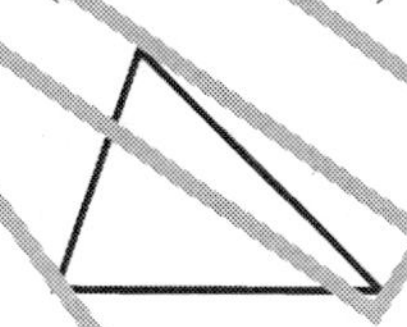

scalene triangle
no sides equal
no angles equal

isosceles triangle
two sides equal
two angles equal

21.2 Interior Angles of a Triangle

The three interior angles of a triangle always add up to 180°.

Example 1: Find the missing angle in the triangle.

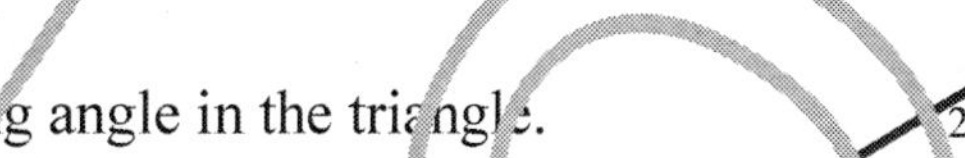

Solution:

$$\begin{aligned} 20^\circ + 125^\circ + x &= 180^\circ \\ -20^\circ\ -125^\circ \quad &\quad -20^\circ - 125^\circ \\ x &= 180^\circ - 20^\circ - 125^\circ \\ x &= 35^\circ \end{aligned}$$

Subtract 20° and 125° from both sides to get x by itself.

The missing angle is 35°.

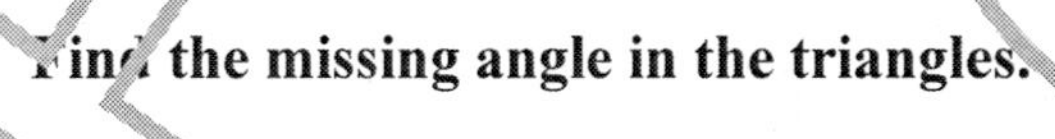

Find the missing angle in the triangles.

1. 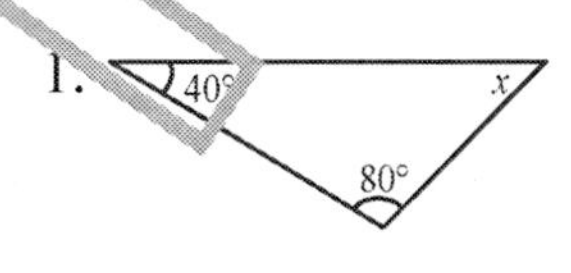

2. 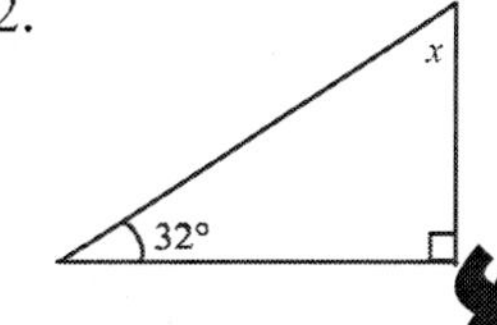

3. 55°, x, 25°

21.3 Similar Triangles

Two triangles are similar if the measurements of the three angles in both triangles are the same. If the three angles are the same, then their corresponding sides are proportional.

Corresponding Sides - The triangles below are similar. Therefore, the two shortest sides from each triangle, c and f, are corresponding. The two longest sides from each triangle, a and d, are corresponding. The two medium length sides, b and e, are corresponding.

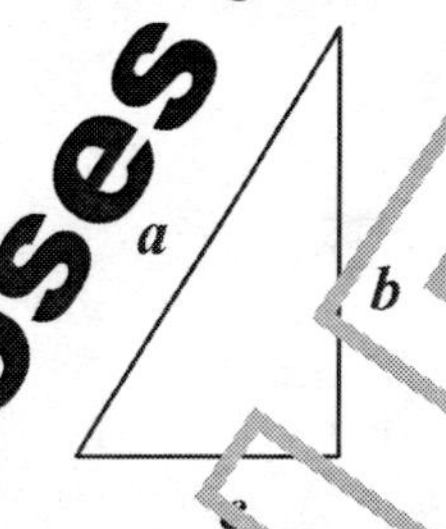

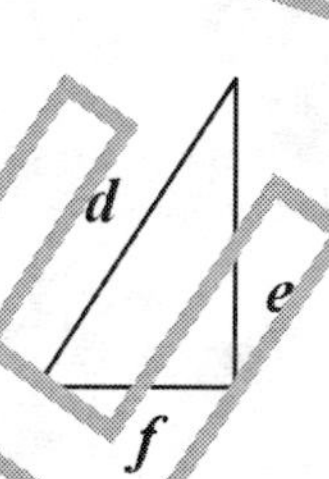

Proportional - The corresponding sides of similar triangles are proportional to each other. This means if we know all the measurements of one triangle, and we only know one measurement of the other triangle, we can figure out the measurements of the two other sides with proportion problems. The two triangles below are similar.

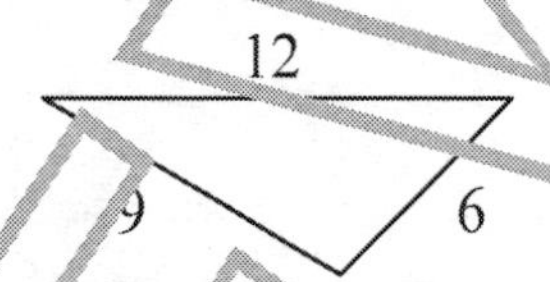

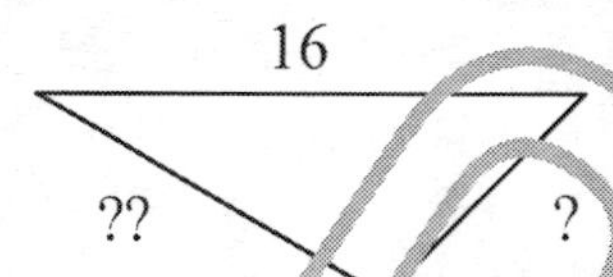

Note: To set up the proportion correctly, it is important to keep the measurements of each triangle on opposite sides of the equal sign.

To find the short side:
Step 1: Set up the proportion

$$\frac{\text{long side}}{\text{short side}} \quad \frac{12}{6} = \frac{16}{?}$$

Step 2: Solve the proportion. Multiply the two numbers diagonal to each other and then divide by the other number.

$16 \times 6 = 96$
$96 \div 12 = 8$

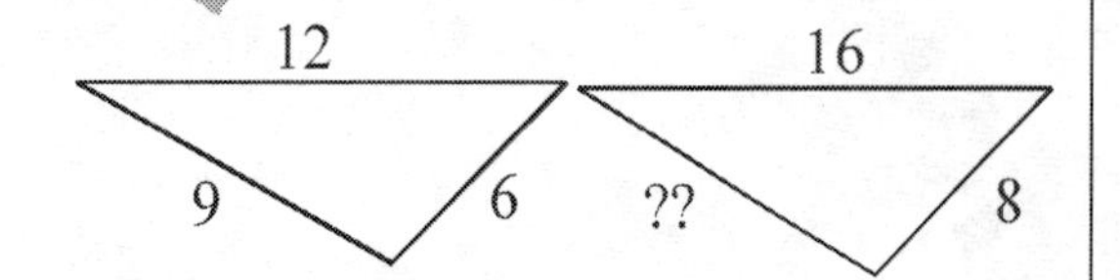

To find the medium length side:
Step 1: Set up the proportion

$$\frac{\text{long side}}{\text{medium}} \quad \frac{12}{9} = \frac{16}{??}$$

Step 2: Solve the proportion. Multiply the two numbers diagonal to each other and then divide by the other number.

$16 \times 9 = 144$
$144 \div 12 = 12$

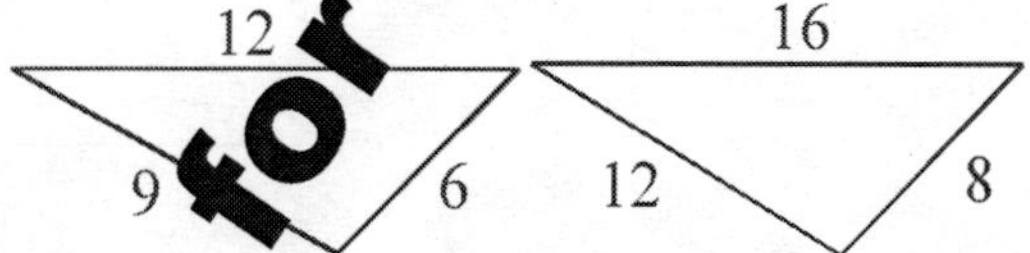

Find the missing side from the following similar triangles.

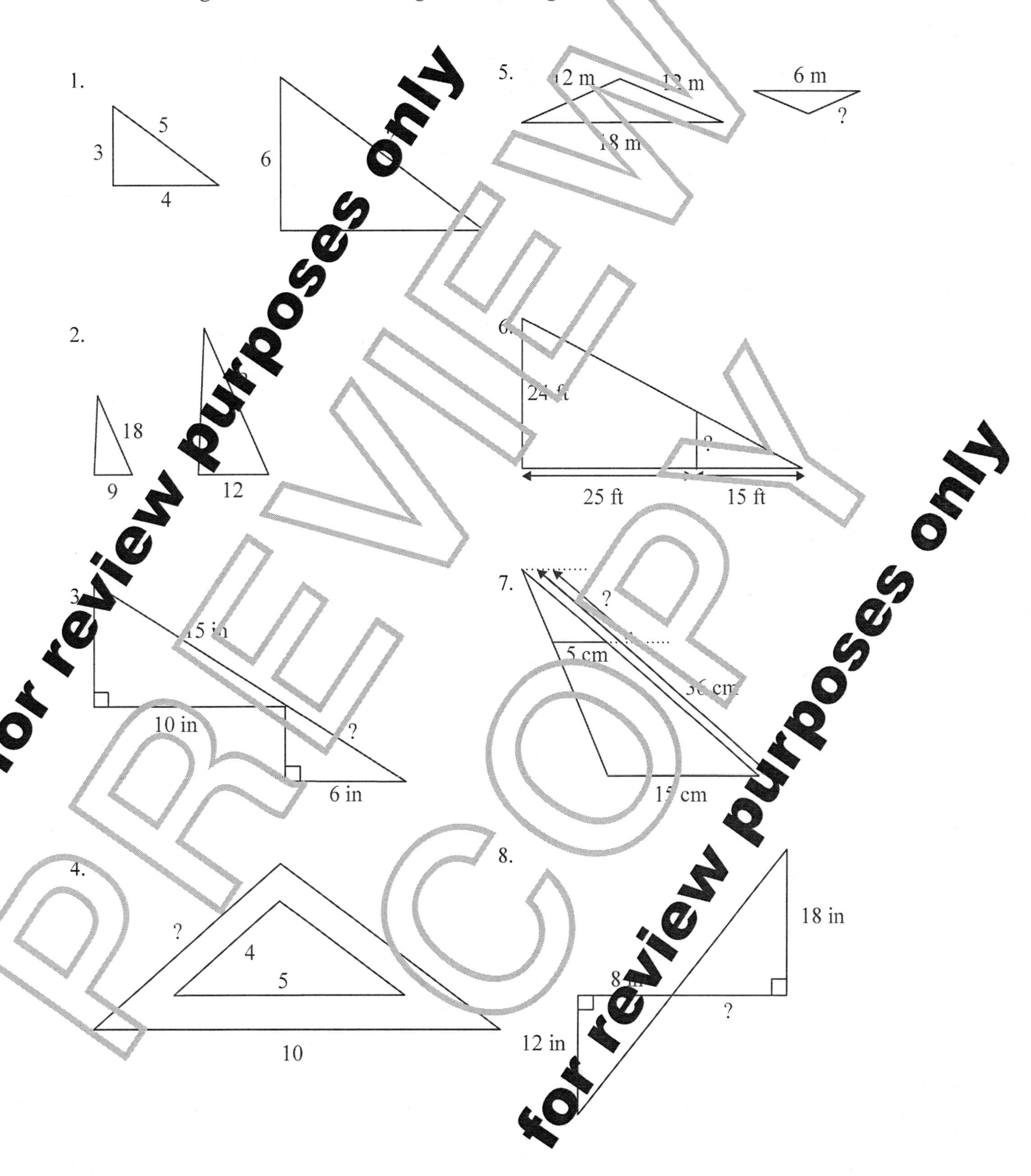

21.4 Side and Angle Relationships

Some triangles are congruent and some are similar. There are specific principles that determine whether are not two triangles are congruent or similar. Two triangles are congruent, denoted by the symbol $\cong$, if all of the corresponding sides and angles of each triangle are equal to each other.

Side-Side-Side (SSS) Congruence Postulate
If three sides of one triangle are congruent to three sides of another triangle then the two triangles are congruent.

If $\overline{AB} \cong \overline{xy}$, and $\overline{BC} \cong \overline{yz}$, and $\overline{AC} \cong \overline{xz}$, then $\triangle ABC \cong \triangle xyz$

Side-Angle-Side (SAS) Congruence Postulate
If two sides and the included angle of one triangle are congruent to two sides and the included angle of the second triangle, then the two triangles are congruent.

If $\overline{AB} \cong \overline{xy}$, and $\angle A \cong \angle x$, and $\overline{AC} \cong \overline{xz}$, then $\triangle ABC \cong \triangle xyz$

Angle-Side-Angle (ASA) Congruence Postulate
If two angles and the included side of one triangle are congruent to two angles and the included side of the second triangle, then the two triangles are congruent.

If $\angle B \cong \angle y$, and $\angle C \cong \angle z$, and $\overline{BC} \cong \overline{yz}$, then $\triangle ABC \cong \triangle xyz$

Angle-Angle-Side (AAS) Congruence Theorem
If two angles and a nonincluded side of one triangle are congruent to two angles and the corresponding side of a second triangle, then the two triangles are congruent.

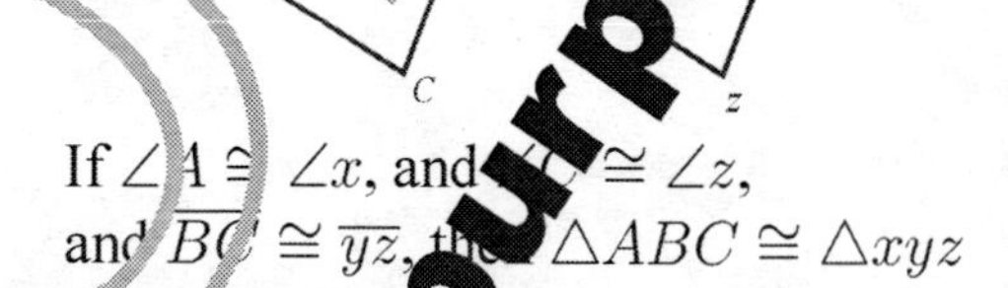

If $\angle A \cong \angle x$, and $\angle C \cong \angle z$, and $\overline{BC} \cong \overline{yz}$, then $\triangle ABC \cong \triangle xyz$

A triangle is an **equilateral triangle** if all of its sides and angles are equal. A triangle is an **isosceles triangle** if two of its sides and the angles opposite those sides are equal. A triangle is a **right triangle** if one of its angles equals 90°.

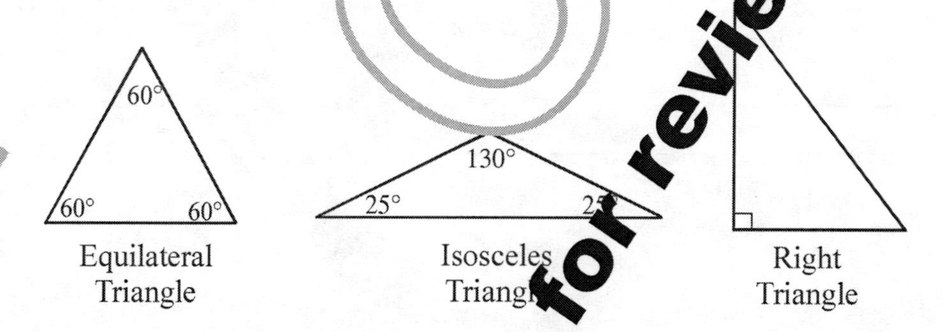

Equilateral Triangle

Isosceles Triangle

Right Triangle

21.5 Pythagorean Theorem

Pythagoras was a Greek mathematician and philosopher who lived around 600 B.C. He started a math club among Greek aristocrats called the Pythagoreans. Pythagoras formulated the **Pythagorean Theorem** which states that in a **right triangle**, the sum of the squares of the legs of the triangle are equal to the square of the hypotenuse. Most often you will see this formula written as $a^2 + b^2 = c^2$. **This relationship is only true for right triangles.**

Example 2: Find the length of side c.

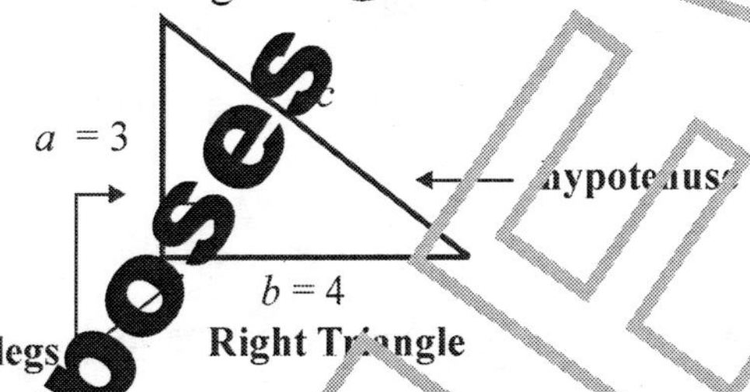

Formula:
$$a^2 + b^2 = c^2$$
$$3^2 + 4^2 = c^2$$
$$9 + 16 = c^2$$
$$25 = c^2$$
$$\sqrt{25} = \sqrt{c^2}$$
$$5 = c$$

Find the hypotenuse of the following triangles. Round the answers to two decimal places.

1.

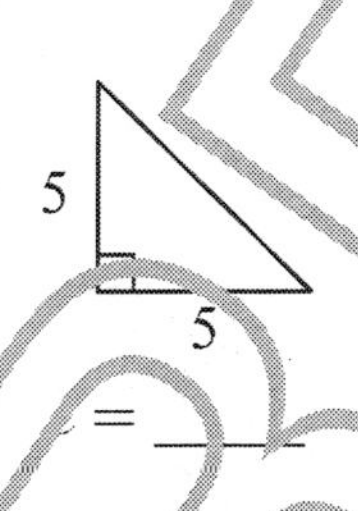

$c =$ _____

4.

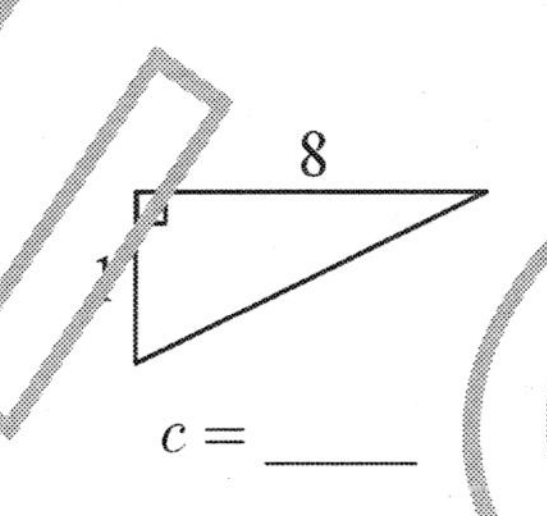

$c =$ _____

7.

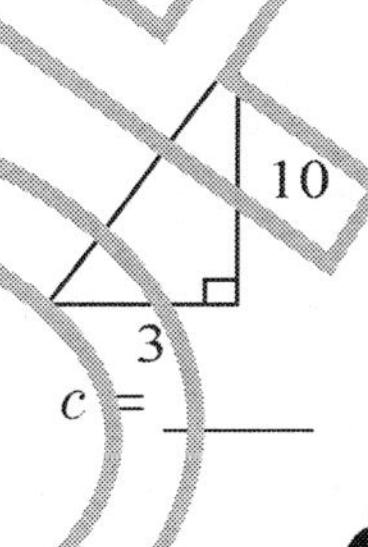

$c =$ _____

2.

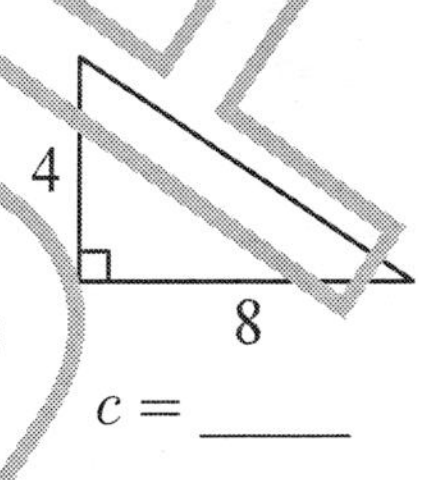

$c =$ _____

5.

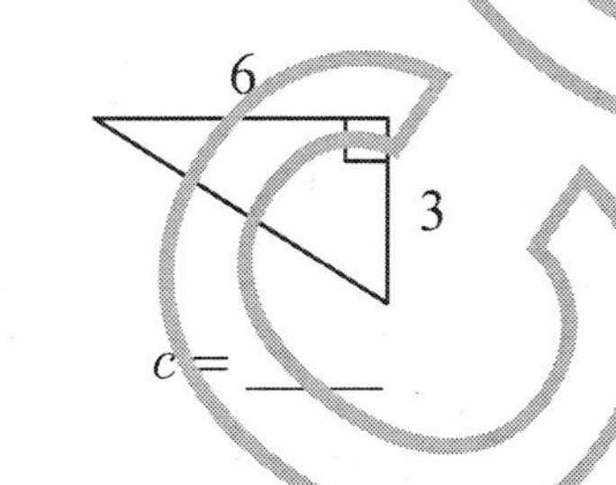

$c =$ _____

8.

$c =$ _____

3.

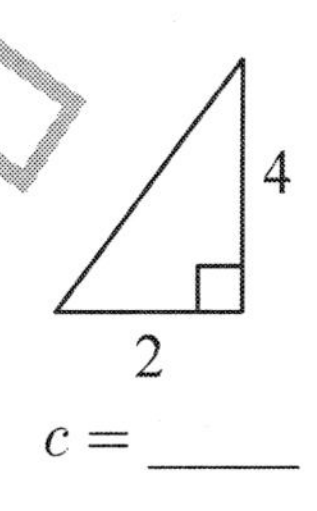

$c =$ _____

6.

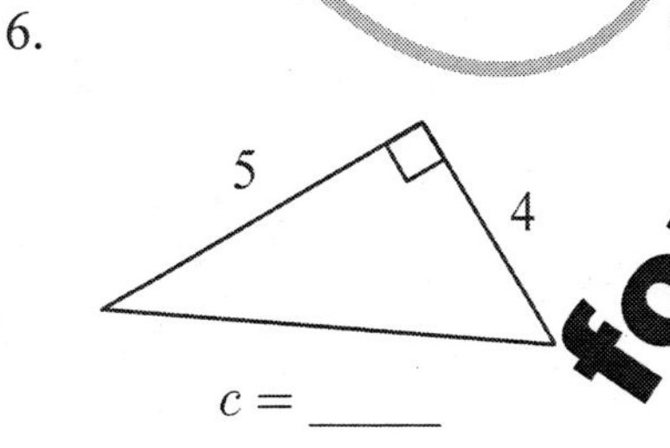

$c =$ _____

9.

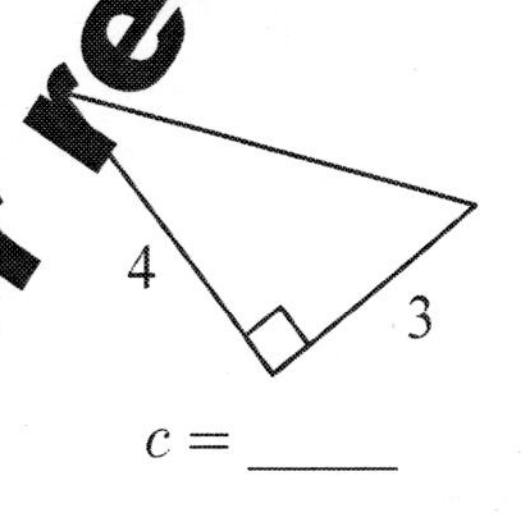

$c =$ _____

21.6 Finding the Missing Leg of a Right Triangle

In some triangles, we know the measurement of the hypotenuse as well as one of the legs. To find the measurement of the other leg, use the Pythagorean theorem by filling in the known measurements, and then solve for the unknown side.

Example 3: Find the measure of b.

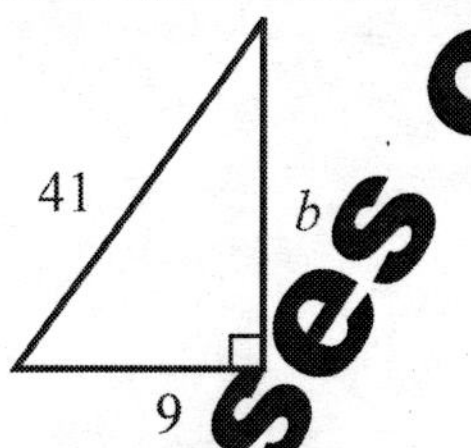

In the formula, $a^2 + b^2 = c^2$, a and b are the legs and c is always the hypotenuse.

$9^2 + b^2 = 41^2$
$81 + b^2 = 1681$
$b^2 = 1681 - 81$
$b^2 = 1600$
$\sqrt{b^2} = \sqrt{1600}$
$b = 40$

Practice finding the measure of the missing leg in each right triangle below. Simplify square roots.

1.

____

2.

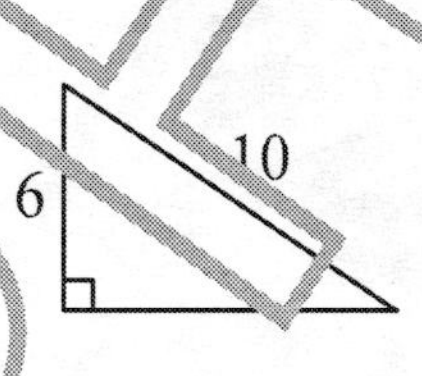

____

3.

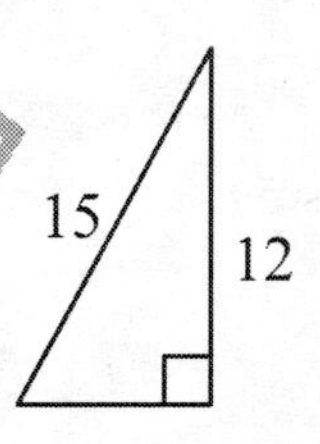

____

4.

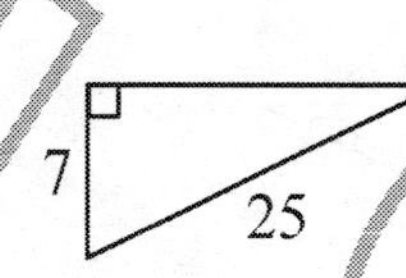

____

5.

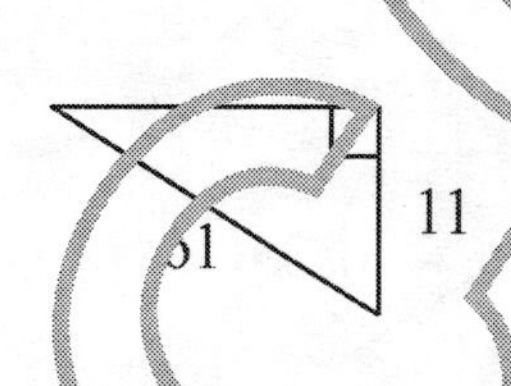

____

6.

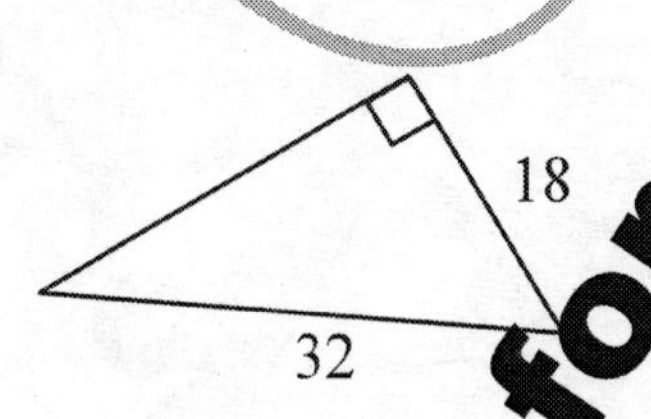

____

7.

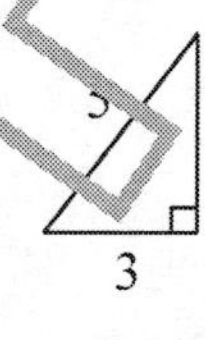

____

8.

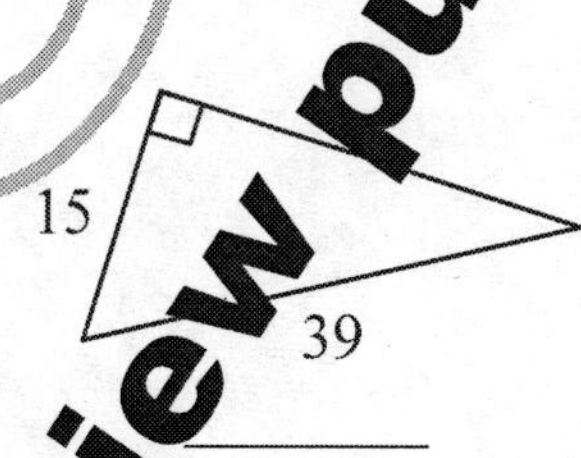

____

9.

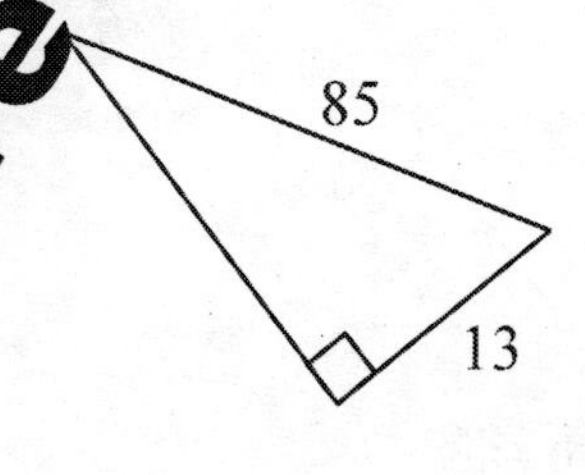

____

21.7 Applications of the Pythagorean Theorem

The Pythagorean Theorem can be used to determine the distance between two points in some situations. Recall that the formula is written $a^2 + b^2 = c^2$.

Example 4: Find the distance between point B and point A given that the length of each square is 1 inch long and 1 inch wide.

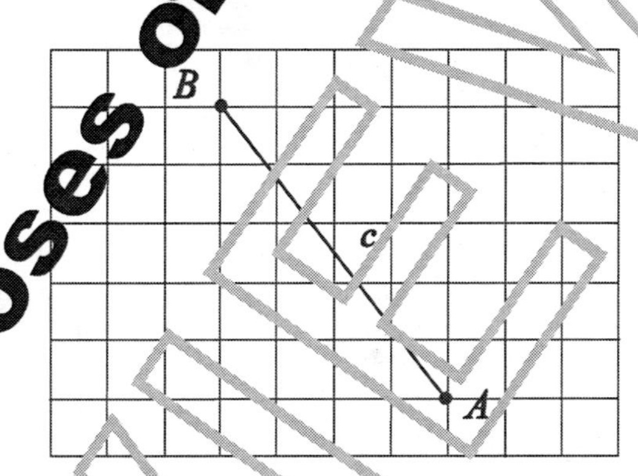

Step 1: Draw a straight line between the two points. We will call this side c.

Step 2: Draw two more lines, one from point B and one from point A. These lines should make a $90°$ angle. The two new lines will be labeled a and b. Now we can use the Pythagorean Theorem to find the distance from Point B to Point A.

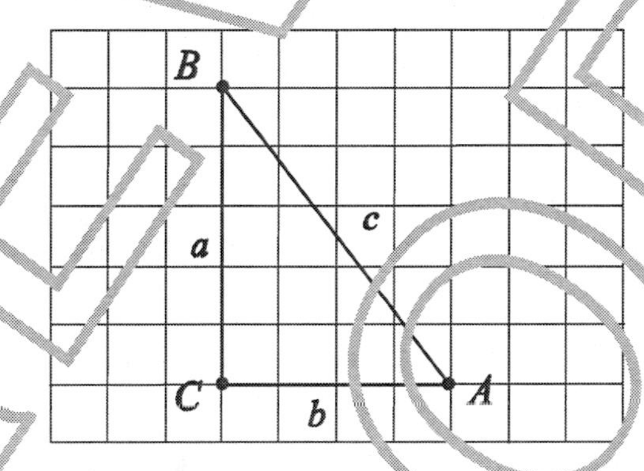

Step 3: Find the length of a and b by counting the number of squares each line has. We find that $a = 5$ inches and $b = 4$ inches. Now, substitute the values found into the Pythagorean Theorem.

$$
\begin{aligned}
a^2 + b^2 &= c^2 \\
5^2 + 4^2 &= c^2 \\
25 + 16 &= c^2 \\
41 &= c^2 \\
\sqrt{41} &= \sqrt{c^2} \\
\sqrt{41} &= c
\end{aligned}
$$

Use the Pythagorean Theorem to find the distances asked. Round your answers to two decimal points.

Below is a diagram of the mall. Use the grid to help answer questions 1 and 2. Each square is 25 feet × 25 feet.

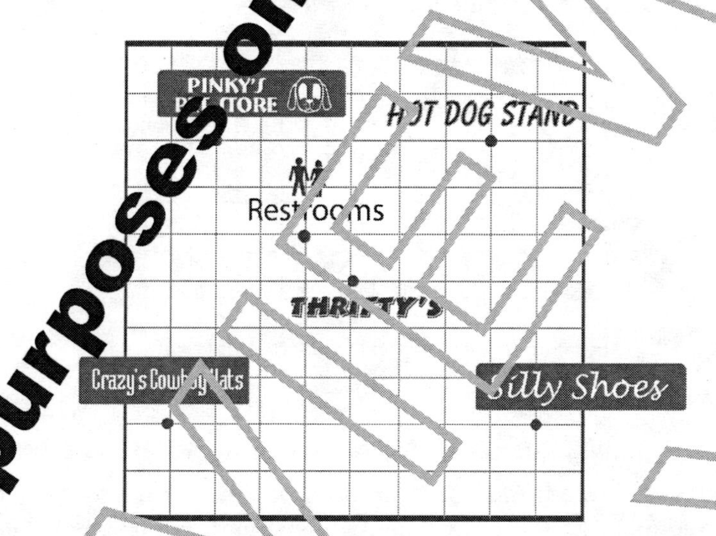

1. Marty walks from Pinky's Pet Store to the restroom to wash his hands. How far did he walk?
2. Betty needs to meet her friend at Silly Shoes, but she wants to get a hot dog first. If Betty is at Thrifty's, how far will she walk to meet her friend?

Below is a diagram of a football field. Use the grid on the football field to help find the answers to questions 3 and 4. Each square is 10 yards × 10 yards.

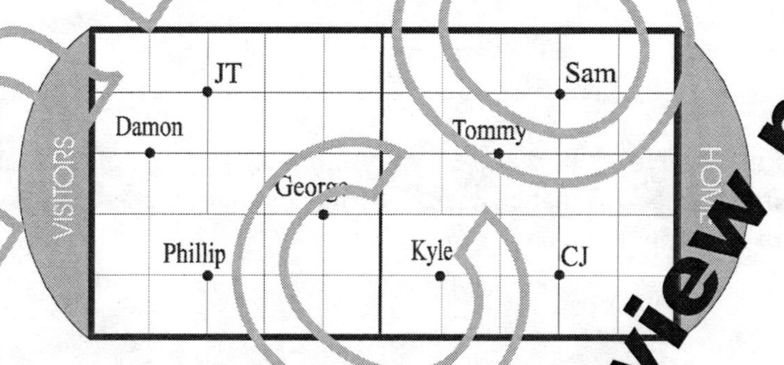

3. George must throw the football to a teammate before he is tackled. If CJ is the only person open, how far must George be able to throw the ball?
4. Damon has the football and is about to make a touchdown. If Phillip tries to stop him, how far must he run to reach Damon?

21.8 Special Right Triangles

Two right triangles are special right triangles if they have fixed ratios among their sides.

45-45-90 Triangles

In a 45-45-90 triangle, the two sides opposite the 45° angles will always be equal. The length of the hypotenuse is $\sqrt{2}$ times the length of one of the sides opposite a 45° angle.

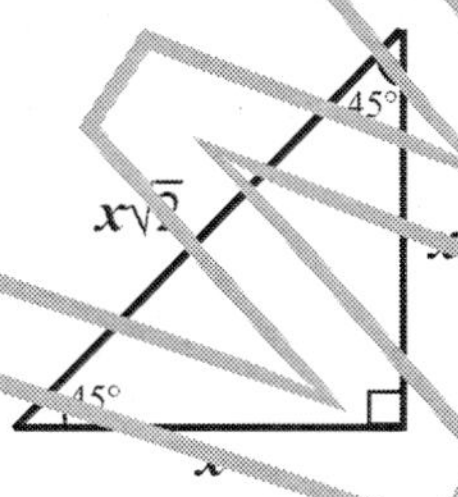

Example 5: What are the lengths of sides a and b?

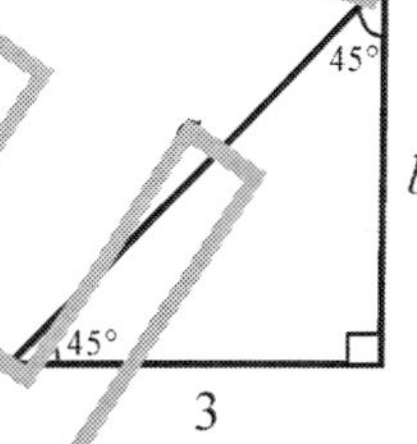

Step 1: The two sides opposite the 45° angles are equal. Therefore, side $b = 3$.

Step 2: The hypotenuse is $\sqrt{2}$ times the length of a side opposite a 45° angle.
Therefore, $a = 3 \times \sqrt{2}$
Simplify: $a = 3\sqrt{2}$

30-60-90 Triangles

In a 30-60-90 triangle, the side opposite the 30° angle is the shortest leg. The side opposite the 60° angle is $\sqrt{3}$ times as long as the shortest leg, and the hypotenuse is twice as long as the shortest leg.

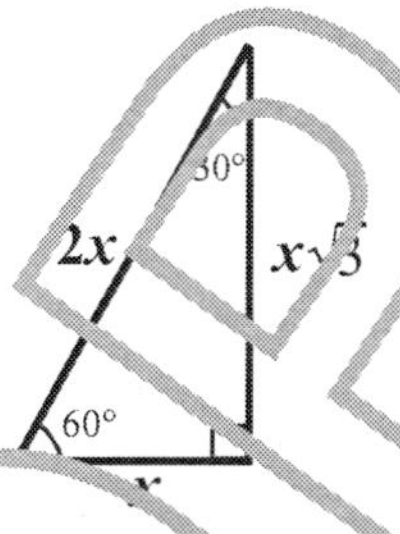

Example 6: What are the lengths of sides a and b?

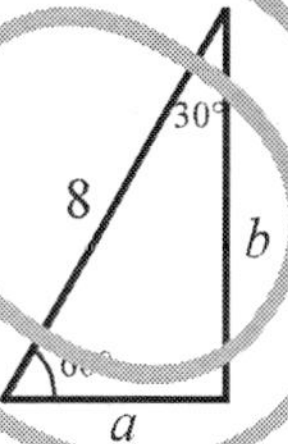

Step 1: The hypotenuse is 2 times the side opposite the 30° angle.
Write the above statement using algebra and then solve.

$8 = 2a$

$\frac{8}{2} = \frac{\not{2}a}{\not{2}}$

$4 = a$

Step 2: Now that it is known that the shortest leg has a length of 4, the side opposite the 60° angle can be calculated.

$b = a \times \sqrt{3}$

$b = 4 \times \sqrt{3}$

$b = 4\sqrt{3}$

Find the missing leg of each of the special right triangles. Simplify your answers.

Find the lengths of sides *a* and *b* in each of the special right triangles.

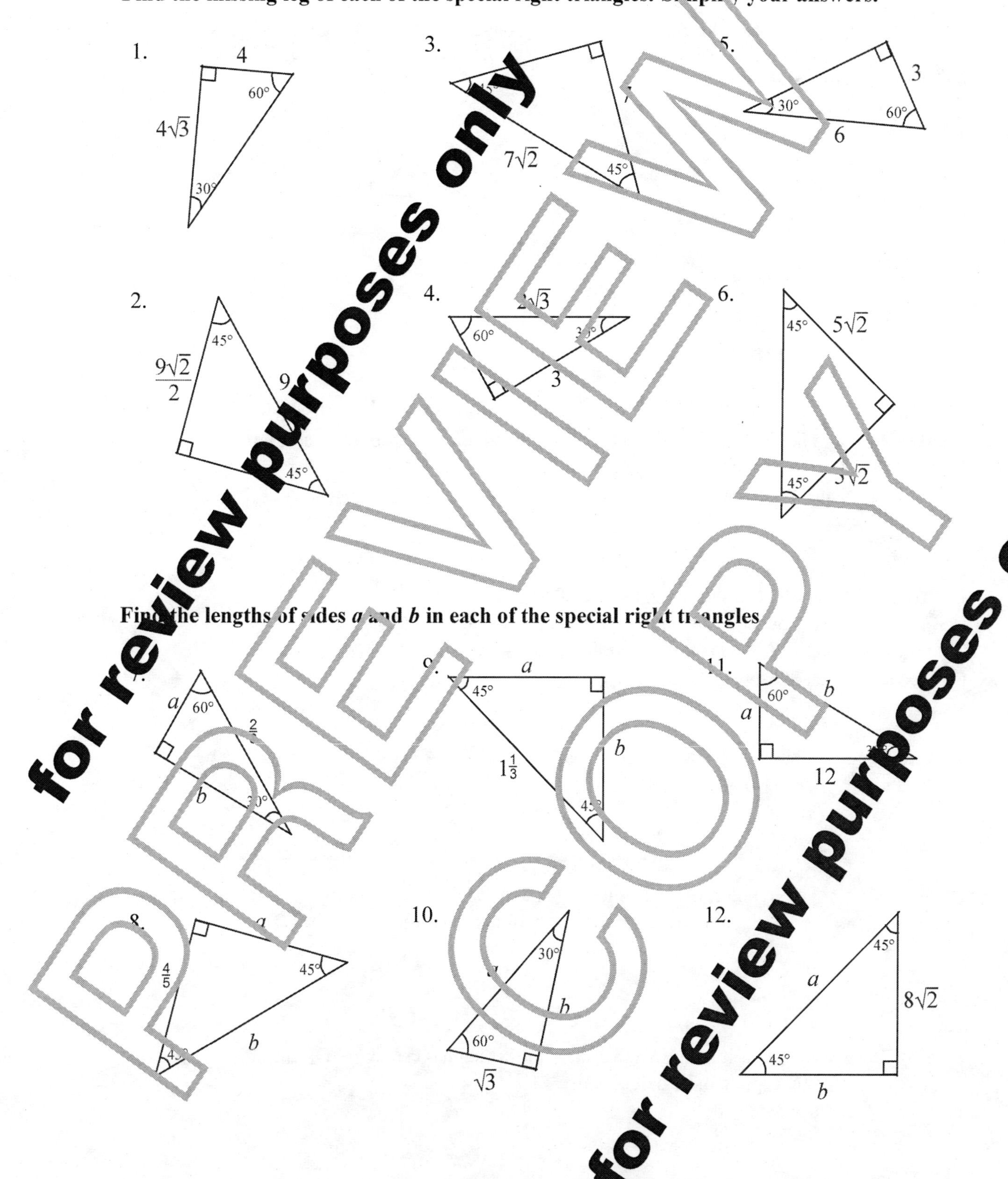

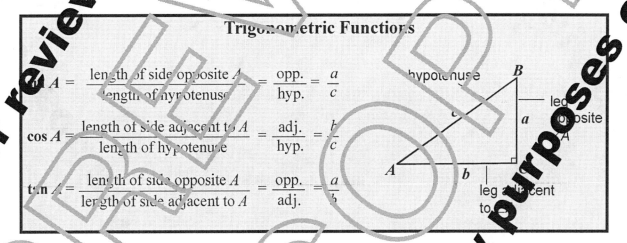

21.9 Introduction to Trigonometric Ratios

Trigonometry is a mathematical topic that applies the relationships between sides and angles in right triangles. Recall that a right triangle has one 90° angle and two acute angles. Consider the right triangle shown below. Note that the angles are labeled with capital letters. The sides are labeled with lowercase letters that correspond to the angles opposite them.

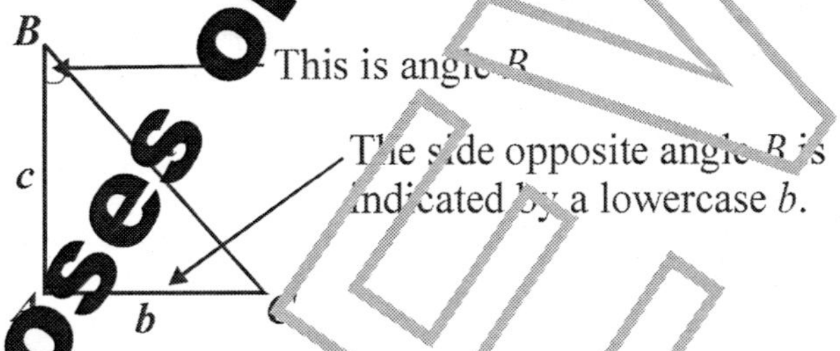

Trigonometric ratios are ratios of the measures of two sides of a right triangle and are related to the acute angles of a right triangle, not the right angle. The value of a trigonometric ratio is dependent only on the size of the acute angle and is not affected by the lengths of the sides of the triangle.

We will consider the three basic trigonometric ratios in this section: **sine, cosine, and tangent**.

Definitions and descriptions of the sine, cosine, and tangent functions are presented below.

Trigonometric Functions

$$\sin A = \frac{\text{length of side opposite } A}{\text{length of hypotenuse}} = \frac{\text{opp.}}{\text{hyp.}} = \frac{a}{c}$$

$$\cos A = \frac{\text{length of side adjacent to } A}{\text{length of hypotenuse}} = \frac{\text{adj.}}{\text{hyp.}} = \frac{b}{c}$$

$$\tan A = \frac{\text{length of side opposite } A}{\text{length of side adjacent to } A} = \frac{\text{opp.}}{\text{adj.}} = \frac{a}{b}$$

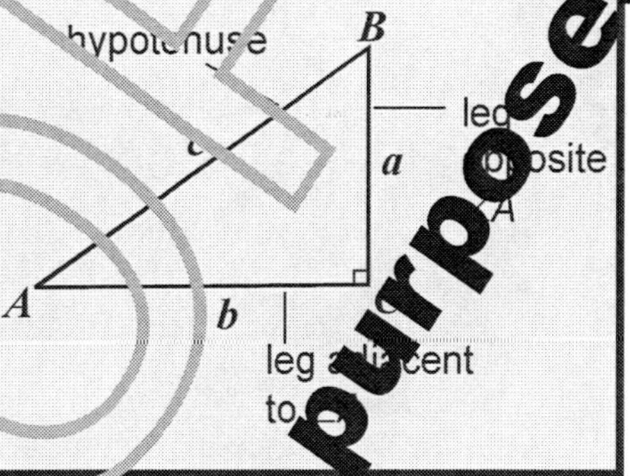

Example 7: For right triangle ABC, find $\sin A$, $\cos A$, $\tan A$, $\sin C$, $\cos C$, and $\tan C$.

$$\sin A = \frac{\text{opp.}}{\text{hyp.}} = \frac{3}{5} = 0.6 \qquad \sin C = \frac{\text{opp.}}{\text{hyp.}} = \frac{4}{5} = 0.8$$

$$\cos A = \frac{\text{adj.}}{\text{hyp.}} = \frac{4}{5} = 0.8 \qquad \cos C = \frac{\text{adj.}}{\text{hyp.}} = \frac{3}{5} = 0.6$$

$$\tan A = \frac{\text{opp.}}{\text{adj.}} = \frac{3}{4} = 0.75 \qquad \tan C = \frac{\text{opp.}}{\text{adj.}} = \frac{4}{3} = 1.\overline{3}$$

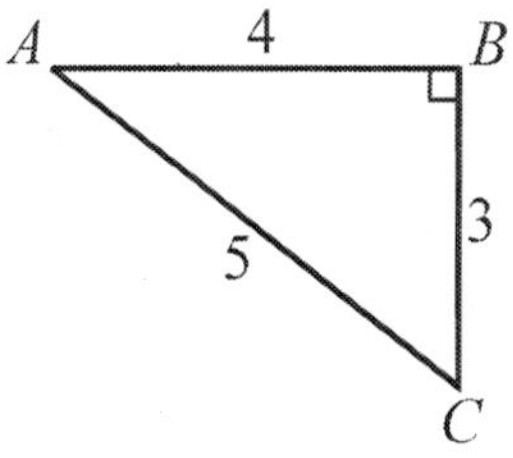

Find $\sin A$, $\cos A$, $\tan A$, $\sin B$, $\cos B$, and $\tan B$ in each of the following right triangles. Express answers as fractions and as decimals rounded to three decimal places.

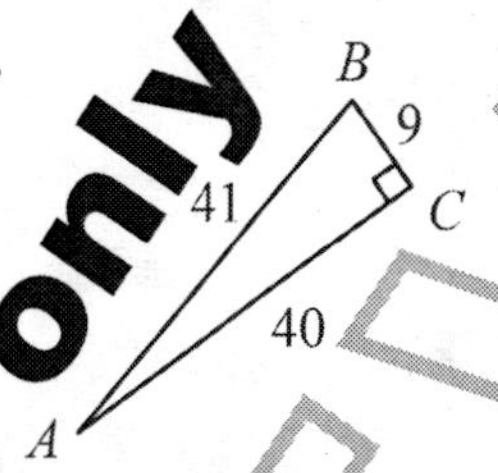

1.

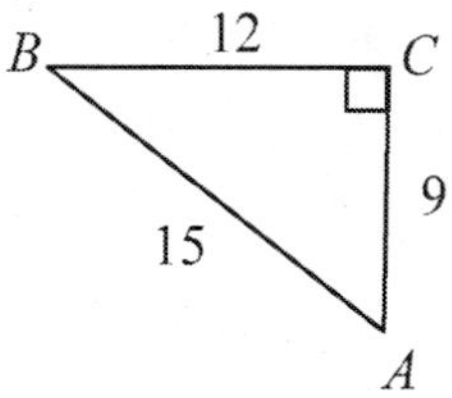

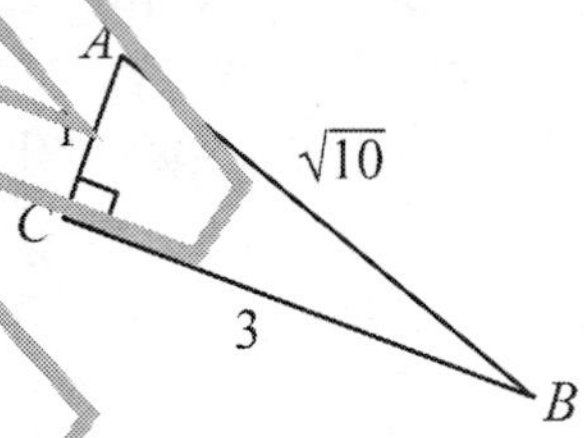

2.

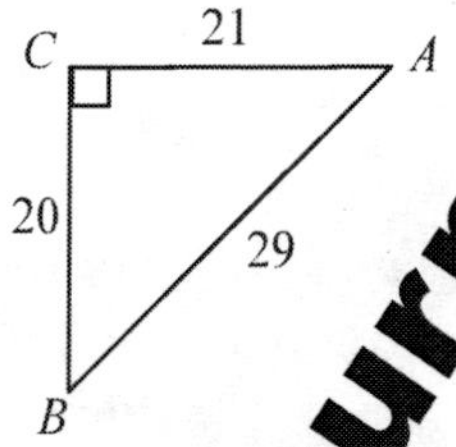

3.

4.

5.

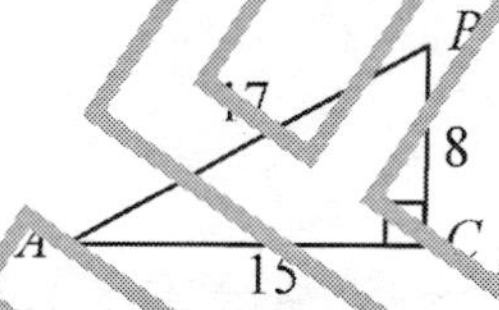

6.

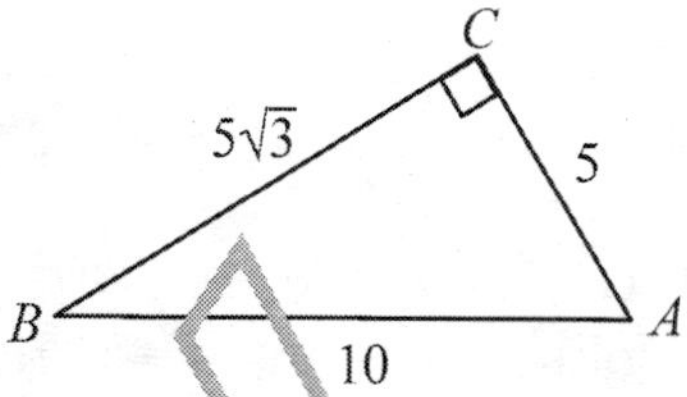

Once the values of the trigonometric ratios are found, then the measures of the angles within the triangle can be found using the arcsine and arccosine. The arcsine and arccosine can also be written as $\sin^{-1}$ and $\cos^{-1}$. The arc functions' identities can be defined as:

$$\begin{aligned} \arcsin(\sin(A)) &= A \\ \arccos(\cos(A)) &= A \\ \arctan(\tan(A)) &= A \end{aligned}$$

Example 8: For right triangle ABC, where $\sin A = 0.6$, find the measures of angles A and C. Round to the nearest whole number.

Step 1: Using your calculator, arcsine is $\sin^{-1}$.**
$\sin A = 0.6$
$\sin^{-1}(\sin A) = \sin^{-1}(0.6)$ Take the arcsine of both sides.
$A = 37°$

Step 2: Since all the angles in a triangles add up to 180°, and $A = 37°$ and $B = 90°$, then
$A + B + C = 180°$
$37° + 90° + C = 180°$
$C = 180° - 37° - 90° = 53°$
Therefore, $A = 37°$, $B = 90°$, and $C = 53°$

** To find the arcsine or arccosine using a TI-83 calculator, you must press the 2nd button, then press the trig function, SIN, COS, or TAN. After this, $\sin^{-1}$, $\cos^{-1}$, or $\tan^{-1}$ will appear on the screen and you can enter the trig ratio, such as 0.6 from the example above. When finding an angle using a TI-83, you must always remember to be in degree mode. To check this, press MODE and make sure that Degree is highlighted, not Radian. If Degree is not highlighted, then go down and over to Degree and press ENTER. This will highlight Degree. To get out of this menu, hit 2nd, then MODE. In other simpler scientific calculators, you might have to type the trig ratio, 0.6, first, then type 2nd SIN, 2nd COS, or 2nd TAN. If there isn't a 2nd button, you will have to the inverse button and then the SIN, COS, or TAN button. The inverse button is usually abbreviated INV.

Find the measures of the angles given the trigonometric function. Round your answers to the nearest degree.

1. $\sin A = 0.4$
2. $\tan x = 1$
3. $\sin b = 0.7$
4. $\cos C = \frac{\sqrt{2}}{2}$
5. $\tan A = -1.5$
6. $\cos y = -1$
7. $\sin B = -0.6$
8. $\cos A = 0$
9. $\tan z = 2.6$
10. $\tan c = 59$
11. $\sin x = \frac{\sqrt{2}}{2}$
12. $\cos x = 0.1$
13. $\tan y = 0$
14. $\cos a = -0.4$
15. $\sin C = 1$

Example 9: Find the values of the sine, cosine, and tangent functions of both acute angles in the right triangle ABC shown below.

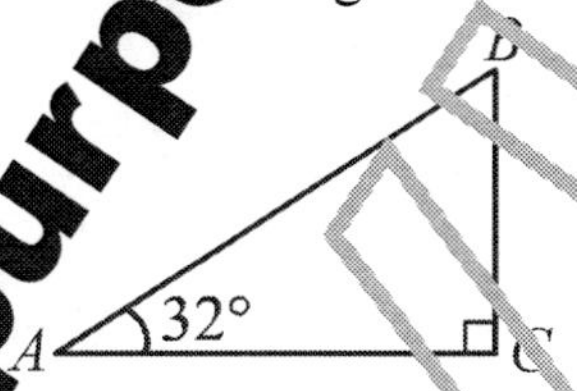

Step 1: Find the third angle.
$A + B + C = 180°$
$32° + B + 90° = 180°$
$B + 90° = 180° - 32° - 90° = 58°$

Step 2: Plug the angle values into $\sin A$, $\cos A$, $\tan A$, $\sin B$, $\cos B$, and $\tan B$.

$\sin A = \sin 32° = \mathbf{0.5299}$ $\quad$ $\sin B = \sin 58° = \mathbf{0.8480}$
$\cos A = \cos 32° = \mathbf{0.8480}$ $\quad$ $\cos B = \cos 58° = \mathbf{0.5299}$
$\tan A = \tan 32° = \mathbf{0.6249}$ $\quad$ $\tan B = \tan 58° = \mathbf{1.6003}$

Find $\sin A$, $\cos A$, $\tan A$, $\sin B$, $\cos B$, and $\tan B$ in each of the following right triangles. Express answers as decimals rounded to three decimal places.

1.
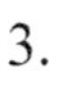

3.
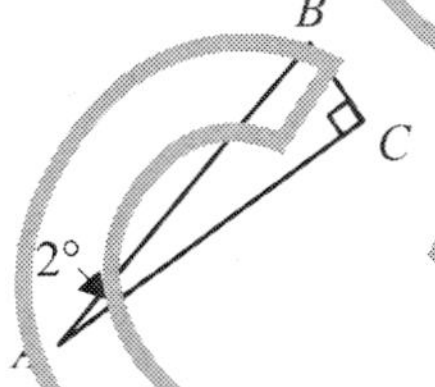

5.

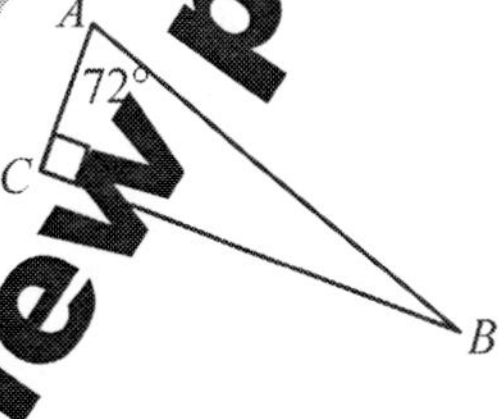

2.

4.
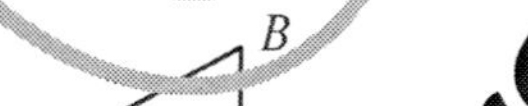
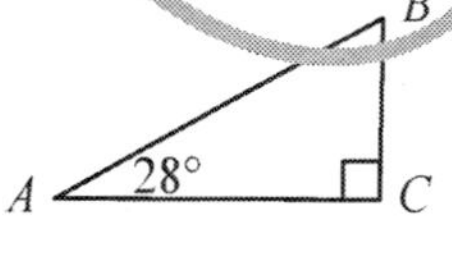

6.
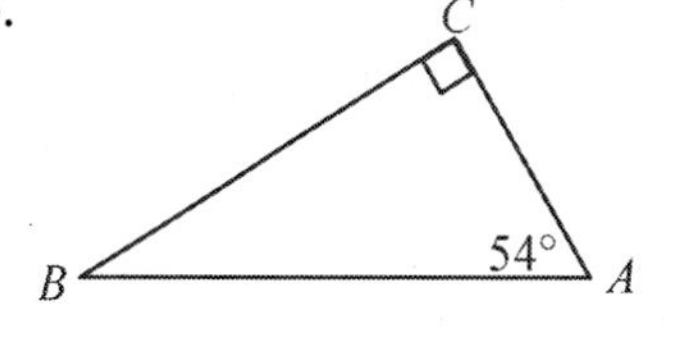

When given one acute angle and one side of a right triangle, the other angle and two sides can be found using trigonometric functions.

Example 10: Find the third angle and the other two sides of the triangle.

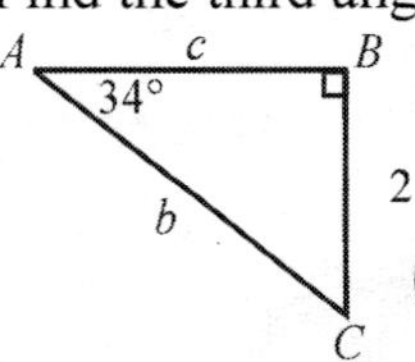

Step 1: Find the third angle. Since all the angles in a triangle add up to 180°, and $A = 34°$ and $B = 90°$, then
$A + B + C = 180°$
$34° + 90° + C = 180°$
$C = 180° - 34° - 90° = 56°$

Step 2: Find the missing sides. This can be done several different ways: using sine, cosine, or tangent. We are going to use sine to find b and tangent to find c.

$$\begin{aligned} \sin A &= \frac{\text{opp.}}{\text{hyp.}} & \tan A &= \frac{\text{opp.}}{\text{adj.}} \\ \sin 34° &= \frac{2}{b} & \tan 34° &= \frac{2}{c} \\ 0.5592 &= \frac{2}{b} & 0.6745 &= \frac{2}{c} \\ 0.5592b &= 2 & 0.6745c &= 2 \\ \frac{0.5592b}{0.5592} &= \frac{2}{0.5592} & \frac{0.6745c}{0.6745} &= \frac{2}{0.6745} \\ b &= 3.58 & c &= 2.97 \end{aligned}$$

$C = 56°$, $b = 3.58$, and $c = 2.97$

Note: After you have calculated the second side using one of the trigonometric ratios, you can use the Pythagorean Theorem to find the third side.
$3.58^2 = 2^2 + c^2 \longrightarrow c^2 = 3.58^2 - 2^2 \longrightarrow c^2 = 8.8164 \longrightarrow c = 2.97$

Find the missing sides and angles using the information given.

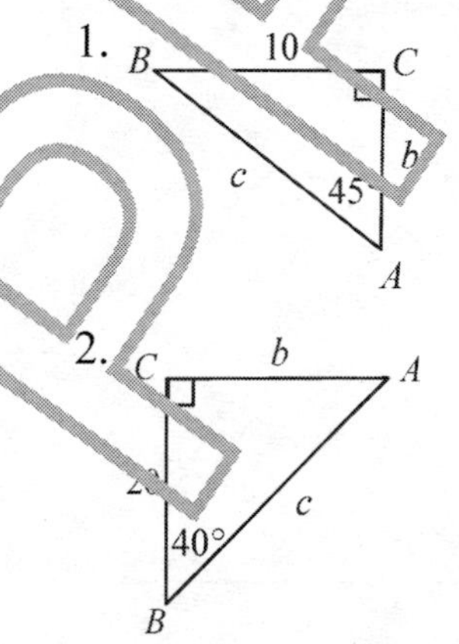

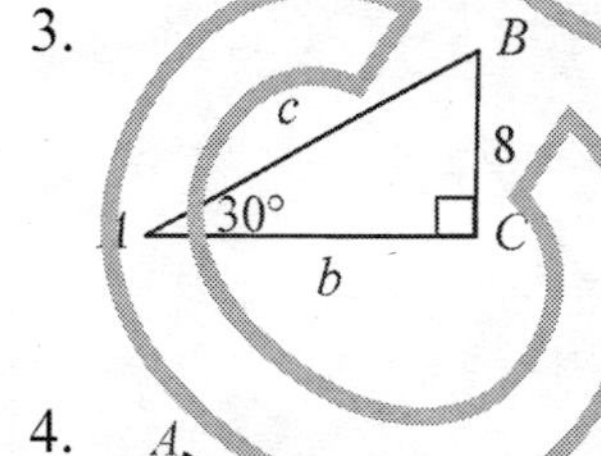

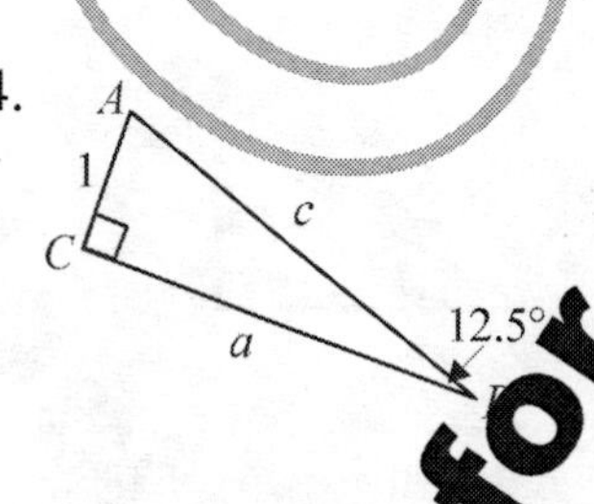

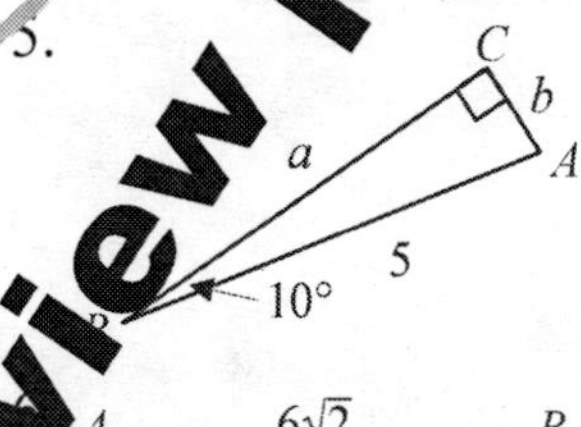

A $6\sqrt{2}$ B
50°
b a
C

Use the pictures to help solve the problems.

1. An F-22 is flying over two control towers. There is a point above the two towers where the fighter pilot can get a clear signal to both the towers. If he is 120 feet from tower one and is making a 59° angle with the two towers, find the distance, x, the F-22 is from the second tower and find the distance, y, between the two towers.

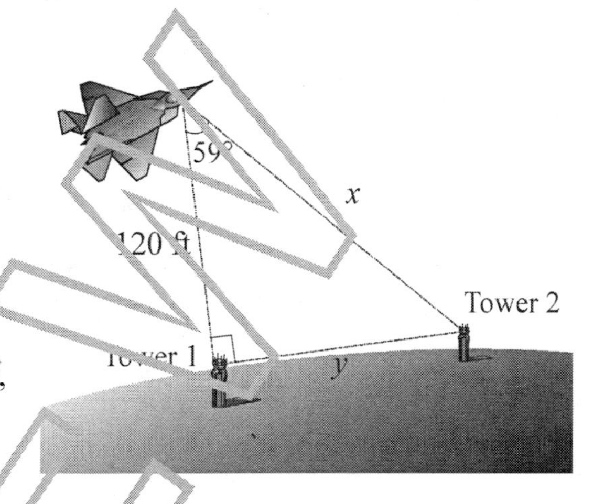

2. Sandra is trying to use her old cell phone to call her best friend. She can be no more than 3 miles from a tower in order to get a signal for her phone. If the telephone tower is 252 feet tall, find the angle of elevation, m, when Sandra's phone is the maximum 3 miles from the tower.
   HINT: Convert 3 miles to feet first.
   1 mile = 5,280 feet

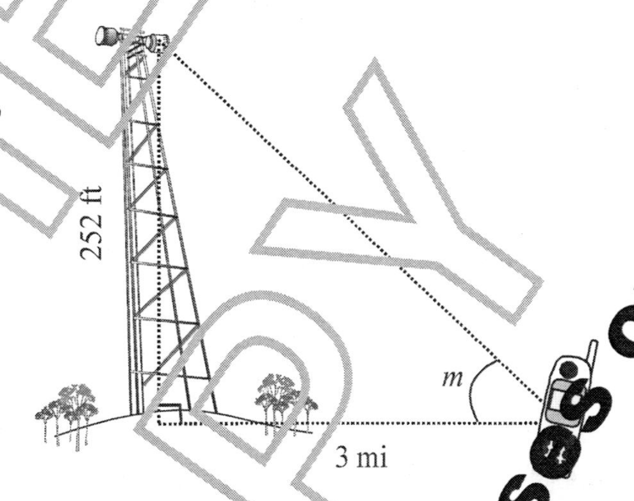

3. Sir Stephen is returning from fighting a war. His first concern on his homecoming journey is to see if his family's banner still flies above their castle. If the flag rises 95 feet above his head, and he emerges from the forest 185 feet from the tower, at what angle, m, is his line of sight to the banner?

Chapter 21 Review

1. Find the missing angle.

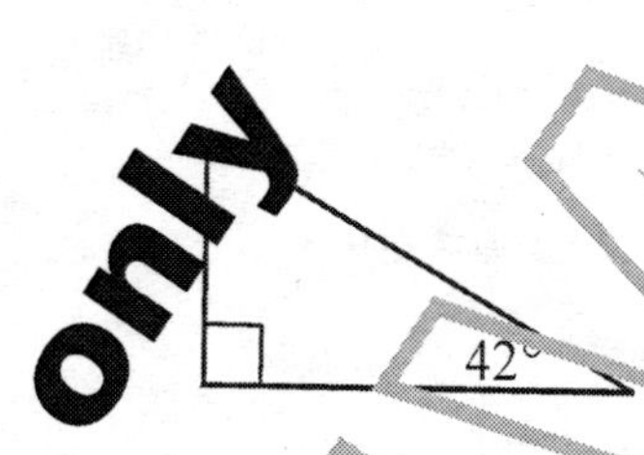

2. What is the length of line segment $\overline{WY}$?

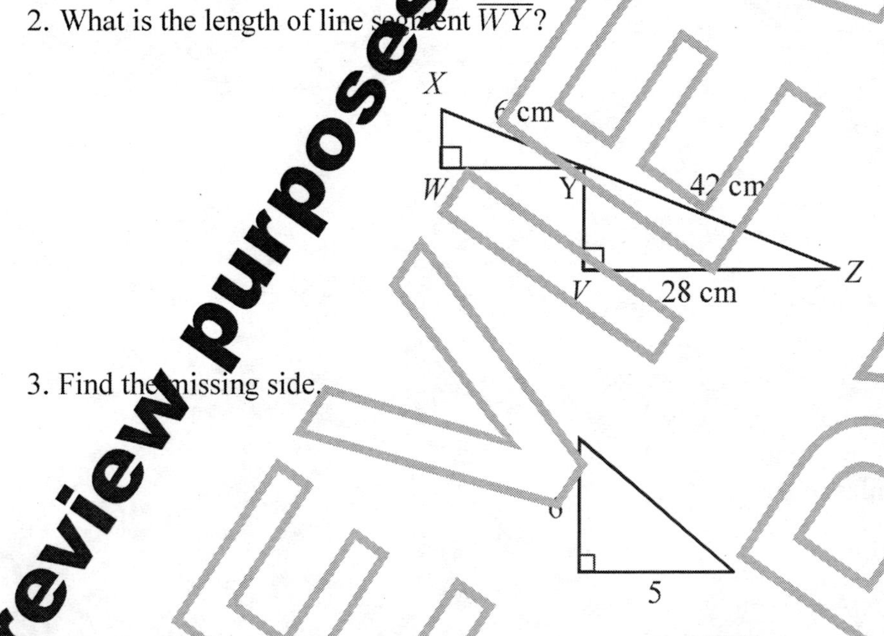

3. Find the missing side.

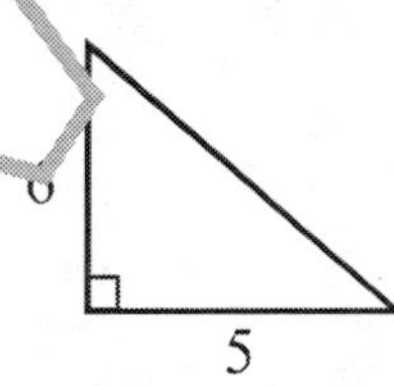

4. Find the measure of the missing leg of the right triangle below.

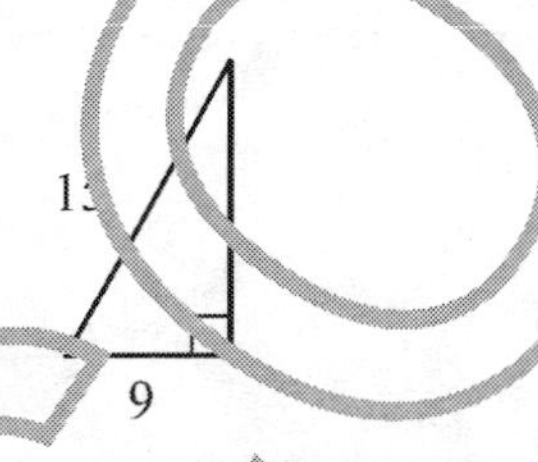

5. The following two triangles are similar. Find the length of the missing side.

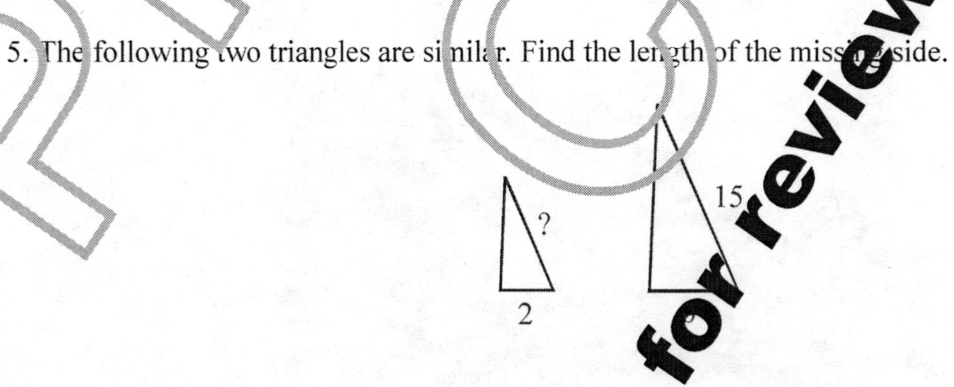

For questions 6 and 7, find the value of x.

6. $\sin x = 0.5$

7. $\tan x = -1$

For questions 8 and 9, find the missing angle and sides.

8. (right triangle ABC: angle $A = 39°$, right angle at C, $BC = 3$, $AC = b$, $AB = c$)

9.

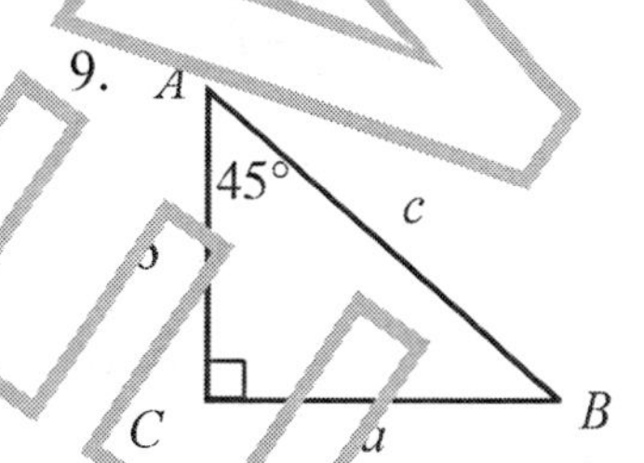

10. Chris walked east from his house to the gas station, which was 1.2 miles away. Then, he walked south from the gas station to his piano teacher's house. His piano teacher lives 2,112 feet from the gas station.
   (A) Use the Pythagorean theorem to find the direct distance in miles from Chris's house to his piano teacher's house.
   (B) Use a trigonometric ratio to find the angle measure between the direct path from Chris's house to the gas station and the direct path from Chris's house to his piano teacher's house.

11.

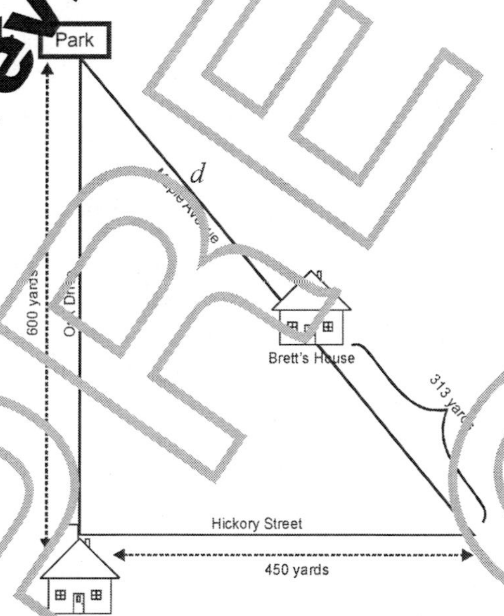

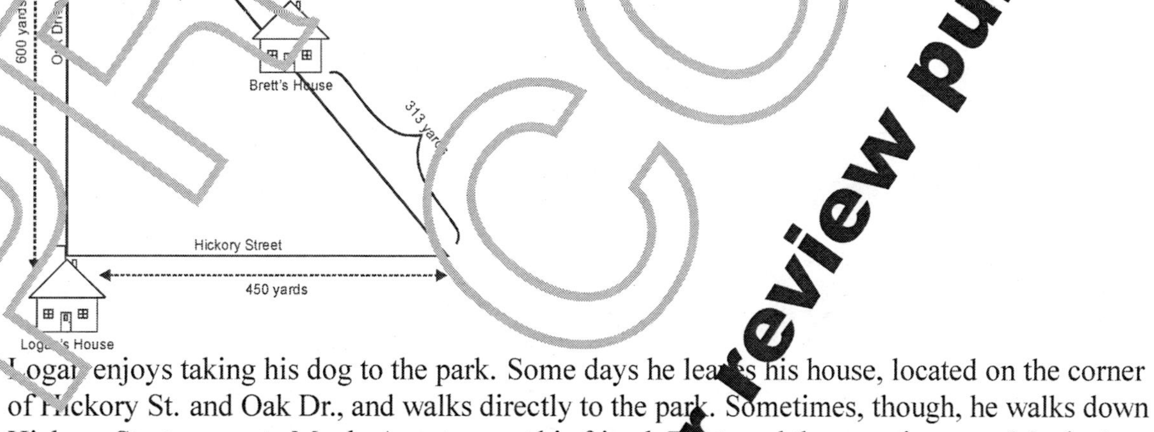

Logan enjoys taking his dog to the park. Some days he leaves his house, located on the corner of Hickory St. and Oak Dr., and walks directly to the park. Sometimes, though, he walks down Hickory St., turns onto Maple Ave. to meet his friend, Brett, and then continues on Maple Ave. to the park. What is the approximate distance (d) from Brett's house to the park?

Chapter 22
Plane and Solid Geometry

22.1 Lines and Line Segments

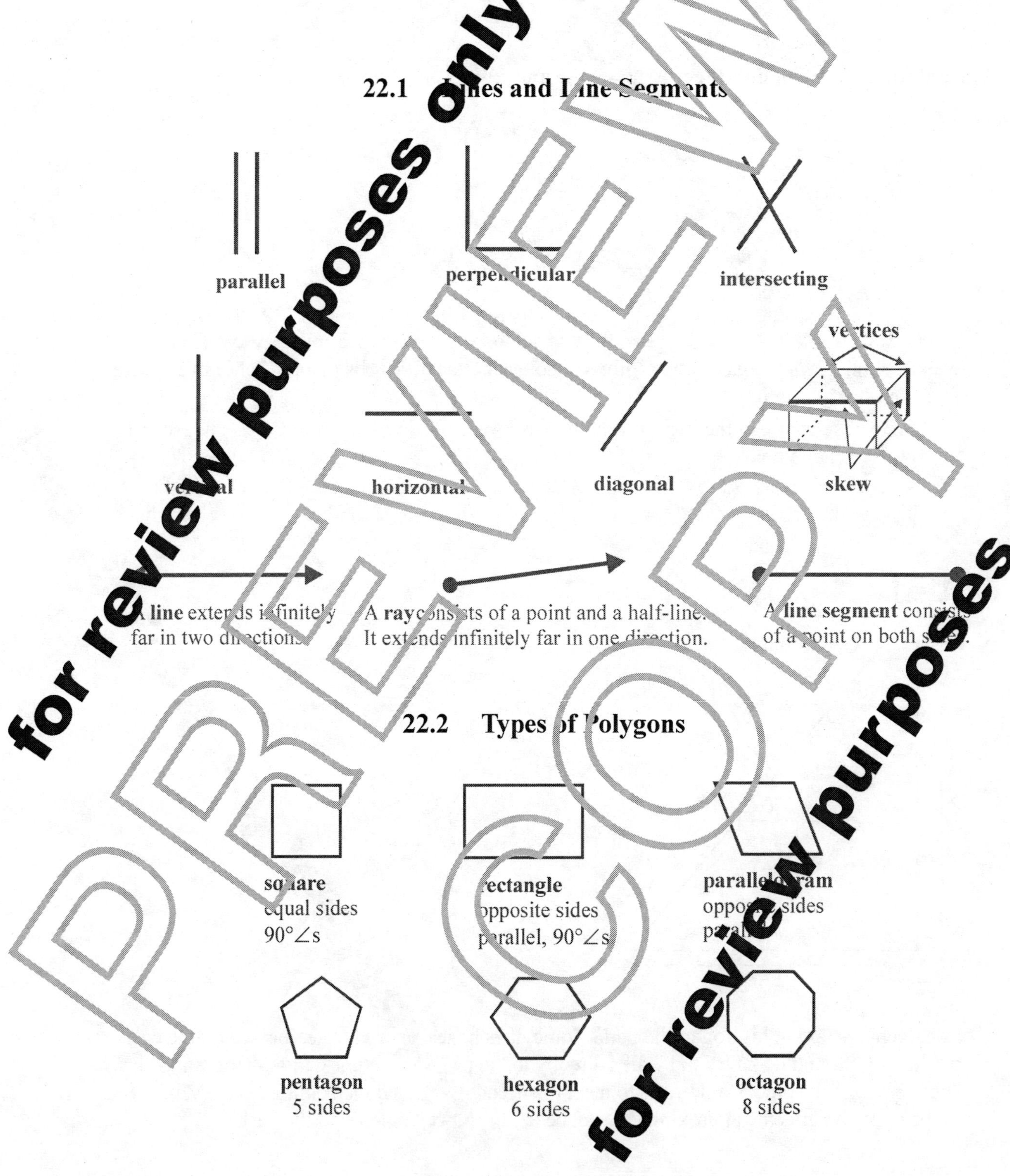

22.3 Perimeter

The **perimeter** is the distance around a polygon. To find the perimeter, add the lengths of the sides.

Examples:

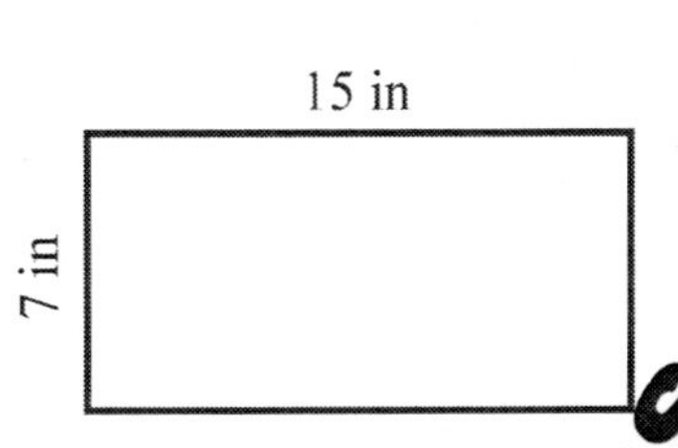

$P = 7 + 15 + 7 + 15$
$P = 44$ in

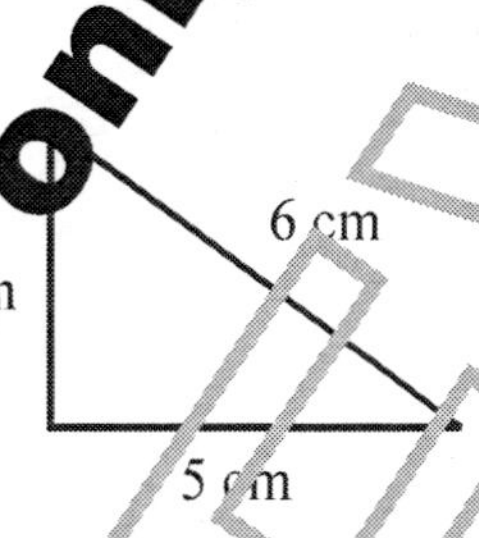

$P = 4 + 6 + 5$
$P = 15$ cm

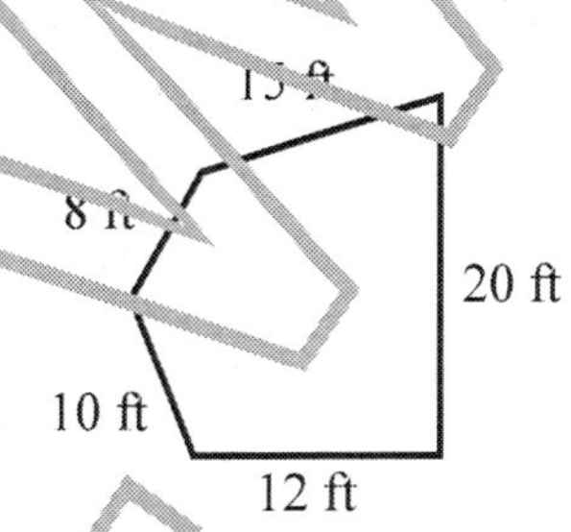

$P = 8 + 15 + 20 + 12 + 10$
$P = 65$ ft

Find the perimeter of the following polygons.

1.

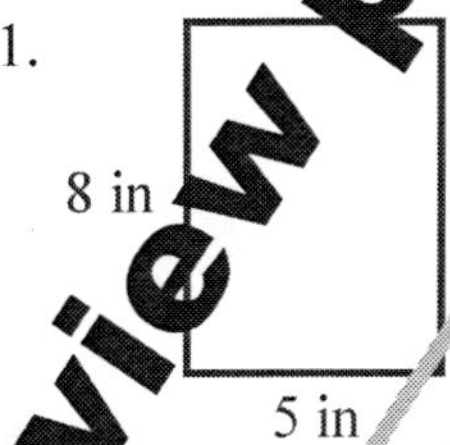

2.

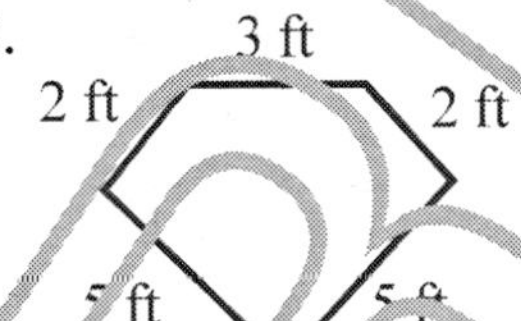

3.

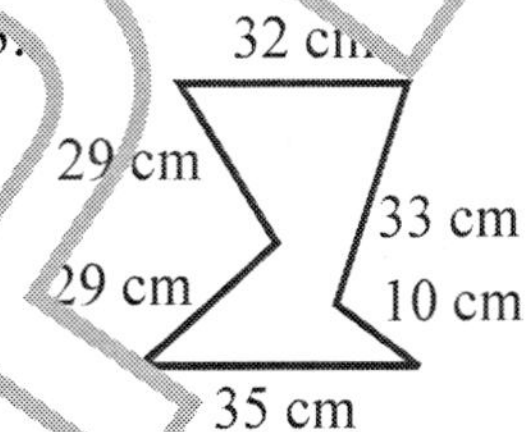

4.

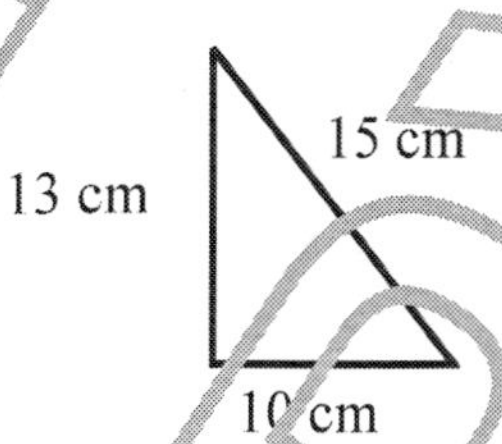

5.

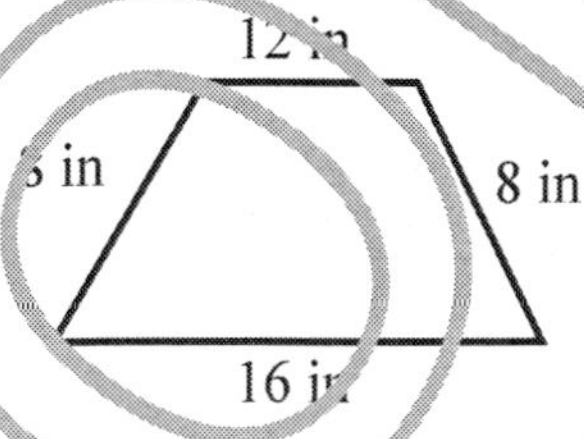

6.

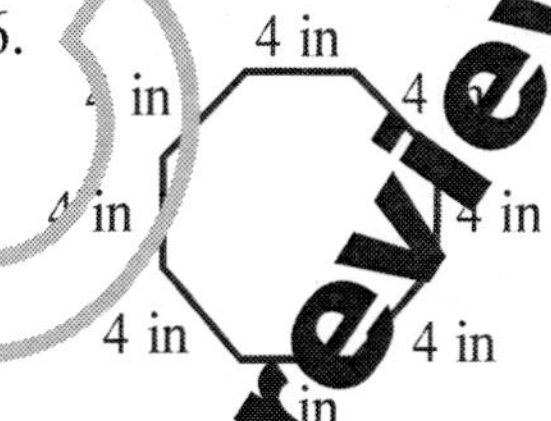

22.4 Area of Squares and Rectangles

Area - area is always expressed in square units, such as in^2, m^2, and ft^2.

The area, (A), of squares and rectangles equals length (l) times width (w). $A = l \times w$.

Example 1:

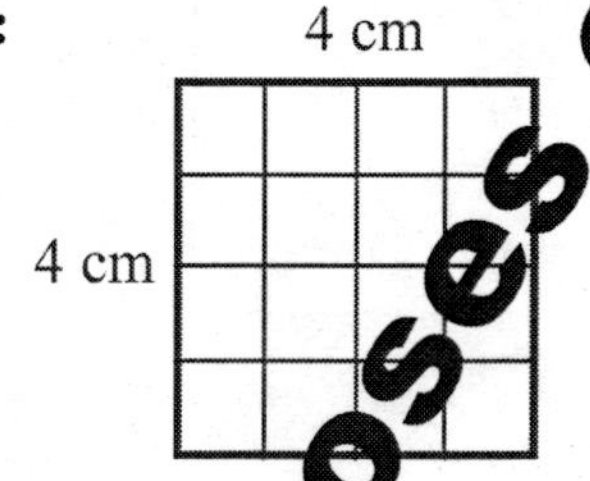

$A = lw$
$A = 4 \times 4$
$A = 16\ \text{cm}^2$

If a square has an area of $16\ \text{cm}^2$, it means that it will take 16 squares that are 1 cm on each side to cover the area that is 4 cm on each side.

Find the area of the following squares and rectangles using the formula $A = lw$.

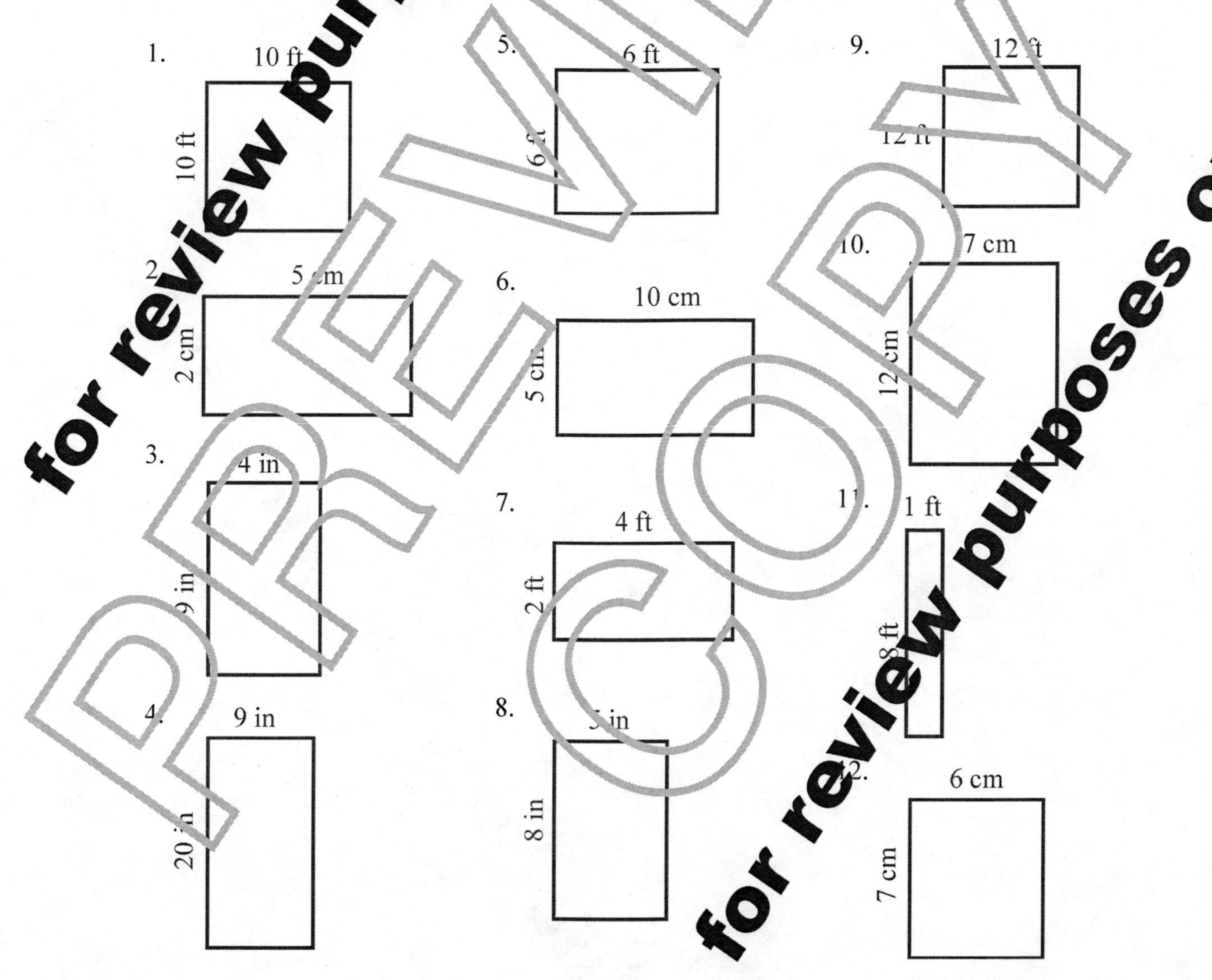

22.5 Area of Triangles

Example 2: Find the area of the following triangle.
The formula for the area of a triangle is as follows:

$$A = \frac{1}{2} \times b \times h$$

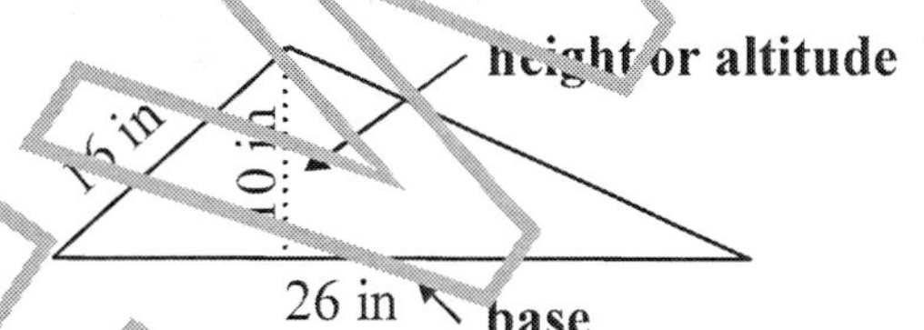

$A =$ area
$b =$ base
$h =$ height or altitude

Step 1: Insert the measurements from the triangle into the formula: $A = \frac{1}{2} \times 26 \times 10$

Step 2: Cancel and multiply. $A = \frac{1}{\cancel{2}_1} \times \frac{\cancel{26}^{13}}{1} \times \frac{10}{1} = 130 \text{ in}^2$

Note: **Area is always expressed in square units such as in^2, ft^2, or m^2.**

Find the area of the following triangles. Remember to include units.

1.
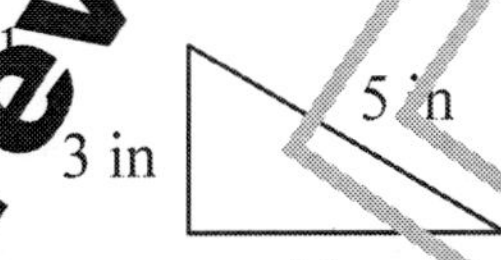

2.
7 cm
6 cm
height
12 cm

3.
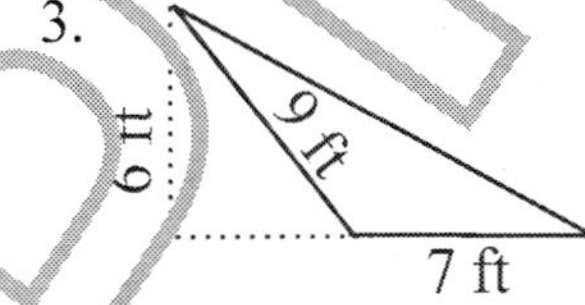

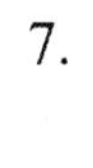

4.
12 cm
12 cm

5.
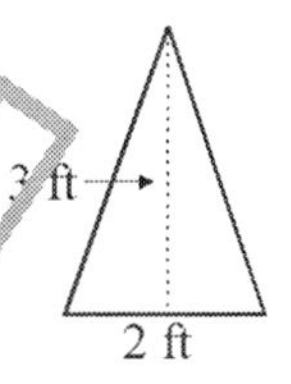

6.
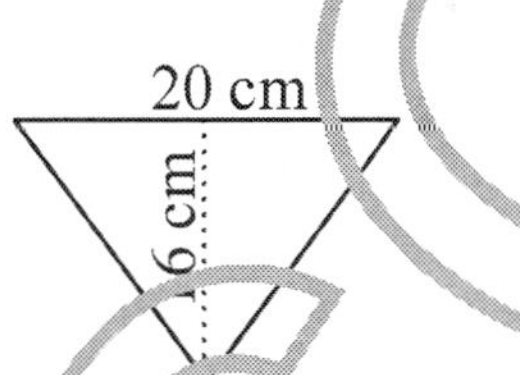

7.
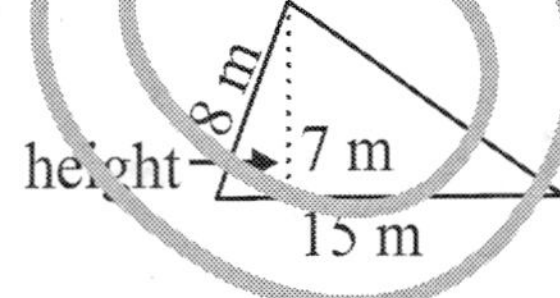

8.
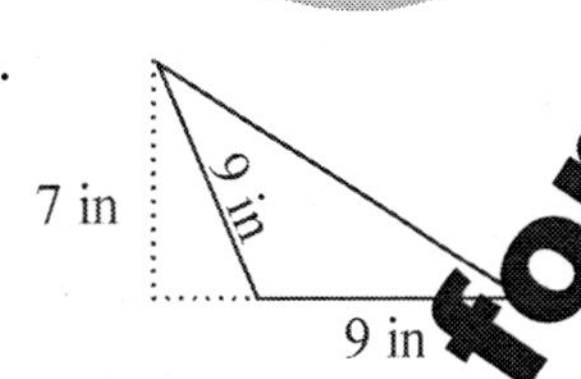

9.
2 ft
2 ft

10.
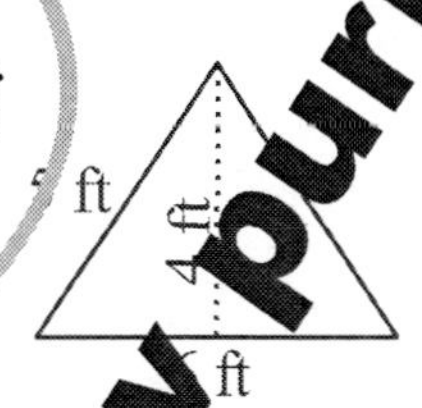

11.
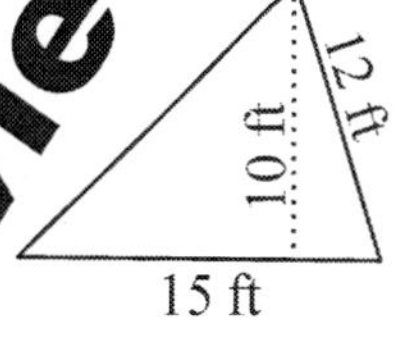

12.
5 m
3 m
10 m

22.6 Parts of a Circle

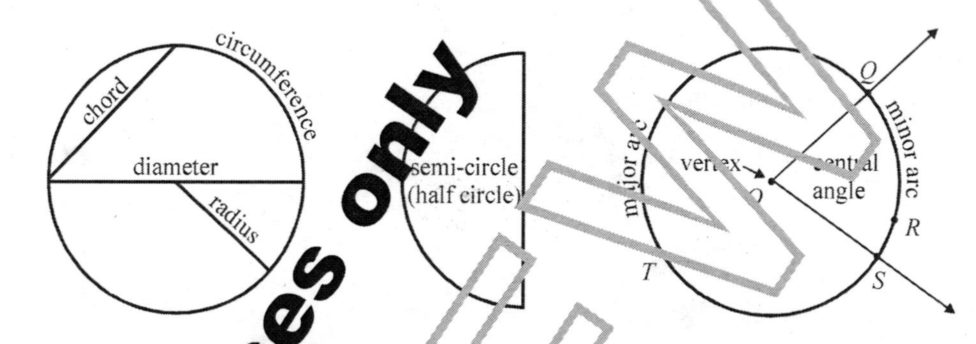

A **central angle** of a circle has the center of the circle as its vertex. The rays of a central angle each contain a radius of the circle. $\angle QOS$ is a central angle.

The points Q and S separate the circle into **arcs**. The arc lies on the circle itself. It does not include any points inside or outside the circle. $\widehat{QRS}$ or $\widehat{QS}$ is a **minor arc** because it is less than a semicircle. A minor arc can be named by 2 or 3 points. $\widehat{QTS}$ is a **major arc** because it is more than a semicircle. A major arc must be named by 3 points.

An **inscribed angle** is an angle whose vertex lies on the circle and whose sides contain **chords** of the circle. $\angle ABC$ in Figure 1 is an inscribed angle. A line is **tangent** to a circle if it only touches the circle at one point, which is called the point of tangency. See Figure 2 for an example. A **secant**, shown in figure 3, is a line that intersects with a circle at two points. Every secant forms a chord. In Figure 3, secant $\overleftrightarrow{AB}$ forms chord $\overline{AB}$.

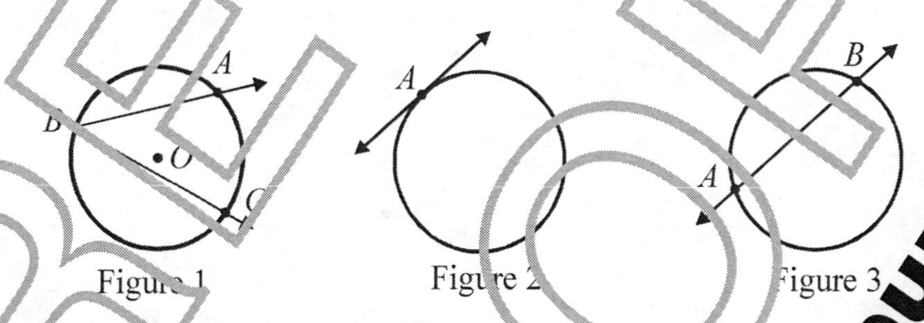

Figure 1 Figure 2 Figure 3

Refer to the figure on the right, and answer the following questions.

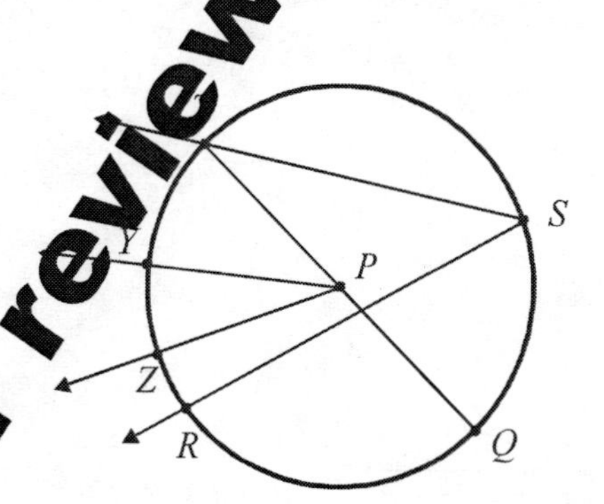

1. Identify the 2 line segments that are chords of the circle but not diameters.
2. Identify the largest major arc of the circle that contains point S.
3. Identify the vertex of the circle.
4. Identify the inscribed angle.
5. Identify the central angle.

22.7 Circumference

Circumference, C - the distance around the outside of a circle
Diameter, d - a line segment passing through the center of a circle from one side to the other
Radius, r - a line segment from the center of a circle to the edge of a circle
Pi, π- the ratio of a circumference of a circle to its diameter $\pi = 3.14$ or $\pi = \frac{22}{7}$

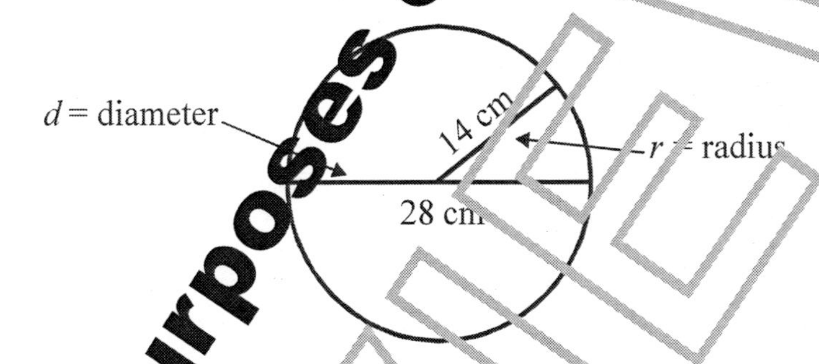

The formula for the circumference of a circle is $C = 2\pi r$ or $C = \pi d$. (The formulas are equal because the diameter is equal to twice the radius, $d = 2r$.)

Example 3: Find the circumference of the circle above.

$C = \pi d$ Use $\pi = 3.14$
$C = 3.14 \times 28$
$C = 87.92$ cm

$C = 2\pi r$
$C = 2 \times 3.14 \times 14$
$C = 87.92$ cm

Use the formulas given above to find the circumferences of the following circles. Use $\pi = 3.14$.

1. 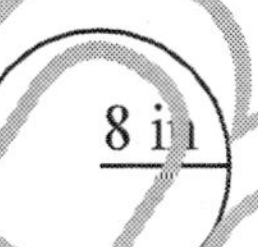

$C =$ ______

2. 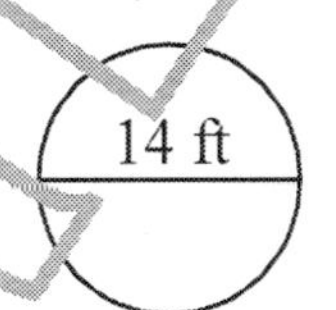

$C =$ ______

3. 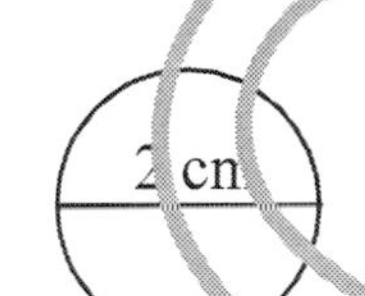

$C =$ ______

4. 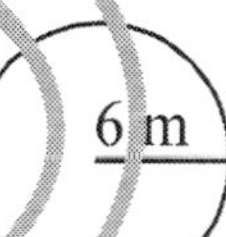

$C =$ ______

5. 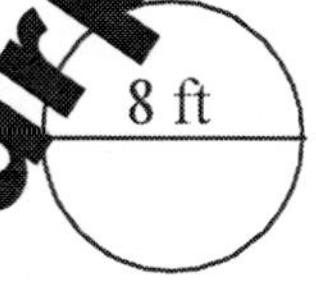

$C =$ ______

Use the formulas given above to find the circumferences of the following circles. Use $\pi = \frac{22}{7}$.

6. 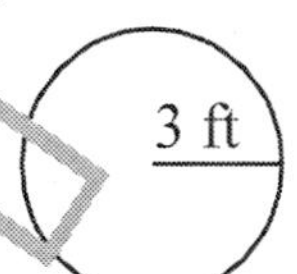

$C =$ ______

7. 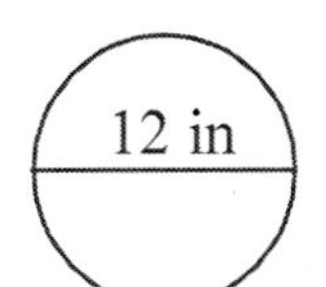

$C =$ ______

8. 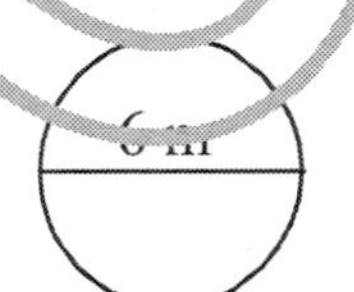
$C =$ ______

9. 

$C =$ ______

10. 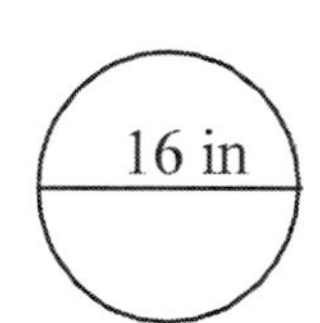

$C =$ ______

22.8 Area of a Circle

The formula for the area of a circle is $A = \pi r^2$. The area is how many square units of measure would fit inside a circle.

Example 4: Find the area of the circle, using both values for π.

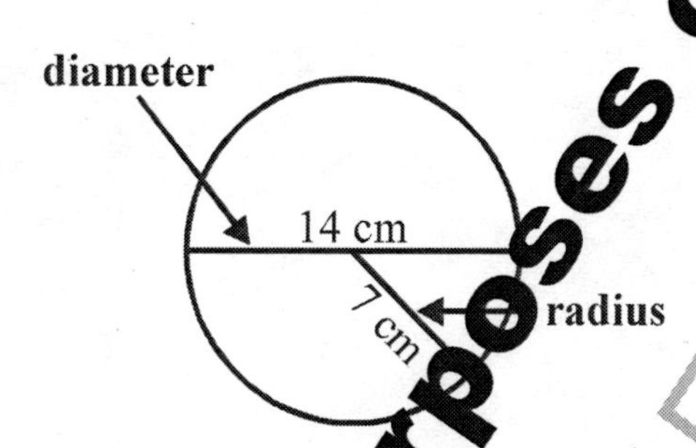

Let $\pi = \frac{22}{7}$
$A = \pi r^2$
$A = \frac{22}{7} \times 7^2$
$A = \frac{22}{\cancel{7}\,1} \times \frac{\cancel{49}\,7}{1}$
$= 154 \text{ cm}^2$

Let $\pi = 3.14$
$A = \pi r^2$
$A = 3.14 \times 7^2$
$A = 3.14 \times 49$
$= 153.86 \text{ cm}^2 \approx 154 \text{ cm}^2$

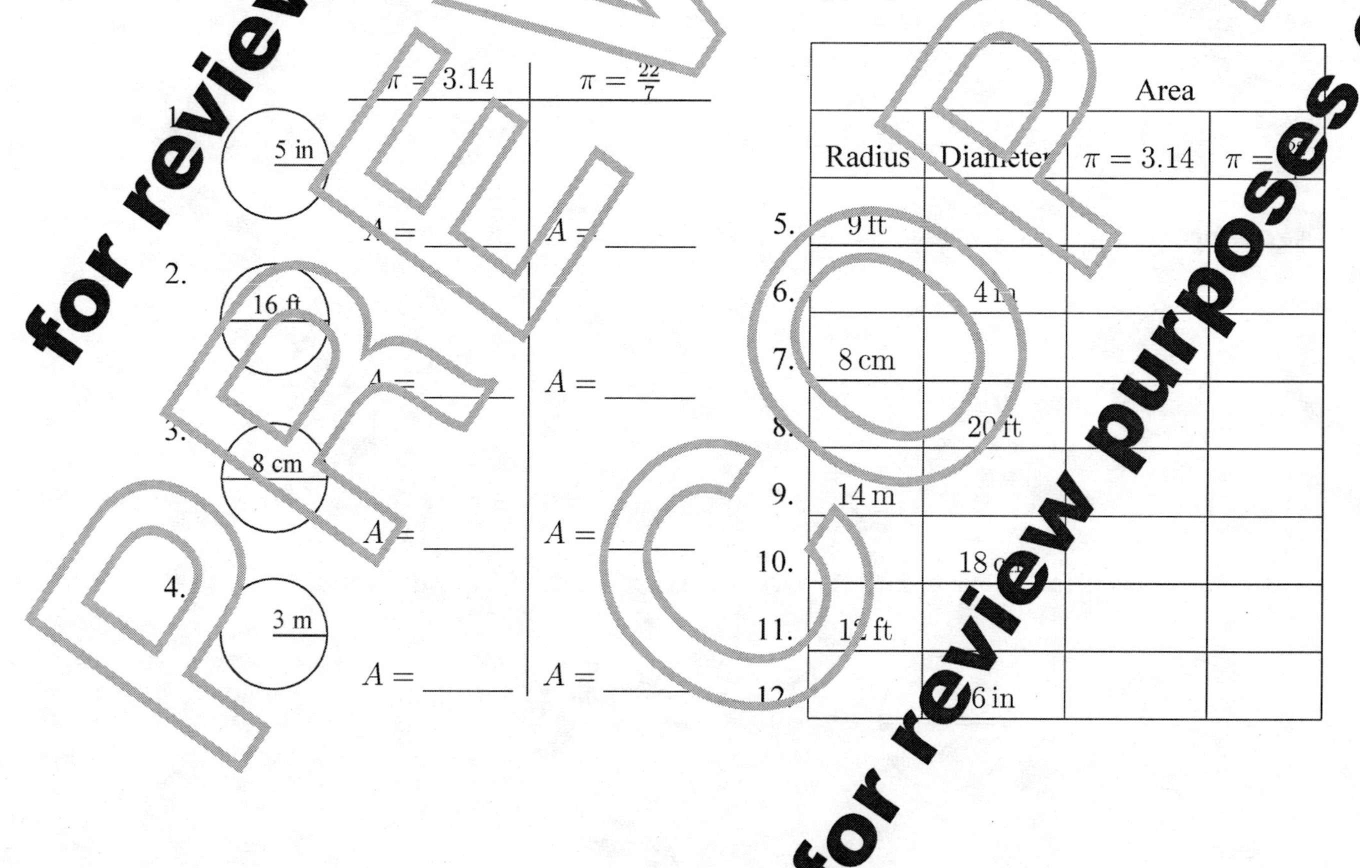

Find the area of the following circles. Remember to include units.

	$\pi = 3.14$	$\pi = \frac{22}{7}$
1. (5 in)	$A =$ ______	$A =$ ______
2. (16 ft)	$A =$ ______	$A =$ ______
3. (8 cm)	$A =$ ______	$A =$ ______
4. (3 m)	$A =$ ______	$A =$ ______

Fill in the chart below. Include appropriate units.

			Area	
	Radius	Diameter	$\pi = 3.14$	$\pi = \frac{22}{7}$
5.	9 ft			
6.		4 in		
7.	8 cm			
8.		20 ft		
9.	14 m			
10.		18 cm		
11.	12 ft			
12.		6 in		

22.9 Two-Step Area Problems

Solving the problems below will require two steps. You will need to find the area of two figures, and then either add or subtract the two areas to find the answer. **Carefully read the examples.**

Example 5:
Find the area of the living room below.

Figure 1

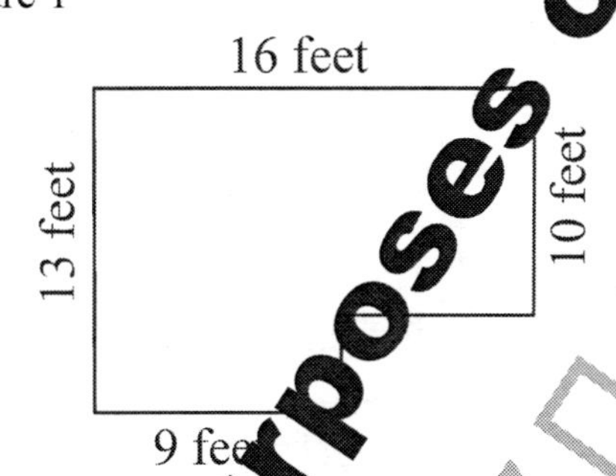

Step 1: Complete the rectangle as in Figure 2, and compute the area as if it were a complete rectangle.

Figure 2

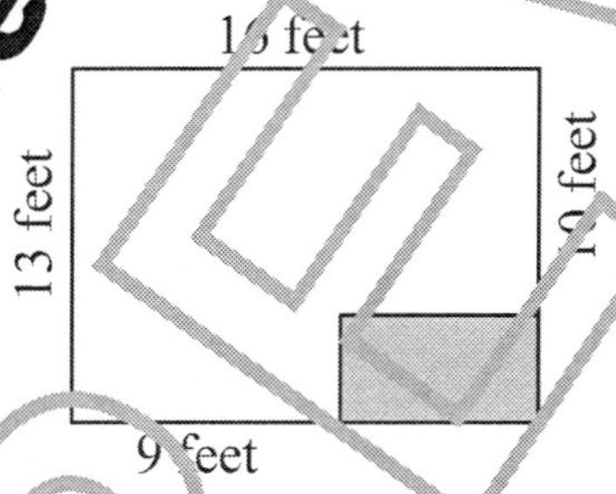

$A = \text{length} \times \text{width}$

$A = 16 \times 13$

$A = 208\ \text{ft}^2$

Step 2: Figure the area of the shaded part.

7 feet

3 feet

$7 \times 3 = 21\ \text{ft}^2$

Step 3: Subtract the area of the shaded part from the area of the complete rectangle

$208 - 21 = 187\ \text{ft}^2$

Example 6:
Find the area of the shaded sidewalk.

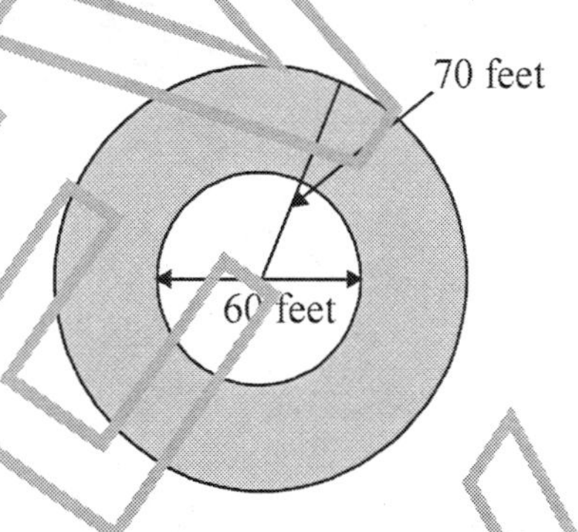

Step 1: Find the area of the outside circle.

$\pi = 3.14$

$A = 3.14 \times 70 \times 70$

$A = 15,386\ \text{ft}^2$

Step 2: Find the area of the inside circle.

$\pi = 3.14$

$A = 3.14 \times 30 \times 30$

$A = 2826\ \text{ft}^2$

Step 3: Subtract the area of the inside circle from the area of the outside circle.

$15,386 - 2826 = 12,560\ \text{ft}^2$

Find the area of the following figures.

1.

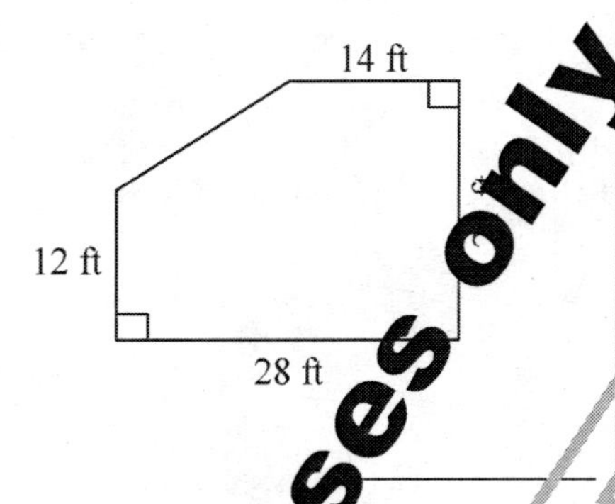

2.

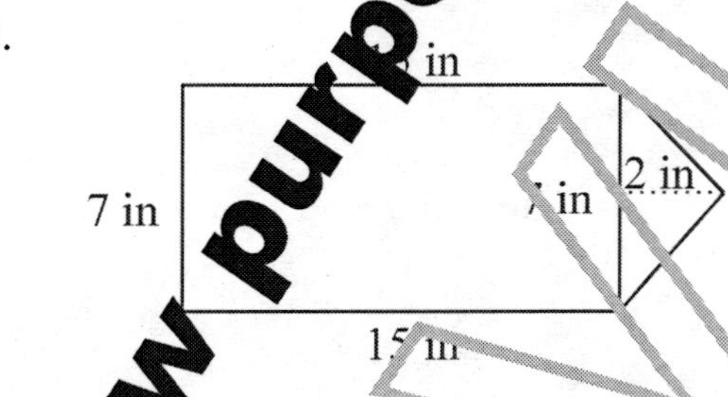

3. What is the area of the shaded circle? Use $\pi = 3.14$, and round the answer to the nearest whole number.

4.

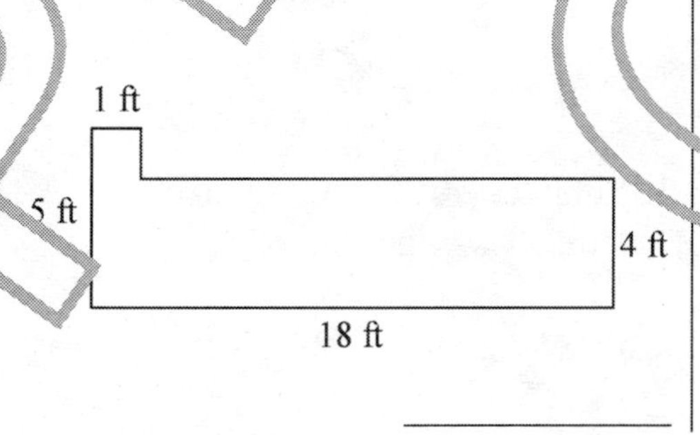

5.

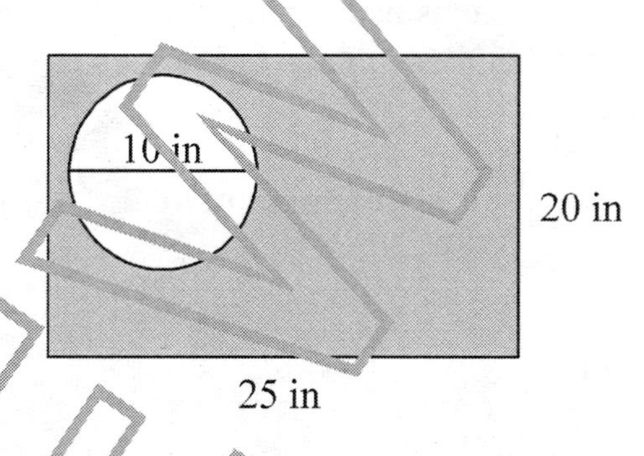

6. What is the area of the shaded part?

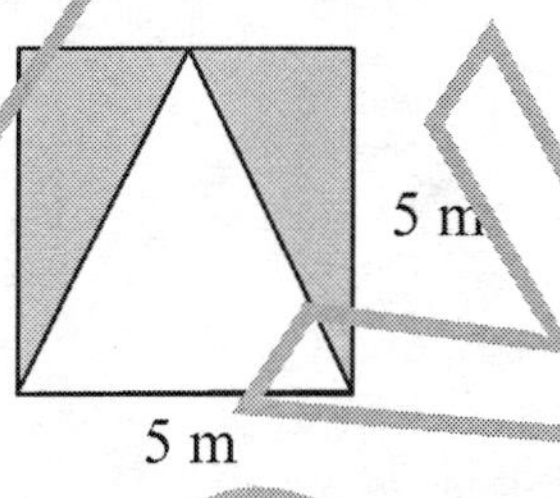

7. What is the area of the shaded part?

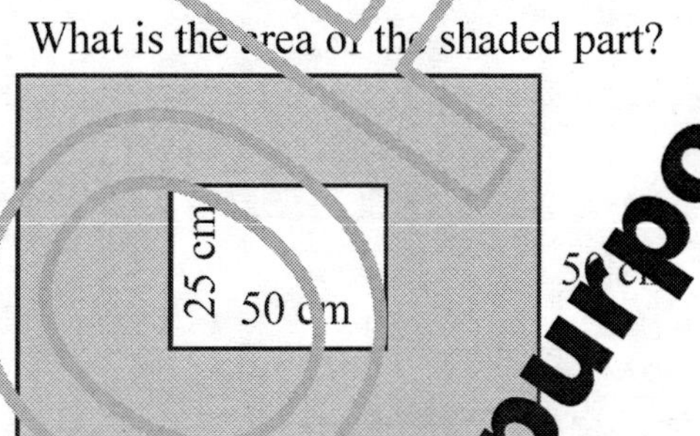

8.

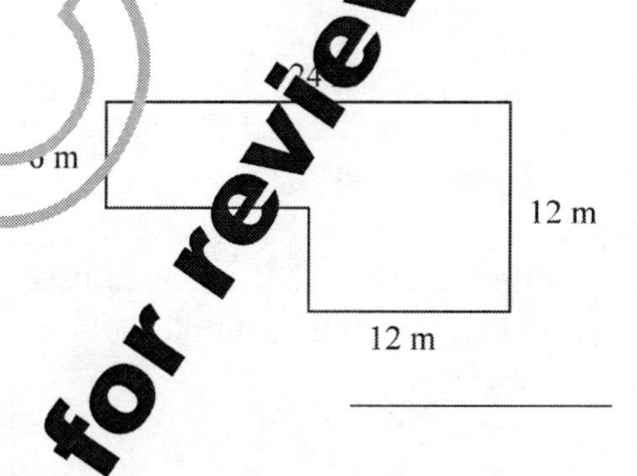

22.10 Geometric Relationships of Plane Figures

This section illustrates what happens to the area of a figure when one or more of the dimensions is doubled or tripled.

Example 7: Sam drew a square that was 2 inches on each side for art class. His teacher said the square needed to be twice as big. When Sam doubled each side to 4 inches, what happened to the area?

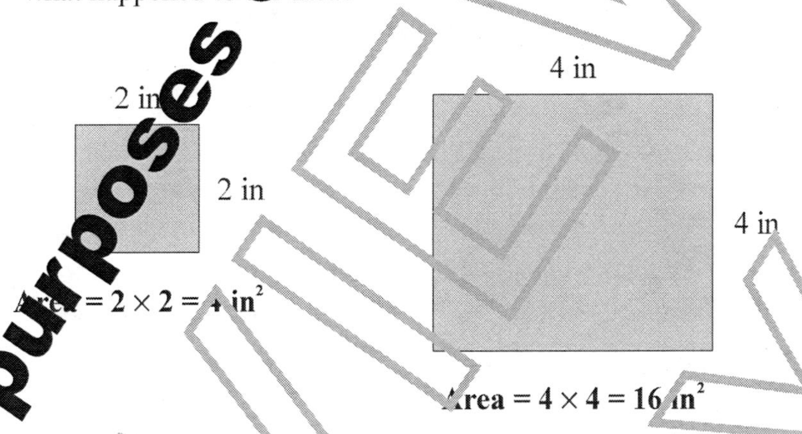

The area of the second square is 4 times larger than the first.

Example 8: Sonya drew a circle which had a radius of 3 inches for a school project. She also needed to make a larger circle which had a radius of 9 inches. When Sonya drew the bigger circle, what was the difference in area?

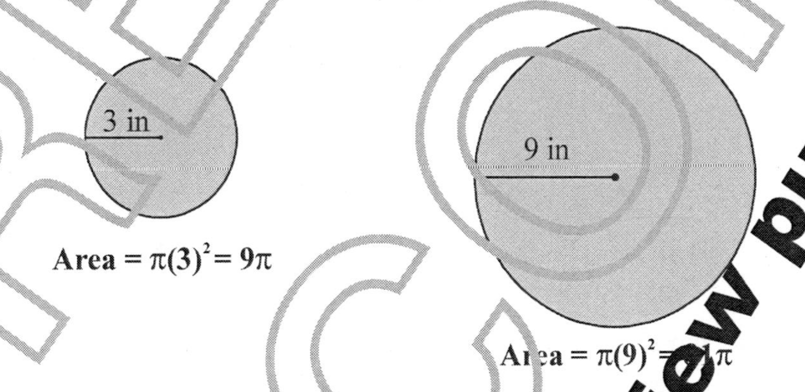

The area of the second circle is 9 times larger than the first.

From these two examples, we can determine that for every doubling or tripling of both sides or of the radius of a planar object, the total area increases by a squared value. In other words, when both sides of the square doubled, the area was 2^2 or 4 times larger. When the radius of the circle became 3 times larger, the area became 3^2 or 9 times larger.

Carefully read each of the problems below and solve.

1. Ken draws a circle with a radius of 5 cm. He then draws a circle with a radius of 10 cm. How many times larger is the area of the second circle?
2. Kobe draws a square with each side measuring 6 inches. He then draws a rectangle with a width of 6 inches and a length of 12 inches. How many times larger is the area of the rectangle than the area of the square? (**Hint:** The increase is *not* equal in both directions.)
3. Toshi draws a square 3 inches on each side. Then he draws a bigger square that is 6 inches on each side. How many times larger is the area of the second square than the area of the first square?
4. Leslie draws a triangle with a base of 5 inches and a height of 3 inches. To use her triangle pattern for a bulletin board design, it needs to be 3 times bigger. If she increases the base and the height by multiplying each by 3, how much will the area of the triangle increase?
5. Heather is using 100 tiles that measure 1 foot by 1 foot to cover a 10 feet by 10 feet floor. If she used tiles that measure 2 feet by 2 feet, how many tiles would she need?
6. The area of circle B is 9 times larger than the area of circle A. If the radius of circle A is represented by x, how would you represent the radius of circle B?

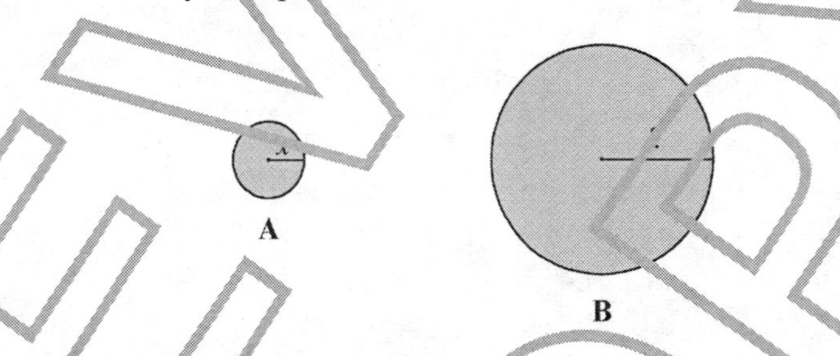

7. How many squares will it take to fill the rectangle below?

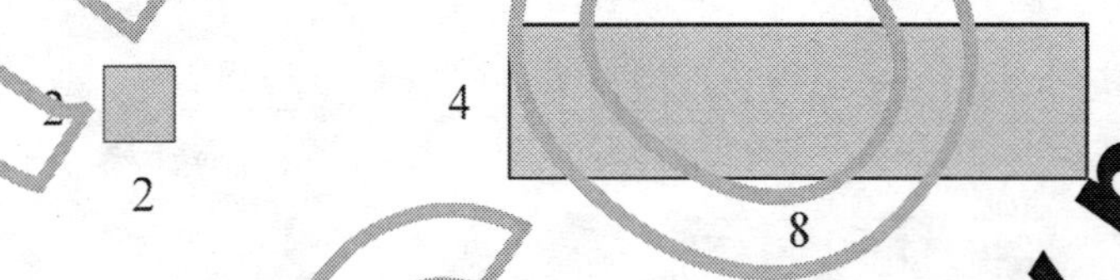

8. If the area of diamond B is one-fourth the area of diamond A, what are the dimensions of diamond B?

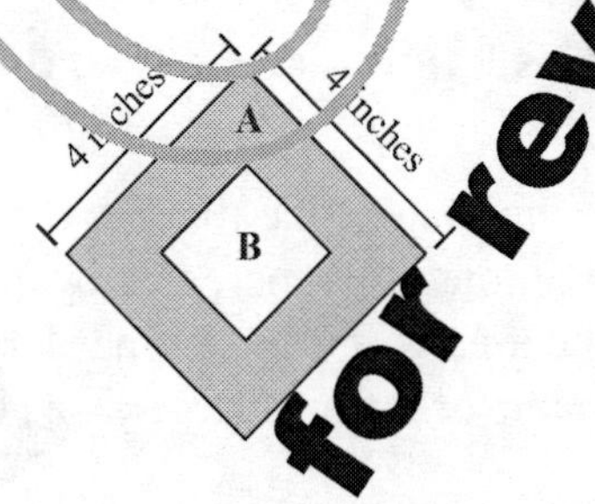

22.11 Solids

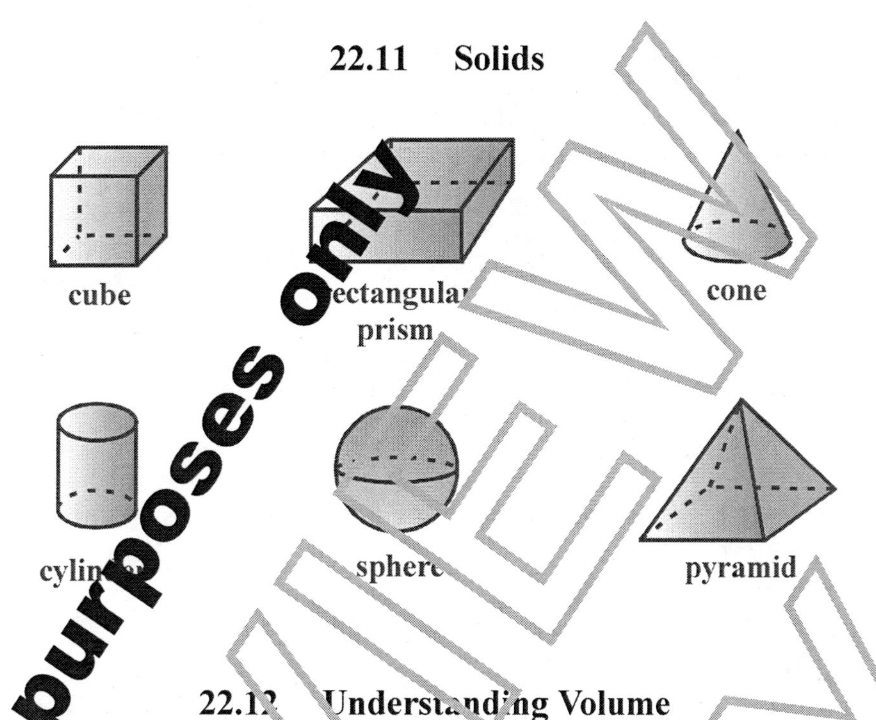

22.12 Understanding Volume

Volume - Measurement of volume is expressed in cubic units such as in^3, ft^3, m^3, cm^3, or mm^3. The volume of a solid is the number of cubic units that can be contained in the solid.

First, let's look at rectangular solids.

Example 1: How many 1 cubic centimeter cubes will it take to fill up the figure below?

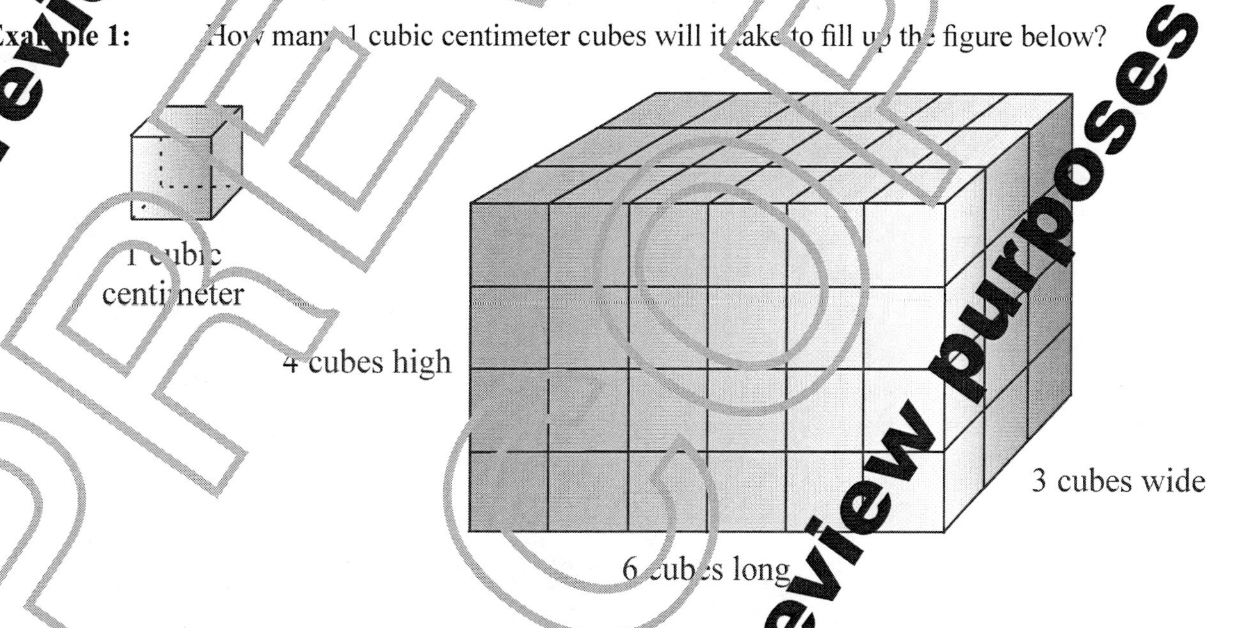

To find the volume, you need to multiply the length times the width times the height.

Volume of a rectangular solid = length × width × height $(V = lwh)$.

$$V = 6 \times 3 \times 4 = 72 \text{ cm}^3$$

22.13 Volume of Rectangular Prisms

You can calculate the volume (V) of a rectangular prism (box) by multiplying the length (l) by the width (w) by the height (h), as expressed in the formula $V = (lwh)$.

Example 2: Find the volume of the box pictured here:

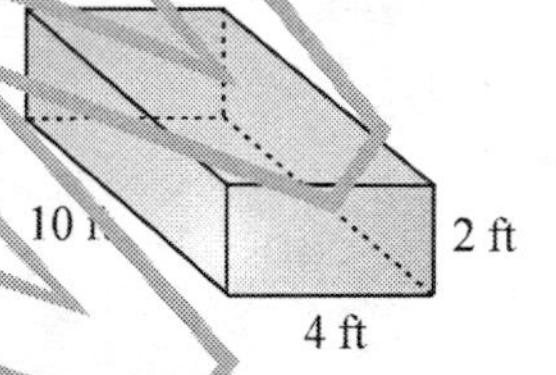

Step 1: Insert measurements from the figure into the formula.

Step 2: Multiply to solve. $10 \times 4 \times 2 = 80 \text{ ft}^3$

Note: **Volume is always expressed in cubic units such as in^3, ft^3, m^3, cm^3, or mm^3.**

Find the volume of the following rectangular prisms (boxes).

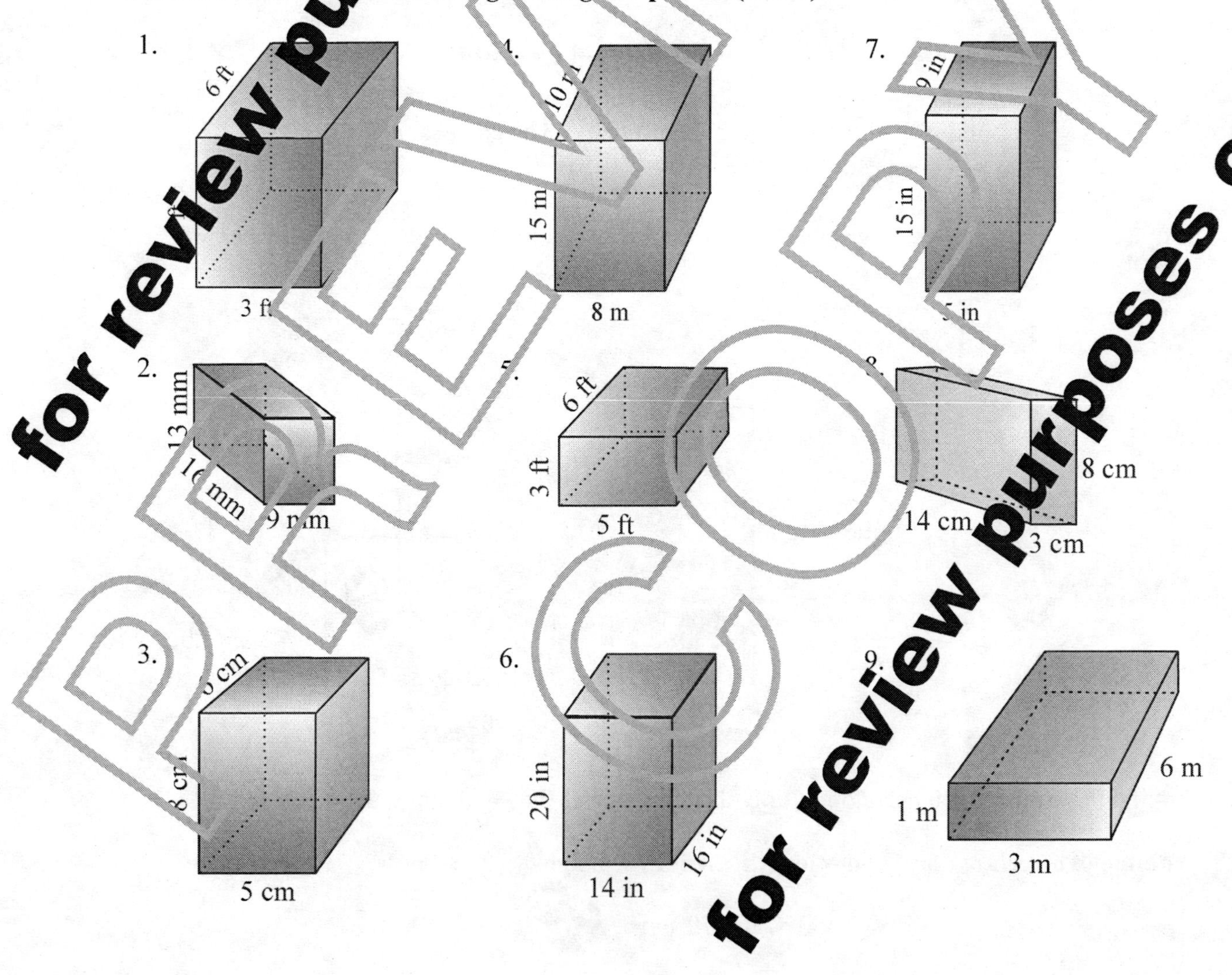

22.14 Volume of Cubes

A **cube** is a special kind of rectangular prism (box). Each side of a cube has the same measure. So, the formula for the volume of a cube is $V = s^3$ $(s \times s \times s)$.

Example 3: Find the volume of the cube at right:

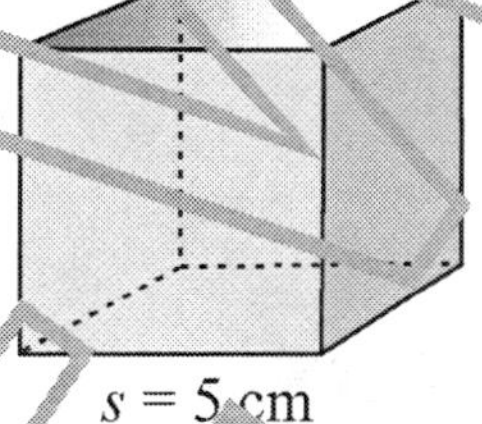
$s = 5$ cm

Step 1: Insert measurements from the figure into the formula.

Step 2: Multiply to solve. $5 \times 5 \times 5 = 125$ cm^3

Note: Volume is always expressed in cubic units such as in^3, ft^3, m^3, cm^3, **or** mm^3**.**

Answer each of the following questions about cubes.

1. If a cube is 3 centimeters on each edge, what is the volume of the cube?

2. If the measure of the edge is doubled to 6 centimeters on each edge, what is the volume of the cube?

3. If the edge of a 3-centimeter cube is tripled to become 9 centimeters on each edge, what will the volume be?

4. How many cubes with edges measuring 3 centimeters would you need to stack together to make a solid 12-centimeter cube?

5. What is the volume of a 2-centimeter cube?

6. Jerry built a 2-inch cube to hold his marble collection. He wants to build a cube with a volume 8-times larger. How much will each edge measure?

Find the volume of the following cubes.

7.

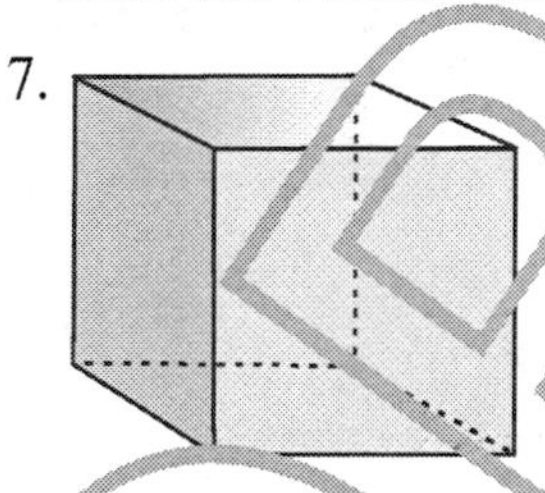
$s = 7$ in.

8.

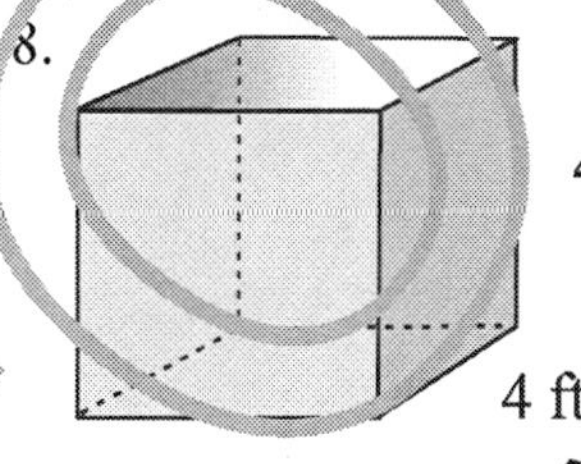

9. 12 inches = 1 foot

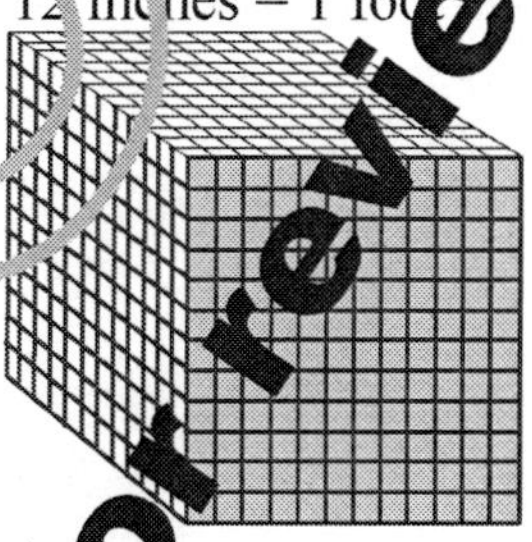
$s = 1$ foot

How many cubic inches are in a cubic foot?

22.15 Volume of Cylinders

To find the volume of a solid, insert the measurements given for the solid into the correct formula and solve. Remember, volumes are expressed in cubic units such as in^3, ft^3, m^3, cm^3, or mm^3.

Example 1: Find the volume of the cylinder below.

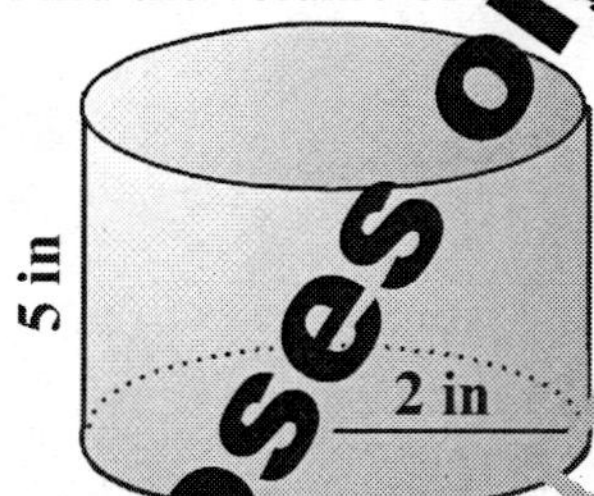

Volume of a cylinder $= V = \pi r^2 h$
Plug in the appropriate values in for the variables.
$V = \pi r^2 h$ where $\pi = \frac{22}{7}$
$V = \frac{22}{7} \times 4 \times 5$
$V = 62\frac{6}{7}\ \text{in}^3$

Find the volume of the following shapes. Use $\pi = 3.14$.

1.
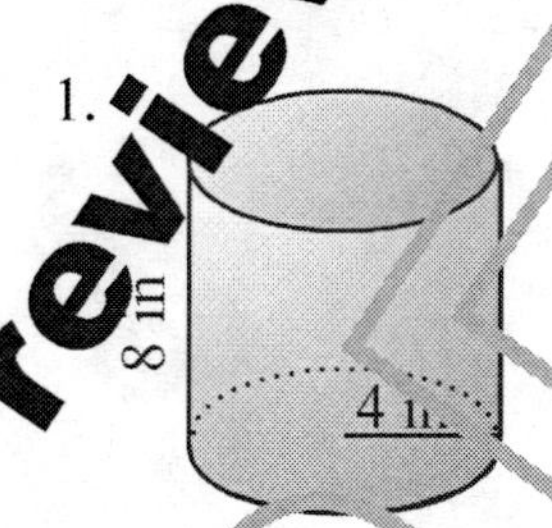

3.
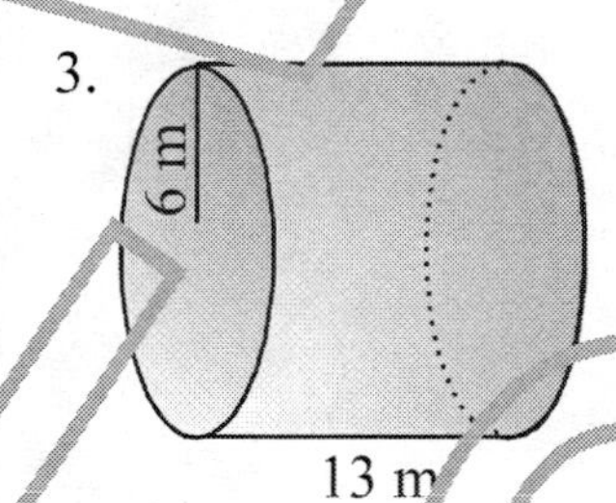

5.
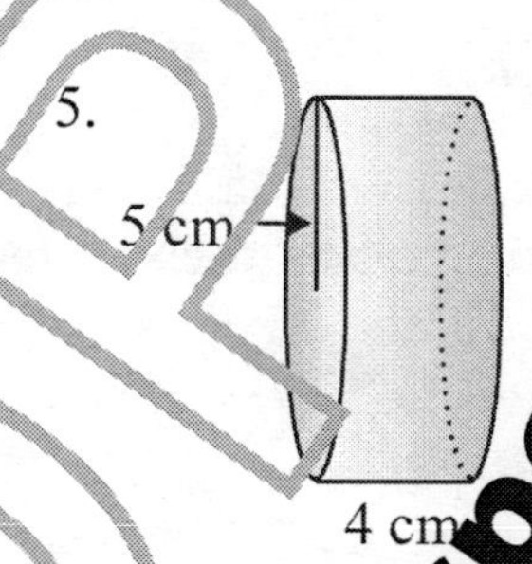

2.
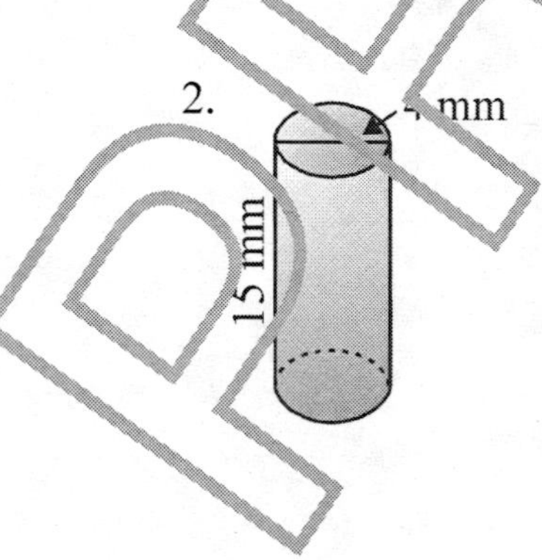

4.
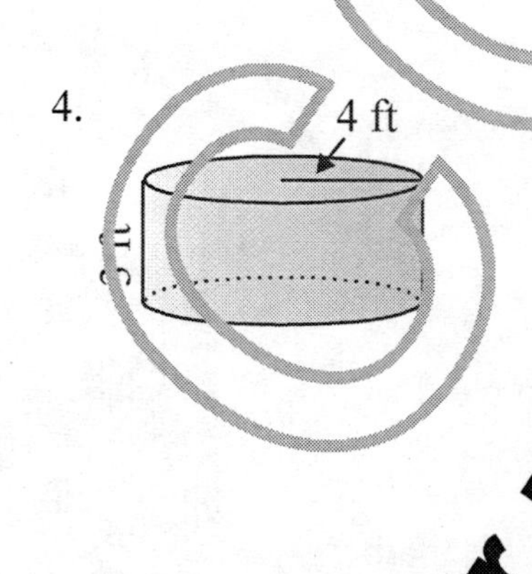

6. 3 cm

Chapter 22 Review

1. Find the area of the shaded region of the figure below.

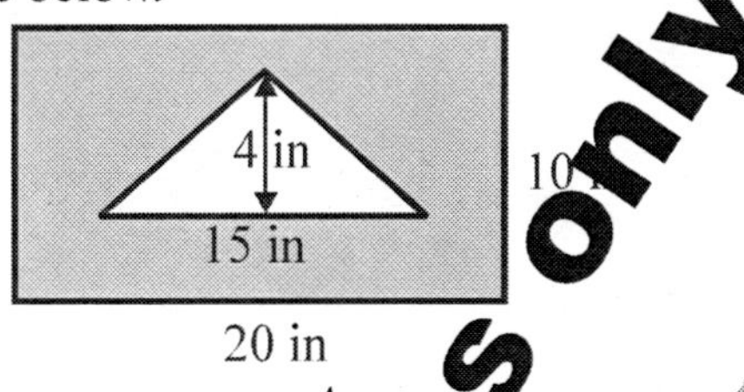

$A =$ ____________

2. Calculate the perimeter.

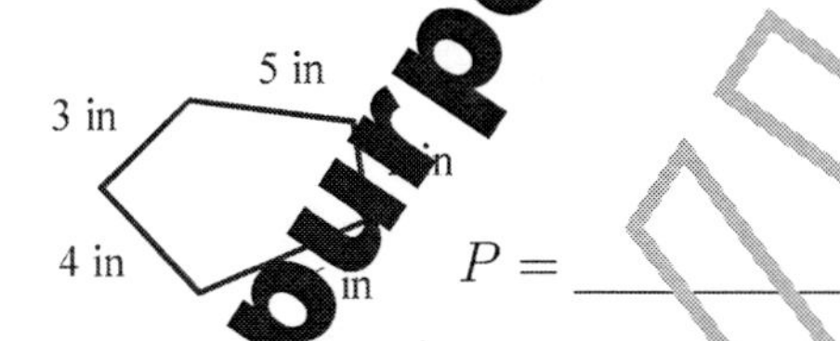

$P =$ ____________

3. Calculate the perimeter and area of the following figure.

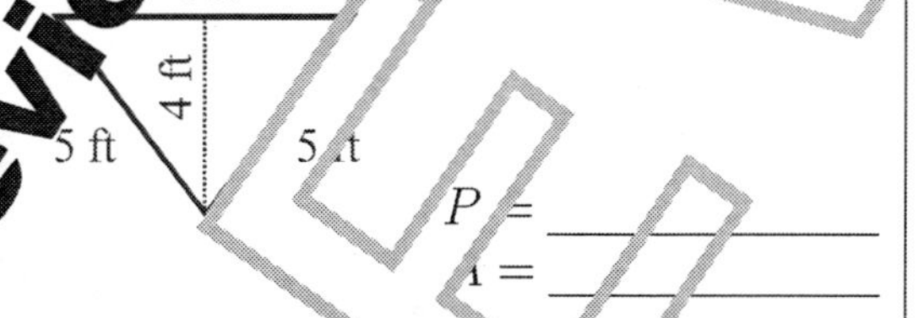

$P =$ ____________
$A =$ ____________

4. Calculate the circumference and the area of the following circle. Use $\pi = 3.14$.

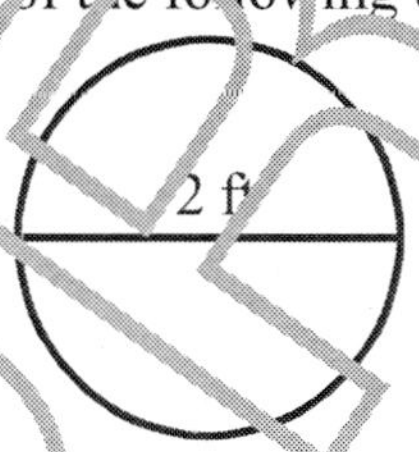

$C =$ ____________
$A =$ ____________

5. Find the area of the shaded part.

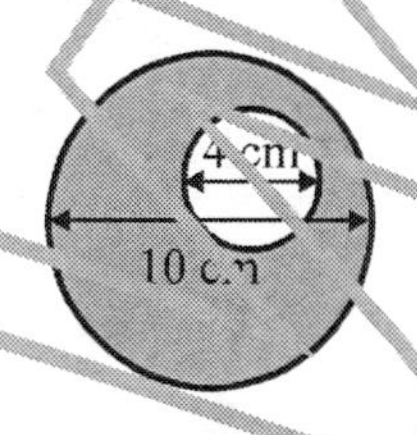

$A =$ ____________

6. Calculate the circumference and the area of the following circle. Use $\pi = \frac{22}{7}$.

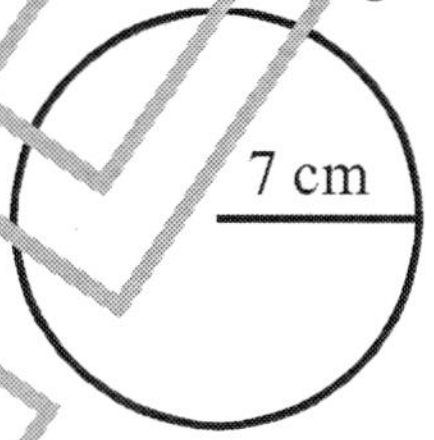

$C =$ ____________
$A =$ ____________

7. If you double the the width of a square, how much does the area of the square increase?

8. What is the area of a square which measures 8 inches on each side?

Find the volume of the following solids.

9.

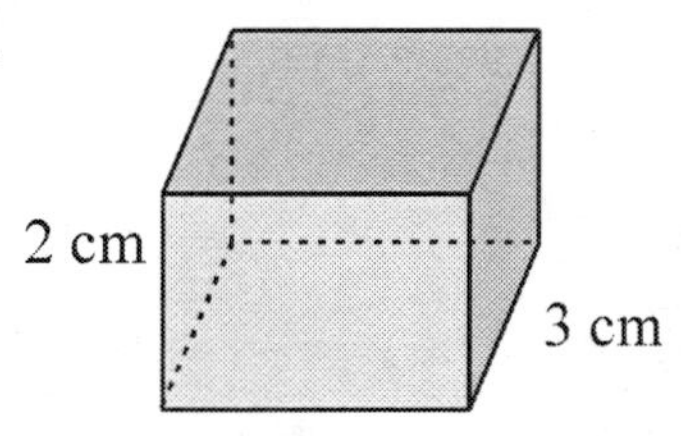

3 cm

$V =$ _________

10.

Use $\pi = \frac{22}{7}$.
$V =$ _________

11. The sandbox at the local elementary school is 60 inches wide and 100 inches long. The sand in the box is 6 inches deep. How many cubic inches of sand are in the sandbox?

12. A grain silo is in the shape of a cylinder. If the silo has an inside diameter of 10 feet and a height of 35 feet, what is the maximum volume inside the silo? Use $\pi = \frac{22}{7}$.

13. Find the volume of the figure below. Each side of each cube measures 4 feet.

Chapter 23
Transformations

23.1 Drawing Geometric Figures on a Cartesian Coordinate Plane

You can use a Cartesian coordinate plane to draw geometric figures by plotting **vertices** and connecting them with line segments.

Example 1: What are the coordinates of each vertex of quadrilateral $ABCD$ below?

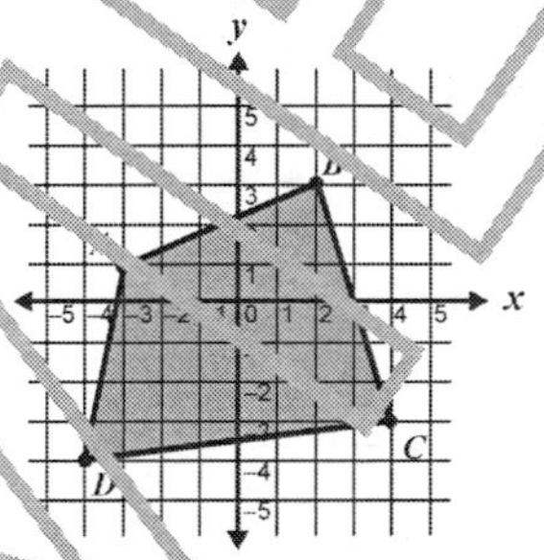

Step 1: To find the coordinates of point A, count over -3 on the x-axis and up 1 on the y axis. point $A = (-3, 1)$.

Step 2: The coordinates of point B are located to the right two units on the x-axis and up 3 units on the y axis. point $B = (2, 3)$.

Step 3: Point C is located 4 units to the right on the x-axis and down -3 on the y-axis. point $C = (4, -3)$.

Step 4: Point D is -4 units left on the x-axis and down -4 units on the y-axis. point $D = (-4, -4)$.

Example 2: Plot the following points. Then construct and identify the geometric figure that you plotted.

$A = (-2, -5)$ $\quad C = (3, 1)$
$B = (-2, 1)$ $\quad D = (3, -5)$

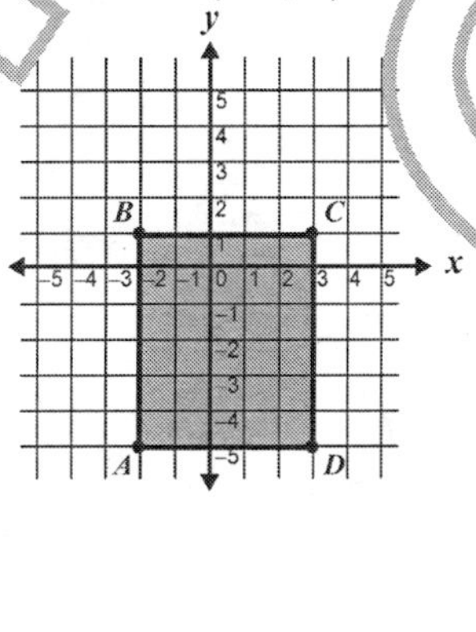

Figure $ABCD$ is a rectangle.

Find the coordinates of the geometric figures graphed below.

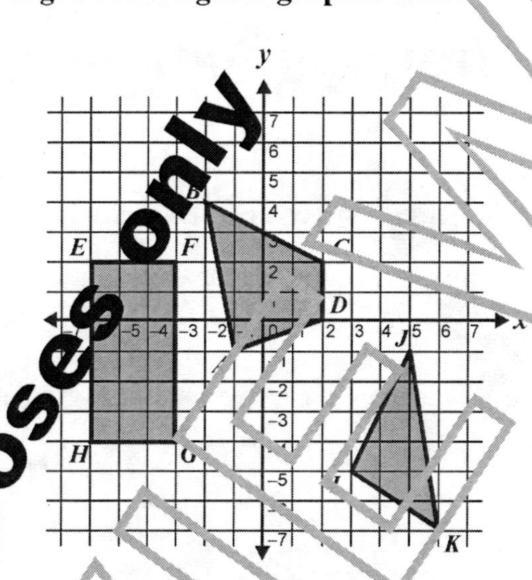

1. Quadrilateral $ABCD$
   $A =$ ________
   $B =$ ________
   $C =$ ________
   $D =$ ________

2. Rectangle $EFGH$
   $E =$ ________
   $F =$ ________
   $G =$ ________
   $H =$ ________

3. Triangle IJK
   $I =$ ________
   $J =$ ________
   $K =$ ________

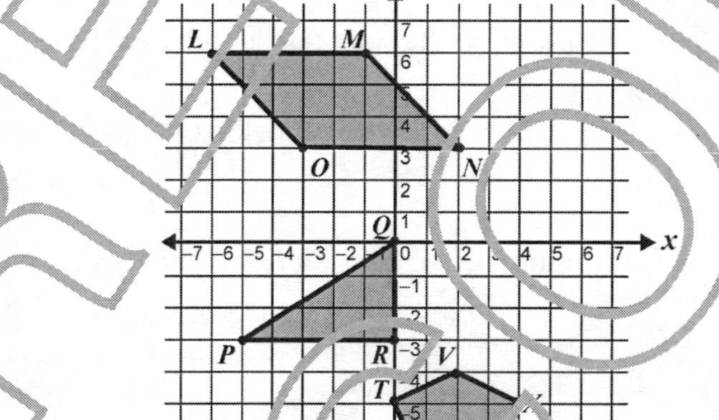

4. Parallelogram $LMNO$
   $L =$ ________
   $M =$ ________
   $N =$ ________
   $O =$ ________

5. Right Triangle PQR
   $P =$ ________
   $Q =$ ________
   $R =$ ________

6. Pentagon $STVXY$
   $S =$ ________
   $T =$ ________
   $V =$ ________
   $X =$ ________
   $Y =$ ________

Plot and label the following points. Then construct and identify the geometric figure you plotted. Question 1 is done for you.

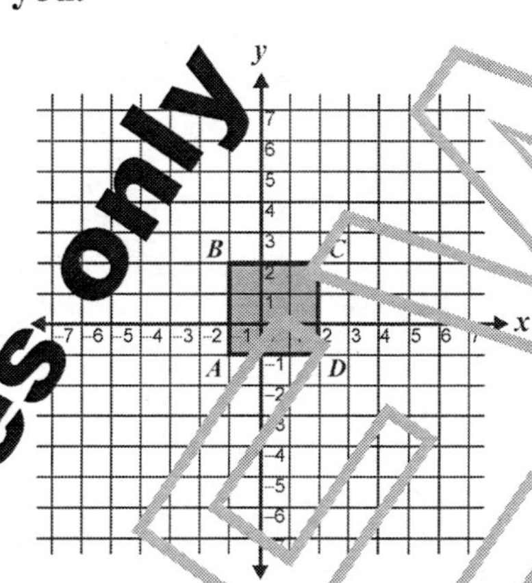

		Figure			Figure
1.	Point $A = (-1, -1)$ Point $B = (-1, 2)$ Point $C = (2, 2)$ Point $D = (2, -1)$	square	3.	Point $H = (-4, 0)$ Point $I = (-6, 0)$ Point $J = (-4, 4)$ Point $K = (-2, 4)$	________
2.	Point $E = (3, -2)$ Point $F = (5, 1)$ Point $G = (7, -2)$	________	4.	Point $L = (-1, -3)$ Point $M = (4, -6)$ Point $N = (-1, -6)$	________

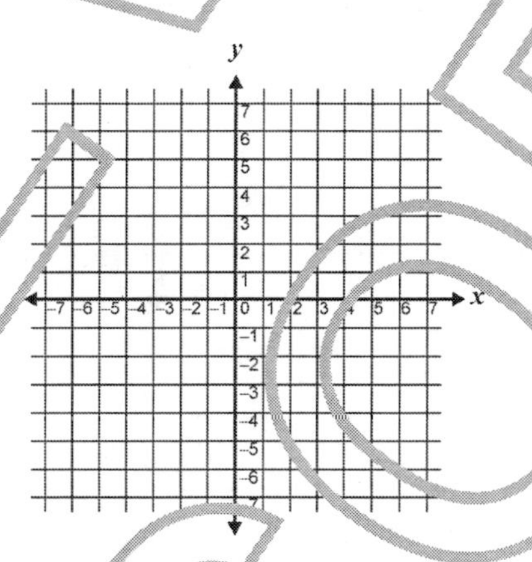

		Figure			Figure
5.	Point $A = (-2, -3)$ Point $B = (-3, 5)$ Point $C = (-1, 6)$ Point $D = (1, 5)$ Point $E = (0, 3)$	________	7.	Point $J = (-1, 2)$ Point $K = (-1, -1)$ Point $L = (3, -2)$	________
6.	Point $F = (-1, -3)$ Point $G = (-3, -5)$ Point $H = (-1, -7)$ Point $I = (1, -5)$	________	8.	Point $M = (6, 2)$ Point $N = (6, -4)$ Point $O = (4, -4)$ Point $P = (4, 2)$	________

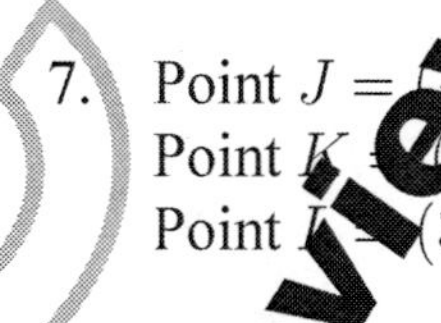

23.2 Reflections

A **reflection** of a geometric figure is a mirror image of the object. Placing a mirror on the **line of reflection** will give you the position of the reflected image.

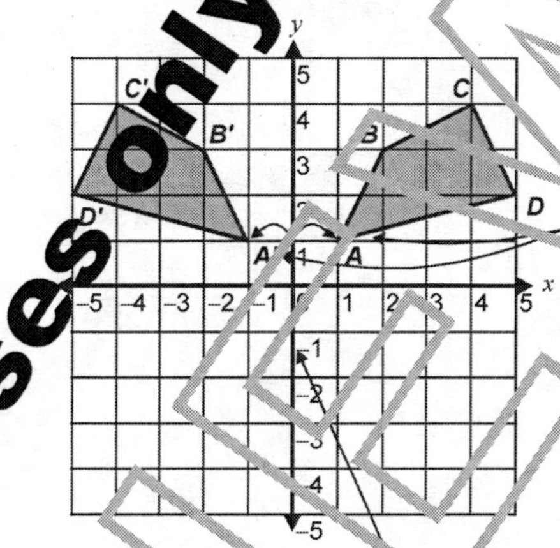

Line of reflection: y-axis

Quadrilateral $ABCD$ is reflected across the y-axis to form quadrilateral $A'B'C'D'$. The y-axis is the line of reflection. Point A' (read as A prime) is the reflection of point A, point B' corresponds to point B, C' to C, and D' to D.

Point A is $+1$ space from the y-axis. Point A's mirror image, point A', is -1 space from the y-axis.
Point B is $+2$ spaces from the y-axis. Point B' is -2 spaces from the y-axis.
Point C is $+4$ spaces from the y-axis and point C' is -4 spaces from the y-axis.
Point D is $+5$ spaces from the y-axis and point D' is -5 spaces from the y-axis.

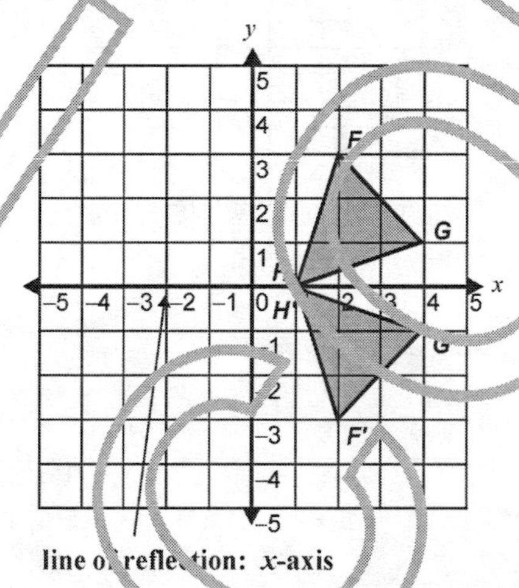

line of reflection: x-axis

Triangle FGH is reflected across the x-axis to form triangle $F'G'H'$. The x-axis is the line of reflection. Point F' reflects point F. Point G' corresponds to point G and H' mirrors H.

Point F is $+3$ spaces from the x-axis. Likewise, point F' is -3 spaces from the x-axis.
Point G is $+1$ space from the x-axis, and point G' is -1 space from the x-axis.
Point H is 0 spaces from the x-axis, so point H' is also 0 spaces from the x-axis.

Reflecting Across a 45° Line

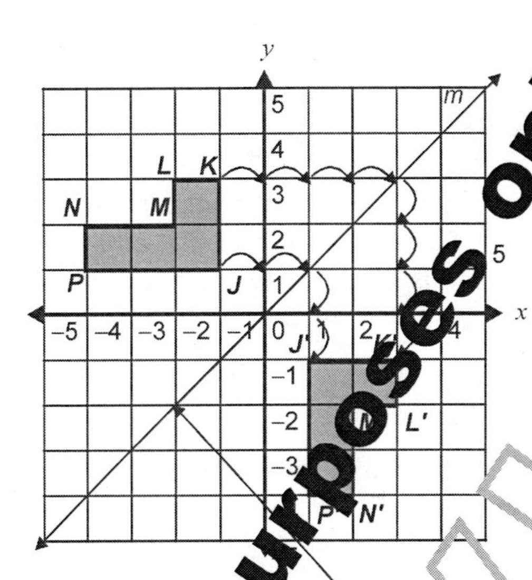

line of reflection, line m

Figure $JKLMNP$ is reflected across line m to form figure $J'K'L'M'N'P'$. Line m is at a 45° angle. Point J corresponds to J', K to K', L to L', M to M', N to N' and P to P'. Line m is the line of reflection. **Pay close attention to how to determine the mirror image of figure $JKLMNP$ across line m described below. This method only works when the line of reflection is at a 45° angle.**

Point J is 2 spaces over from line m, so J' must be 2 spaces down from line m.

Point K is 4 spaces over from line m, so K' is 4 spaces down from line m, and so on.

Draw the following reflections and record the new coordinates of the reflection. The first problem is done for you.

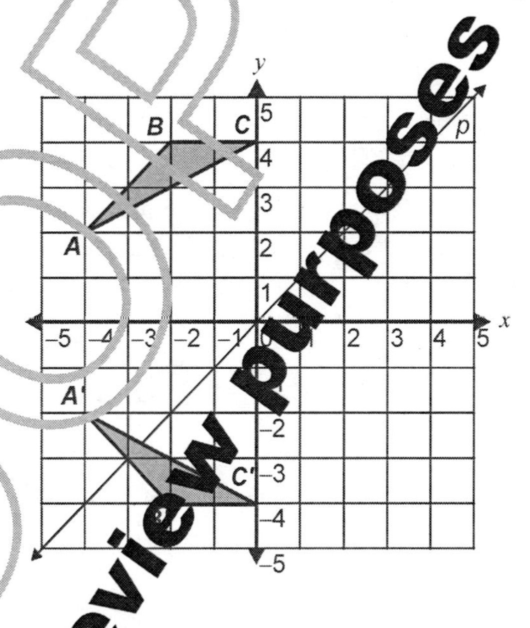

1. Reflect figure ABC across the x-axis. Label vertices $A'B'C'$ so that point A' is the reflection of point A, B' is the reflection of B and C' is the reflection of C.
   $A' =$ (−4, −2) $B' =$ (−2, −4) $C' =$ (0, −4)
2. Reflect figure ABC across the y-axis. Label vertices $A''B''C''$ so that point A'' is the reflection of point A, B'' is the reflection of B, and C'' is the reflection of C.
   $A'' =$ ______ $B'' =$ ______ $C'' =$ ______
3. Reflect figure ABC across line p. Label vertices $A'''B'''C'''$ so that point A''' is the reflection of point A, B''' is the reflection of B, and C''' is the reflection of C.
   $A''' =$ ______ $B''' =$ ______ $C''' =$ ______

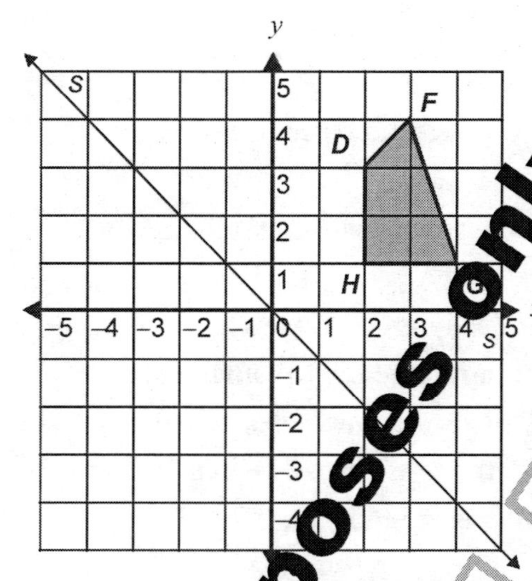

4. Reflect figure $DFGH$ across the y-axis. Label vertices $D'F'G'H'$ so that point D' is the reflection of point D, F' is the reflection of F, G' is the reflection of G, and H' is the reflection of H.

D' = ______ G' = ______
F' = ______ H' = ______

5. Reflect figure $DFGH$ across the x-axis. Label vertices D'', F'', G'', H'' so that point D'' is the reflection of D, F'' is the reflection of F, G'' is the reflection of G, and H'' is the reflection of H.

D'' = ______ G'' = ______
F'' = ______ H'' = ______

6. Reflect figure $DFGH$ across line s. Label vertices $D'''F'''G'''H'''$ so that point D''' is the reflection of D, F''' corresponds to F, G''' to G, and H''' to H.

D''' = ______ G''' = ______
F''' = ______ H''' = ______

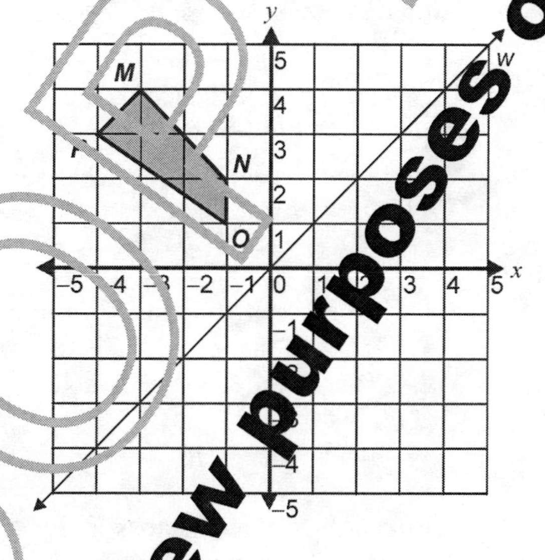

7. Reflect quadrilateral $MNOP$ across the y-axis. Label vertices $M'N'O'P'$ so that point M' is the reflection of point M, N' is the reflection of N, O' is the reflection of O, and P' is the reflection of P.

M' = ______ O' = ______
N' = ______ P' = ______

8. Reflect figure $MNOP$ across the x-axis. Label vertices M'', N'', O'', P'' so that point M'' is the reflection of M, N'' is the reflection of N, O'' is the reflection of O, and P'' is the reflection of P.

M'' = ______ O'' = ______
N'' = ______ P'' = ______

9. Reflect figure $MNOP$ across line w. Label vertices $M'''N'''O'''P'''$ so that point M''' is the reflection of M, N''' corresponds to N, O''' to O, and P''' to P.

M''' = ______ O''' = ______
N''' = ______ P''' = ______

23.3 Translations

To make a translation of a geometric figure, first duplicate the figure and then slide it along a path.

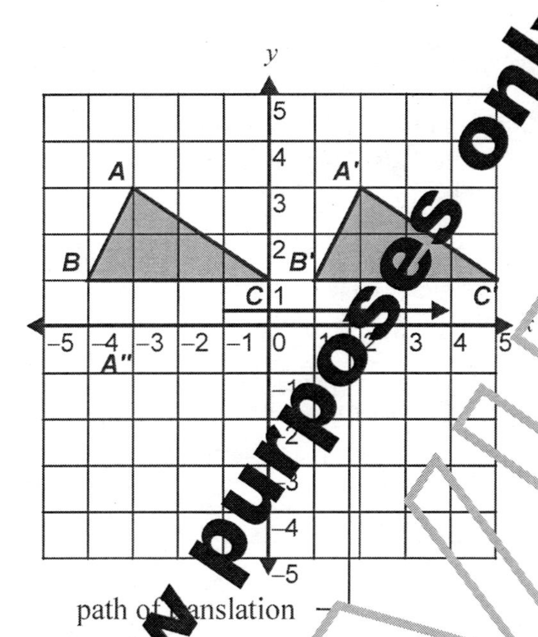

Triangle $A'B'C'$ is a translation of triangle ABC. Each point is translated 5 spaces to the right. In other words, the triangle slid 5 spaces to the right. Look at the path of translation. It gives the same information as above. Count the number of spaces across given by the path of translation, and you will see it represents a move 5 spaces to the right. Each new point is found at $(x + 5, y)$.

Point A is at $(-3, 3)$. Therefore, A' is found at $(-3 + 5, 3)$ or $(2, 3)$.

B is at $(-4, 1)$, so B' is at $(-4 + 5, 1)$ or $(1, 1)$.

C is at $(0, 1)$, so C' is at $(0 + 5, 1)$ or $(5, 1)$.

Quadrilateral $FGHI$ is translated 5 spaces to the right and 3 spaces down. The path of translation shows the same information. It points right 5 spaces and down 3 spaces. Each new point is found at $(x + 5, y - 3)$.

Point F is located at $(-4, 3)$. Point F' is located at $(-4 + 5, 3 - 3)$ or $(1, 0)$.

Point G is at $(-2, 5)$. Point G' is at $(-2 + 5, 5 - 3)$ or $(3, 2)$.

Point H is at $(-1, 4)$. Point H' is at $(-1 + 5, 4 - 3)$ or $(4, 1)$.

Point I is at $(-1, 2)$. Point I' is at $(-1 + 5, 2 - 3)$ or $(4, -1)$.

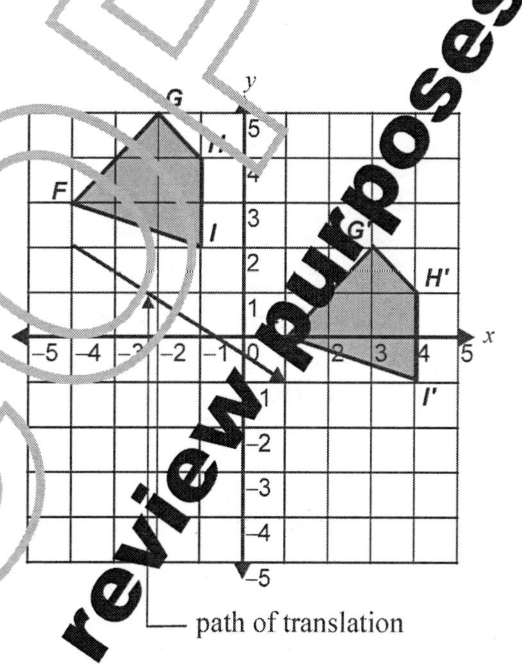

Draw the following translations and record the new coordinates of the translation. The figure for the first problem is drawn for you.

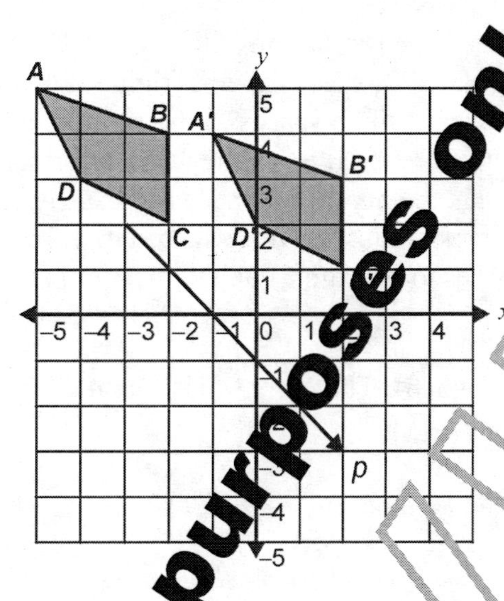

1. Translate figure $ABCD$ 4 spaces to the right and 1 space down. Label the vertices of the translated figure A', B', C', and D' so that point A' corresponds to the translation of point A, B' corresponds to B, C' to C, and D' to D.

   A' = ______ C' = ______
   B' = ______ D' = ______

2. Translate figure $ABCD$ 5 spaces down. Label the vertices of the translated figure A'', B'', C'', and D'' so that point A'' corresponds to the translation of point A, B'' corresponds to B, C'' to C, and D'' to D.

   A'' = ______ C'' = ______
   B'' = ______ D'' = ______

3. Translate figure $ABCD$ along the path of translation, p. Label the vertices of the translated figure A''', B''', C''', and D''' so that point A''' corresponds to the translation of point A, B''' corresponds to B, C''' to C, and D''' to D.

   A''' = ______ C''' = ______
   B''' = ______ D''' = ______

4. Translate triangle FGH 6 spaces to the left and 3 spaces up. Label the vertices of the translated figure F', G', and H' so that point F' corresponds to the translation of point F, G' corresponds to G, and H' to H.

   F' = ______ G' = ______ H' = ______

5. Translate triangle FGH 4 spaces up and 1 space to the left. Label the vertices of the translated triangle $F''G''H''$ so that point F'' corresponds to the translation of point F, G'' corresponds to G, and H'' to H.

   F'' = ______ G'' = ______ H'' = ______

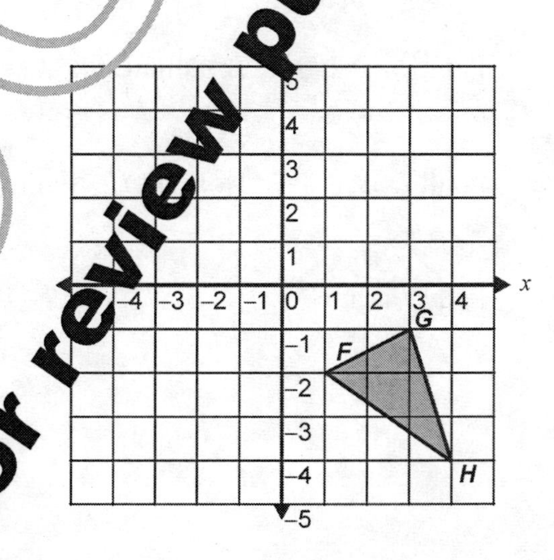

23.4 Rotations

A **rotation** of a geometric figure shows motion around a point.

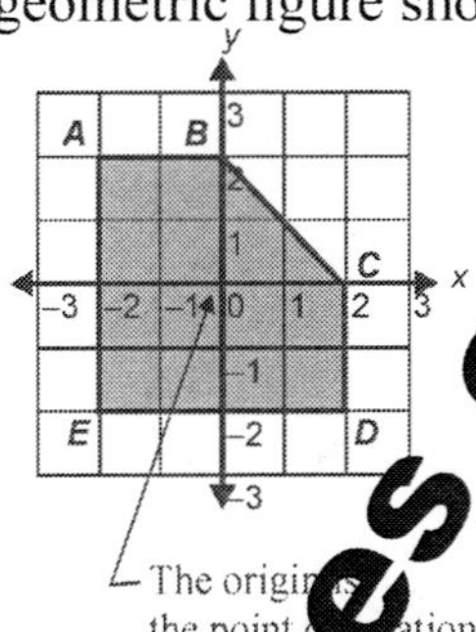

The origin is the point of rotation.

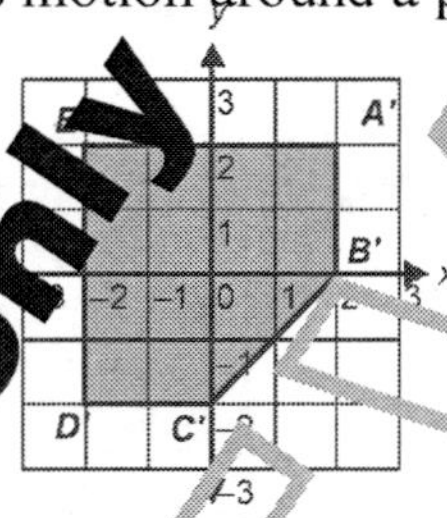

Figure $ABCDE$ has been rotated $\frac{1}{4}$ of a turn clockwise around the origin to form $A'B'C'D'E'$.

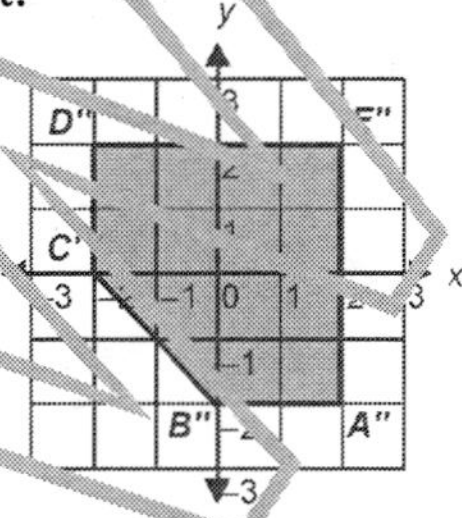

Figure $ABCDE$ has been rotated $\frac{1}{2}$ of a turn around the origin to form $A''B''C''D''E''$.

Draw the following rotations, and record the new coordinates of the rotation. The figure for the first problem is drawn for you.

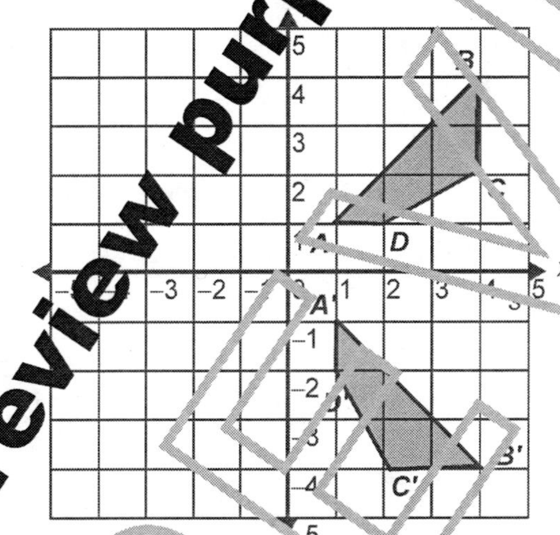

1. Rotate figure $ABCD$ around the origin clockwise $\frac{1}{4}$ turn. Label the vertices A', B', C', and D' so that point A' corresponds to the rotation of point A, B' corresponds to B, C' to C, and D' to D.
   A' = ______ C' = ______
   B' = ______ D' = ______

2. Rotate figure $ABCD$ around the origin clockwise $\frac{1}{2}$ turn. Label the vertices A'', B'', C'', and D'' so that point A'' corresponds to the rotation of point A, B'' corresponds to B, C'' to C, and D'' to D.
   A'' = ______ C'' = ______
   B'' = ______ D'' = ______

3. Rotate figure $ABCD$ around the origin clockwise $\frac{3}{4}$ turn. Label the vertices A''', B''', C''', and D''' so that point A''' corresponds to the rotation of point A, B''' corresponds to B, C''' to C, and D''' to D.
   A''' = ______ C''' = ______
   B''' = ______ D''' = ______

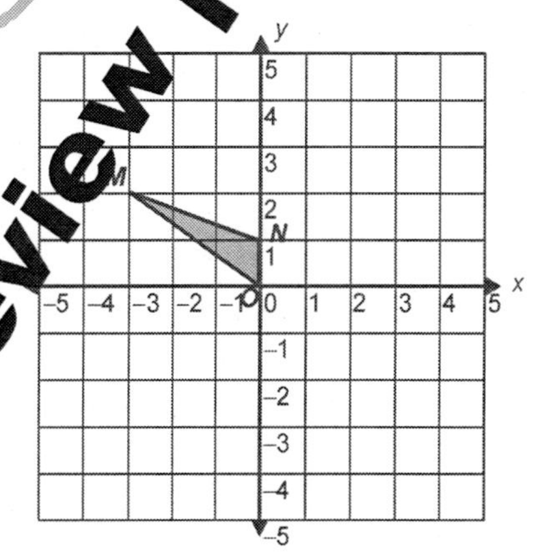

4. Rotate figure MNO around point O clockwise $\frac{1}{4}$ turn. Label the vertices M', N', and O so that point M' corresponds to the rotation of point M and N' corresponds to N.
   M' = ______ N' = ______

5. Rotate figure MNO around point O clockwise $\frac{1}{2}$ turn. Label the vertices M'', N'', and O so that point M'' corresponds to the rotation of point M, and N'' corresponds to N.
   M'' = ______ N'' = ______

6. Rotate figure MNO around point O clockwise $\frac{3}{4}$ turn. Label the vertices M''', N''', and O so that point M''' corresponds to the rotation of point M, and N''' corresponds to N.

   M''' = ______ N''' = ______

23.5 Transformation Practice

Answer the following questions regarding transformations.

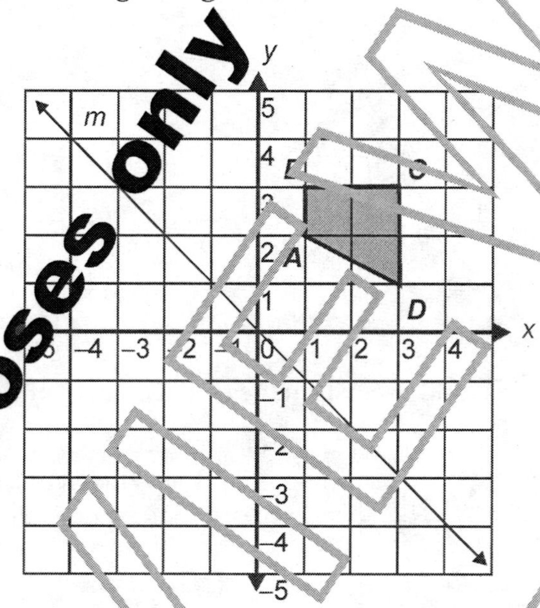

1. Translate quadrilateral $ABCD$ so that point A', which corresponds to point A, is located at coordinates $(-4, 2)$. Label the other vertices B' to correspond to B, C' to C, and D' to D. What are the coordinates of B', C', and D'?

   $A' =$ _____ $C' =$ _____

   $B' =$ _____ $D' =$ _____

2. Reflect quadrilateral $ABCD$ across line m. Label the coordinates A'', B'', C'', and D'', so that point A'' corresponds to the reflection of point A, B'' corresponds to the reflection of B, and C'' corresponds to the reflection of C. What are the coordinates of A'', B'', C'', and D''?

   $A'' =$ _____ $C'' =$ _____

   $B'' =$ _____ $D'' =$ _____

3. Rotate quadrilateral $ABCD$ $\frac{1}{4}$ turn counterclockwise around point D. Label the points $A'''B'''C'''D'''$ so that A''' corresponds to the rotation of point A, B''' corresponds to B, C''' to C and D''' to D. What are the coordinates of A''', B''', C''' and D'''?

   $A''' =$ _____ $C''' =$ _____

   $B''' =$ _____ $D''' =$ _____

23.6 Dilations

A **dilation** of a geometric figure is either an enlargement or a reduction of the figure. The point at which the figure is either reduced or enlarged is called the center of dilation. The dilation of a figure is always the product of the original and a **scale factor**. The scale factor is always a positive number that is multiplied by the coordinates of a shape's vertices, which is usually illustrated in a coordinate plane. If the scale factor is greater than one, then the resulting dilated figure will be an enlargement of the original figure. If the scale factor is less than one, then the resulting dilated figure will be a reduction of the original figure.

Example 3: The triangle ABC has been dilated by a scale factor of $\frac{1}{4}$.

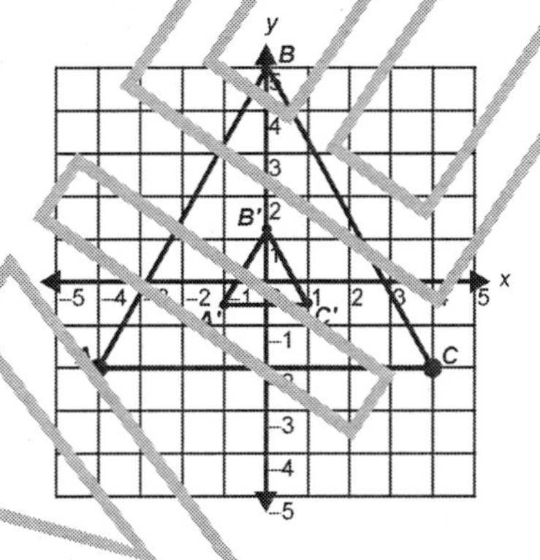

The first step in finding the dilated object is to list all the vertices of the original object, ABC. The next step is to multiply the coordinates of the vertices of ABC by the scale factor, 3, to find the coordinates of the dilated figure. Lastly, draw the dilated object on the coordinate plane as shown above.

$A : (-4, -2)$ $A' : (-1, -\frac{1}{2})$
$B : (0, 5)$ $B' : (0, \frac{5}{4})$
$C : (4, -2)$ $C' : (1, -\frac{1}{2})$

Note: Since the scale factor is less than one, the dilated figure $A'B'C'$ is a reduction of original triangle, ABC.

Circle the coordinate plane that contains the shape that has been dilated.

(A)

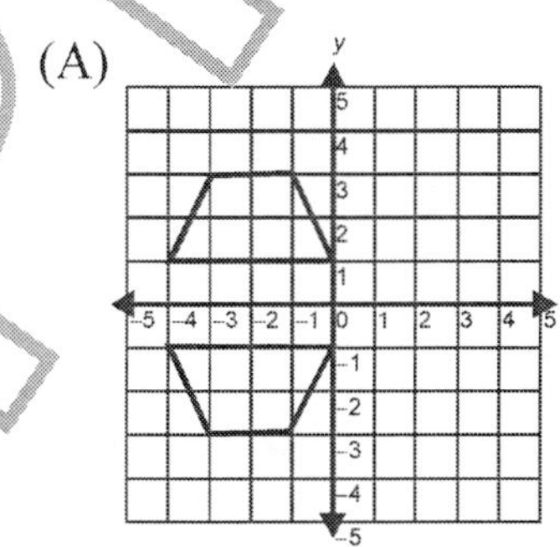

(B)

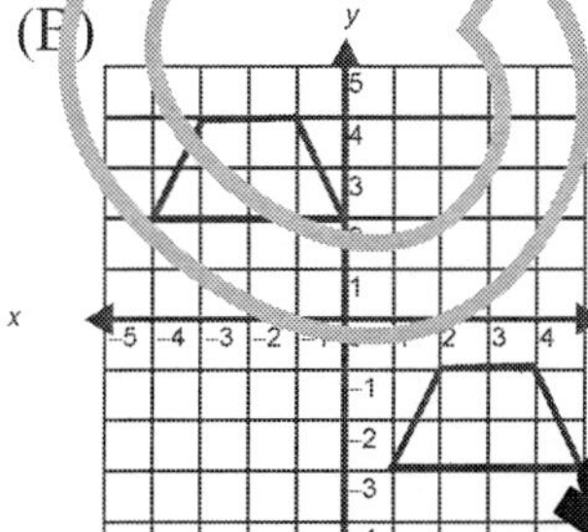

(C)

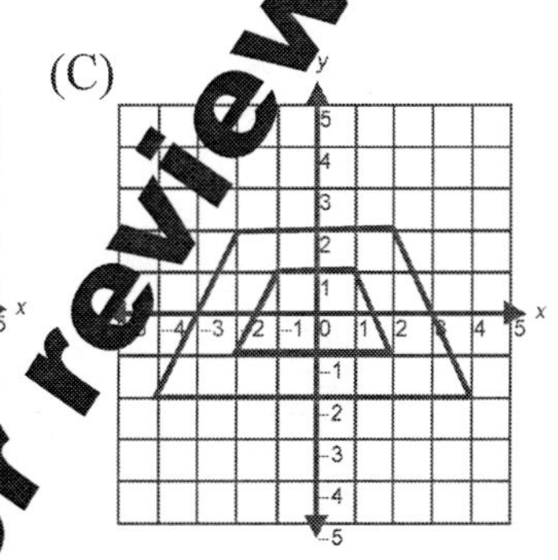

On your own graph paper sketch the dilated and original figures.

For questions 1–6, find the coordinates of the vertices of the dilated figure.

1. A: $(-3,1)$
   B: $(-1,4)$
   C: $(1,4)$
   D: $(3,1)$
   Scale factor: 4

2. A: $(-6,5)$
   B: $(3,5)$
   C: $(3,-4)$
   D: $(-6,-4)$
   Scale factor: $\frac{1}{3}$

3. A: $(-10,0)$
   B: $(0,10)$
   C: $(8,5)$
   Scale factor: $\frac{4}{5}$

4. A: $(-1,7)$
   B: $(1,7)$
   C: $(5,5)$
   D: $(5,\frac{1}{2})$
   E: $(1,-3)$
   F: $(-1,-3)$
   G: $(-5,\frac{1}{2})$
   H: $(-5,5)$
   Scale factor: 2

5. A: $(-8,7)$
   B: $(-4,7)$
   C: $(-2,3)$
   D: $(-6,3)$
   Scale factor: $\frac{3}{[illegible]}$

6. A: $(-4,12)$
   B: $(6,-2)$
   C: $(-14,-2)$
   Scale factor: $\frac{1}{2}$

For questions 7–10, find the scale factor.

7. A: $(-3,2)$ A': $(-10.5,7)$
   B: $(1,2)$ B': $(3.5,7)$
   C: $(1,-3)$ C': $(3.5,-10.5)$
   D: $(-3,-3)$ D': $(-10.5,-10.5)$

8. A: $(-6,9)$ A': $(-2,3)$
   B: $(3,12)$ B': $(1,4)$
   C: $(6,3)$ C': $(2,1)$
   D: $(-9,0)$ D': $(-3,0)$

9. A: $(0,-3)$ A': $(0,-2)$
   B: $(6,0)$ B': $(4,0)$
   C: $(0,3)$ C': $(0,2)$

10. A: $(-2,6)$ A': $(-10,30)$
    B: $(2,6)$ B': $(10,30)$
    C: $(3,3)$ C': $(15,15)$
    D: $(2,0)$ D': $(10,0)$
    E: $(-2,0)$ E': $(-10,0)$
    F: $(-3,3)$ F': $(-15,15)$

For questions 11 and 12, determine whether or not $A'B'C'D'$ is a dilation of $ABCD$.

11. A: $(-2,5)$ A': $(-1,2)$
    B: $(8,8)$ B': $(4,4)$
    C: $(12,0)$ C': $(6,0)$
    D: $(2,-6)$ D': $([illegible],-3)$

12. A: $(0,8)$ A': $(-2,6)$
    B: $(5,8)$ B': $(3,6)$
    C: $(5,-3)$ C': $(3,-1)$
    D: $(0,-3)$ D': $(-2,-1)$

Chapter 23 Review

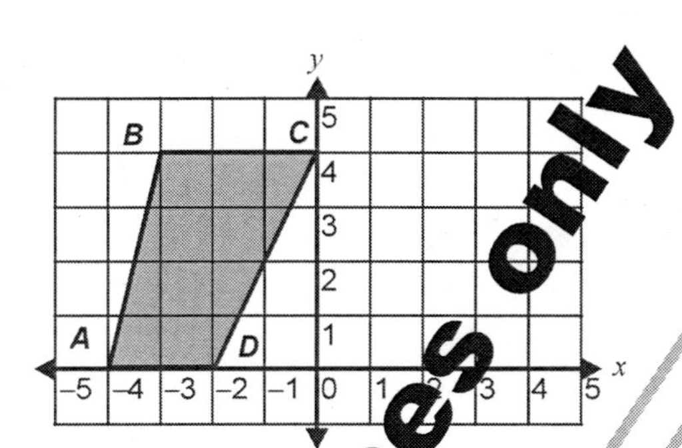

1. Draw the reflection of image $ABCD$ over the y-axis. Label the points A', B', C', and D'. List the coordinates of these points below.

2. A' _____ 4. C' _____

3. B' _____ 5. D' _____

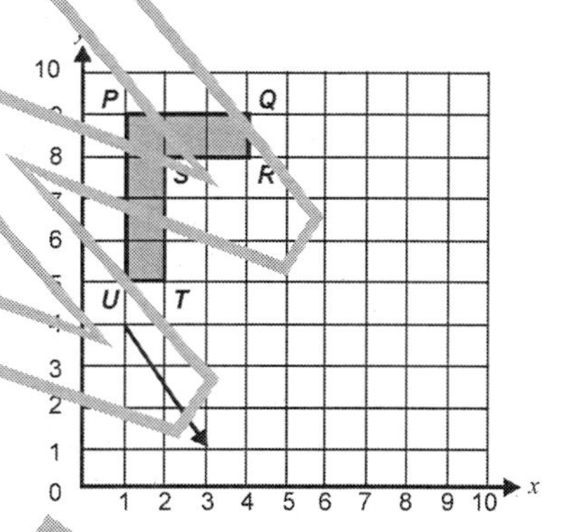

6. Use the translation described by the arrow to translate the polygon above. Label the points P', Q', R', S', T', and U'. List the coordinates of each.

7. P' _____ 10. S' _____

8. Q' _____ 11. T' _____

9. R' _____ 12. U' _____

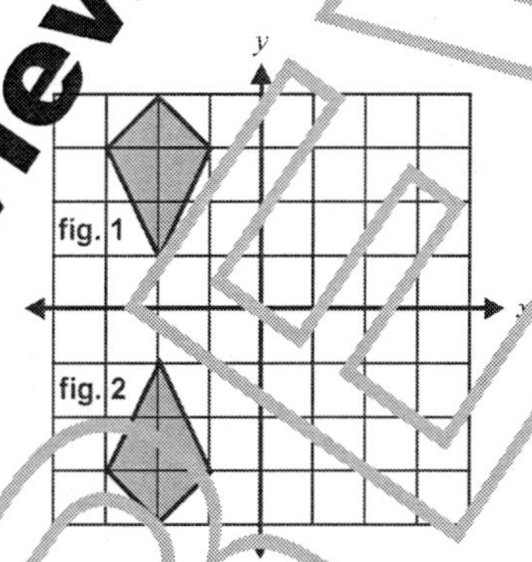

13. Figure 1 goes through a transformation to form figure 2. Which of the following descriptions fits the transformation shown?
A. reflection across the x-axis
B. reflection across the y-axis
C. translation down 5 units
D. translation down 2 units

For questions 14–16, find the coordinates of the vertices of the dilated figures.

14. $(3, 4)$, $(6, 10)$, $(-3, 5)$ scale factor 2

15. $(6, 5)$, $(4, 5)$, $(3, 7)$ scale factor $\frac{1}{2}$

16. $(-8, 7)$, $(-2, 7)$, $(-3, 3)$, $(-6, 3)$ scale factor $\frac{3}{[illegible]}$

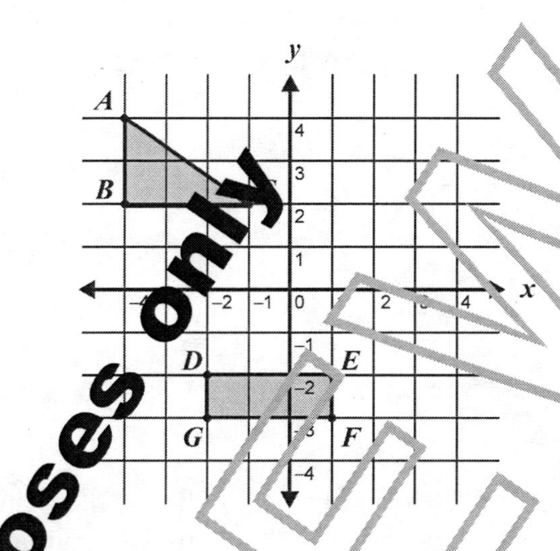

Find the coordinates of the geometric figures graphed above.

17. point A
18. point B
19. point C
20. point D
21. point E
22. point F
23. point G

Plot and label the following points on the same graph.

24. point $H = (1, 1)$
25. point $I = (3, 1)$
26. point $J = (4, -2)$
27. point $K = (2, -2)$
28. What type of figure did you plot?

New SAT Mathematics
Reference Sheet

Notes

1. The use of a calculator is permitted.
2. All numbers used are real numbers.
3. Figures that accompany problems in this test are intended to provide information useful in solving the problems.
   They are drawn as accurately as possible EXCEPT when it is stated in a specific problem that the figure is not drawn to scale. All figures lie in a plane unless otherwise indicated.
4. Unless otherwise specified, the domain of any function f is assumed to be the set of all real numbers x for which $f(x)$ is a real number.

Reference Information

$A = \pi r^2$
$C = 2\pi r$

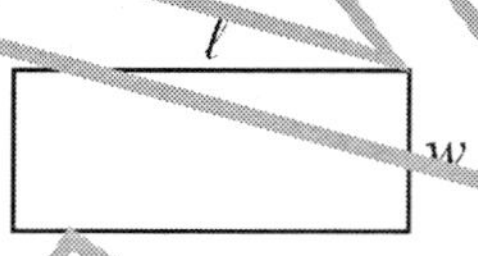

$A = \ell w$

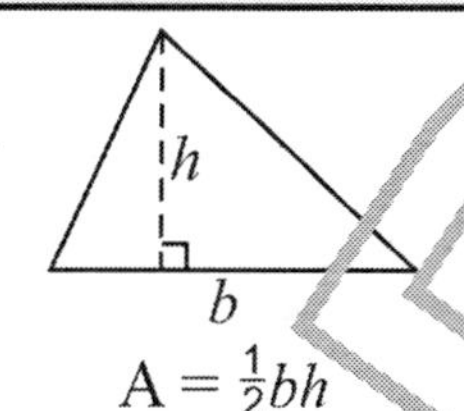

$A = \frac{1}{2}bh$

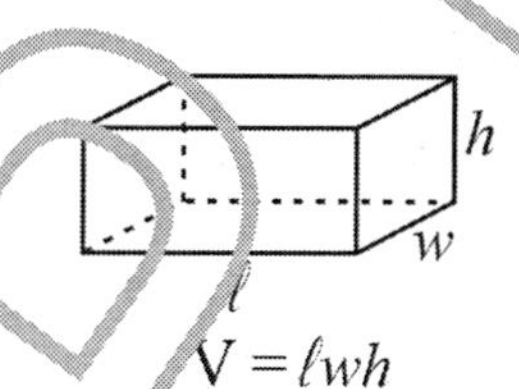

$V = \ell wh$

$V = \pi r^2 h$

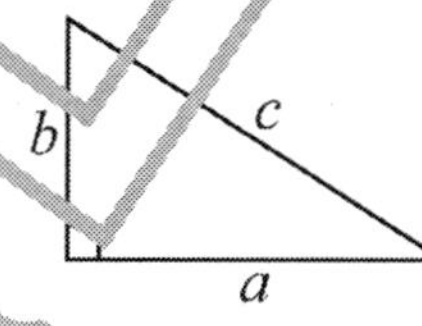

$c^2 = a^2 + b^2$

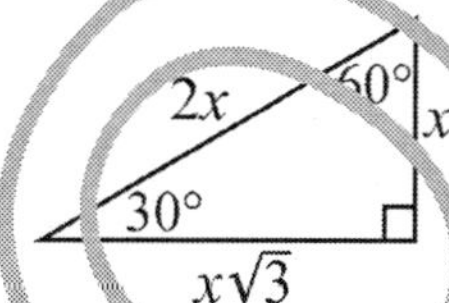

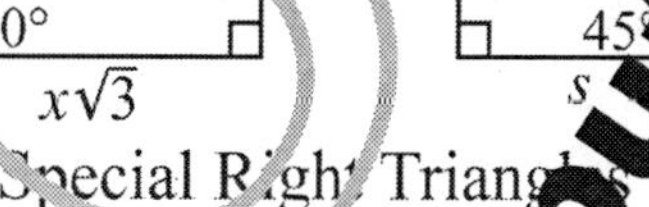

Special Right Triangles

s, $45°$, $s\sqrt{2}$, $45°$, s

The number of degrees of arc in a circle is 360.

The sum of the measures in a degrees of the angles of a triangle is 180.

Practice Test 1

Part 1

Questions 1–8 of part 1 of this test are multiple choice, and questions 9–18 are student-produced response questions. For questions 9–18, use the grids provided by your instructor. You have 25 minutes to complete this part of the test.

Note to instructor: A blank sheet of grids is located at the end of the answer key.

1. $\triangle ABC$ has vertices $A(3,8)$, $B(7,6)$, and $C(5,2)$. What are the coordinates of the vertices of $\triangle A'B'C'$ after the translation $(x-8, y-10)$?

   (A) $A'(-5,18)$, $B'(-3,12)$, $C'(-1,16)$
   (B) $A'(-5,-2)$, $B'(-1,-4)$, $C'(-3,-8)$
   (C) $A'(11,-2)$, $B'(15,-4)$, $C'(13,-8)$
   (D) $A'(11,18)$, $B'(15,16)$, $C'(13,12)$
   (E) $A'(-2,-5)$, $B'(-1,4)$, $C'(-3,-8)$

2. Mrs. Smith bought 7 sweaters of two different styles. The first style cost \$30 each and the second style cost \$20 each. The total price of the sweaters was \$160. How many of each style sweaters did Mrs. Smith buy?

   (A) 2 of the \$30 and 5 of the \$20
   (B) 5 of the \$30 and 2 of the \$20
   (C) 3 of the \$30 and 4 of the \$20
   (D) 1 of the \$30 and 6 of the \$20
   (E) 4 of the \$30 and 3 of the \$20

3. The area of Circle A is given by $A = \pi(9x^2 + 30x + 25)$. What is the radius of Circle A?

   (A) $9x^2 - 25$
   (B) $9x^2 + 25$
   (C) $|3x + 5|$
   (D) $3x^2 - 5$
   (E) $3x^2 + 5$

4. An electric scooter shop charges \$8 to rent a scooter plus \$1.50 for every half hour you ride. If the shop charges you and your friend \$25, how many hours did you each ride?

   (A) 3
   (B) 2.5
   (C) 2
   (D) 1.5
   (E) 1

5. Annie is preparing for her vacation to Italy. She bought 8 rolls of film and 2 camera batteries for \$23.00. The next week she bought 6 more rolls of film and 2 camera batteries and the bill was \$18.00. What is the price of a roll of film and a camera battery?

   (A) film, \$3.00; camera battery, \$1.00
   (B) film, \$2.50; camera battery, \$1.50
   (C) film \$4.00; camera battery, \$3.50
   (D) film, \$3.50; camera battery, \$2.00
   (E) film, \$4.50; camera battery, \$2.50

6. Which line passes through the point $(6,1)$ and has a slope of -3?

   (A) $y = -3x - 19$
   (B) $y = 3x + 19$
   (C) $y = -3x - 21$
   (D) $y = 3x - 19$
   (E) $y = -3x + 19$

7. Solve the system of equations:

$2x + 3y + 4z = 3$
$5x - 9y + 6z = 1$
$\frac{1}{3}x - \frac{1}{2}y + \frac{2}{3}z = \frac{1}{6}$

(A) $(2, 3, 5)$
(B) $\left(\frac{1}{3}, \frac{1}{2}, \frac{1}{4}\right)$
(C) $(1, 3, 2)$
(D) $\left(\frac{1}{2}, \frac{1}{3}, \frac{1}{4}\right)$
(E) $(4, 2, 3)$

8. Last year Eric attended one-half the number of sporting events that Mike did. Chris attended one-third the number that Mike did. If Chris attended 8 sporting events, how many did Eric attend?

(A) 1
(B) 4
(C) 8
(D) 12
(E) [illegible]

9. Find the area of a triangle whose vertices are located at $(-3, 2)$, $(0, -5)$, and $(4, 2)$.

10. How many different ways can the letters of "MATH" be arranged?

11. Suppose that the wealth of a millionaire is increasing linearly. In 1993, he had \$40 million. In 2001, he had \$55 million. How much money will he have in 2009?

12. Solve $y = \arctan\left(\frac{\sqrt{3}}{2}\right)$.

13. Suppose a right triangle has two sides which measure 10 and 13. What is the measure of the hypotenuse?

14. Find the measure of angle complementary to $\angle A = 37°$.

15. If you were going to make a stem-and-leaf plot for the following set of data, how many stems would be needed?

21	51	6[illegible]	29	33
42	5[illegible]	72	15	39
[illegible]	28	36	30	52

16. What is the $m\angle 2$ in the diagram below?

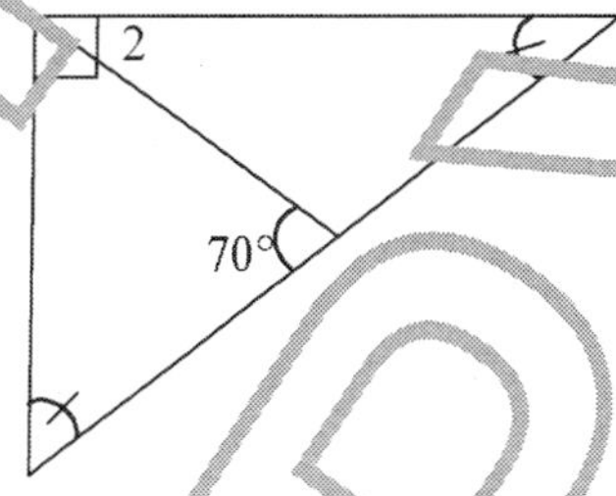

17. Time is measured in seconds and velocity in feet per second. If a car's velocity is described by the function $v(t) = $ [illegible] $+ 3t + t^2$, what is the car's velocity at $t = 4$?

18. A right rectangular prism has a width of 7 meters and a length of [illegible] meters. If the surface area of the prism is $2,142$ square meters, what is its height?

Part 2

Questions 1–20 of part 2 of this test are multiple choice. You have 25 minutes to complete this part of the test.

1. A triangle has two sides that have lengths of 16 inches and 28 inches. Which of the following could not represent the length of the third side?

   (A) 43 inches
   (B) 40 inches
   (C) 33 inches
   (D) 26 inches
   (E) 10 inches

2. In $1\frac{1}{2}$ hours, the minute hand of a clock rotates through an angle of how many degrees?

   (A) 60°
   (B) 180°
   (C) 360°
   (D) 4[illegible]°
   (E) 720°

3. Let $\overrightarrow{v} = (-2, y)$ and $\overrightarrow{w} = (x, 4)$. If $\overrightarrow{v} + \overrightarrow{w} = (6, 11)$, what is the value of x and y?

   (A) $x = 8, y = 8$
   (B) $x = 7, y = 8$
   (C) $x = 4, y = 7$
   (D) $x = 8, y = 7$
   (E) $x = 4, y = 8$

4. Jim owns $\frac{1}{4}$ of a business. He sells half of his shares for $\$12,000$. What is the total value of the business?

   (A) $\$10,000$
   (B) $\$48,000$
   (C) $\$96,000$
   (D) $\$108,000$
   (E) $\$120,000$

5. Which quadrants of the coordinate plane would contain the graph $y = \sqrt{3x - 1} - 2$?

   (A) Quadrant I
   (B) Quadrants I and II
   (C) Quadrants II and III
   (D) Quadrants I and IV
   (E) Quadrant IV

6. If $\frac{1}{x} = \sqrt{0.25}$, then $x^2 = ?$

   (A) 0.25
   (B) 25
   (C) 4
   (D) 400
   (E) 250

7. Points A, B, C, D, and E lie on a line. Point B is between A and C, point D is between C and E, and point E is between A and D. Which of the following cannot be the position of D?

   (A) between A and B
   (B) between A and E
   (C) between B and C
   (D) between B and E
   (E) Cannot be determined by the information given

8. Which function does not have a zero?

   (A) $f(x) =$ [illegible] 3
   (B) $f(x) =$ [illegible]
   (C) $f(x) = 2x + 3$
   (D) $f(x) = -3x + 2$
   (E) $f(x) = -x + 7$

9. The angle of elevation of a ladder leaning against a wall is $55°$. The ladder is 30 feet long. How high up the wall does it reach?

(A) about 17.21 ft
(B) about 24.57 ft
(C) about 33.67 ft
(D) about 42.84 ft
(E) about 52.30 ft

10. Two angles are complementary and one angle has a measure that is 9 times the measure of the other angle. What is the measure of the larger angle?

(A) $9°$
(B) $18°$
(C) $81°$
(D) $90°$
(E) $162°$

11. Find the mode for the stem-and-leaf plot shown below.

Stem	Leaf
2	3, 5, 5, 6, 8
3	
4	0, 1, 2, 2
5	2, 3, 3, 3, 9, 9
6	1, 2, 7

$2 \mid 3 = 23$

(A) 25
(B) 52
(C) 27
(D) 53
(E) 59

12. Suppose a ball is thrown straight up at a speed of 50 feet per second. The time in seconds that it takes for the ball to hit the ground can be found by solving the equation $5 + 50t - 16t^2 = 0$. Approximately how long does it take for the ball to hit the ground?

(A) 1.5 s
(B) 0.1 s
(C) 3.2 s
(D) −0.1 s
(E) 4.8 s

13. Which of the following lines is parallel to $y = \frac{2}{3}x - 4$?

(A) $y = -\frac{2}{3}x + 1$
(B) $y = \frac{3}{2}x - 6$
(C) $y - \frac{2}{3}x = 8$
(D) $y = -\frac{3}{2}x - 9$
(E) $y + \frac{3}{2}x = -1$

14. Name the multiplicative inverse of 5.

(A) $\frac{1}{5}$
(B) 5
(C) $-\frac{1}{5}$
(D) −5
(E) 20

15. Solve $x + 7 < 5$ or $x + 6 > 9$.

(A) $-2 > x > 3$
(B) $x > -2$ or $x < 3$
(C) $-2 < x < 3$
(D) $x < -2$ or $x > 3$
(E) $x > 2$ or $x < -3$

16. If $\frac{a}{b} = \frac{x}{y}$, then which of the following is not necessarily true?

(A) $\frac{a}{x} = \frac{b}{y}$
(B) $\frac{a+b}{b} = \frac{x+y}{y}$
(C) $\frac{b}{a} = \frac{y}{x}$
(D) $\frac{a}{y} = \frac{b}{x}$
(E) $ay = bx$

17. Which function is not linear?

(A) $f(x) = 14x + 7$
(B) $f(x) = -3x - 7$
(C) $f(x) = 8$
(D) $f(x) = \frac{2}{3}x - 5$
(E) $f(x) = 2x^3$

18. The table below shows the land-speed records to the nearest mile per hour. Predict the approximate land-speed record for the year 2005. Assume the rate of increase of speed records remains the same.

Year	Speed Record	Year	Speed Record
1906	128	1933	272
1910	132	1935	301
1911	[illegible]	1937	311
1919	[illegible]	1938	358
1920	[illegible]	1939	369
1926	171	1947	394
1927	204	1963	407
[illegible]	208	1964	[illegible]
[illegible]	231	1965	601
1931	246	1970	622
1932	254	1983	633

(A) 980 mph
(B) 830 mph
(C) 800 mph
(D) 700 mph
(E) 650 mph

19. Consider the formula $F(c) = 1.8c + 32$, where c is the temperature in degrees Celsius and $F(c)$ is the temperature in degrees Fahrenheit. If it is 0°C at the beginning of the day, and it is 11°C at noon, what was the change in temperature in degrees Fahrenheit?

(A) -20.8
(B) 19.8
(C) 51.8
(D) 15.8
(E) 11

20. If $x + y = 15$ and $x - y = -19$, then $xy = ?$

(A) 68
(B) -34
(C) -15
(D) -68
(E) -76

Part 3

Questions 1–16 of part 3 of this test are multiple choice. You have 20 minutes to complete this part of the test.

1. Suppose numbers $\frac{5}{6}$, 1, $\frac{1}{3}$, $\frac{3}{2}$, and $\frac{6}{5}$ are arranged from least to greatest on a number line, with each letter (A, B, C, D, E) corresponding to a number. If the greatest number corresponds to E, which letter corresponds to $\frac{5}{6}$? Assume the letters on the number line start at A and end at E.

   (A) A
   (B) B
   (C) C
   (D) D
   (E) Not enough information

2. Choose the ordered pair that satisfies $7x - 15y <$ [illegible]13.

   (A) [illegible]
   (B) [illegible]
   (C) ([illegible], -2)
   (D) $(2, -3)$
   (E) $(0, 2)$

3. Name the sets of numbers to which $\sqrt{63}$ belongs.

   (A) Naturals, Wholes, Integers, Rationals, and Reals
   (B) Integers, Rationals, and Reals
   (C) Rationals and Reals
   (D) Irrationals and Reals
   (E) Irrationals

4. Simplify: $\sqrt{-256}$.

   (A) ± 16
   (B) No real roots
   (C) -16
   (D) 16
   (E) None of the above

5. Solve the equation $x = \sqrt{4x - 3}$

   (A) $x = 1, 3$
   (B) $x = \pm\sqrt{3}$
   (C) $x = -1, -3$
   (D) $x = 2\sqrt{3}, -2\sqrt{3}$
   (E) $x = \frac{3}{4}$

6. Which integer is between $-\frac{51}{4}$ and $-\frac{27}{2}$?

   (A) $-\frac{27}{4}$
   (B) -13
   (C) $-\frac{51}{2}$
   (D) -12
   (E) $-\frac{2[illegible]}{2}$

7. Simplify: $\dfrac{4a^2 - 9}{10a^2 - 13a - 3}$

   (A) $\dfrac{(2a+3)(2a-3)}{5a+1}$
   (B) $\dfrac{2a+3}{5a+1}$
   (C) $2a - 3$
   (D) $\dfrac{2a-3}{5a+1}$
   (E) $\dfrac{2a+3}{10a^2 - 13a - 3}$

8. The equations $a = -0.38b + 56.6$ and $a = 0.38b + 43.4$ represent the percent of men and women, respectively, receiving bachelor's degrees, where b is the number of years since 1968. In approximately what year did the same percent of men and women receive bachelor's degrees?

(A) 1975
(B) 1980
(C) 1985
(D) 1990
(E) 1995

9. Find the axis of symmetry of the graph of $g(x) = x^2 - 5x + 2$.

(A) $-\frac{5}{2}$
(B) $\frac{5}{2}$
(C) 5
(D) -2
(E) 2

10. The Smiths just opened a carpet store. Their startup costs were $\$5,000$ and their cost of carpet is $\$8$ per square yard. How many square yards of carpet do they need to sell in order to break even if they sell the carpet for $\$18$ per square yard?

(A) 500
(B) 50
(C) 278
(D) 196
(E) 192

11. When is it true that $x^2 + x > -2$?

(A) $\frac{1}{2}$
(B) 2
(C) No solution
(D) 0
(E) All real numbers

12. The point $A = (-3, -10)$ is reflected in the line $y = -1$. What are the coordinates of A'?

(A) $(-3, -10)$
(B) $(-3, -8)$
(C) $(-3, 9)$
(D) $(-3, 8)$
(E) $(-3, 10)$

13. A boat is traveling toward a dock following the line $4x - y = 10$. Another boat is traveling toward the dock following the line $x + y = 5$. What are the coordinates of the dock?

(A) $(3, 2)$
(B) $(2, 3)$
(C) $(-2, 3)$
(D) $(3, -2)$
(E) $(-2, -3)$

14. How many possible negative real zeros are there for the polynomial $g(x) = 2x^4 + x^3 - 4x + 1$?

(A) 5
(B) 4
(C) 3
(D) 2
(E) 1

15. Which of the following is prime?

(A) $5xy^2 + 10y$
(B) $7x^2 - y^2$
(C) $3xy + 13x^2$
(D) $3x^2 - 6y$
(E) None of the above

16. Find $4[h(x)]$ if $h(x) = x^2 - 3x + 2$.

(A) $x^2 - 3x + 2$
(B) $4x^2 - 12x + 8$
(C) $16x^2 - 12x + 2$
(D) $4x^2 - 3x + 2$
(E) $4x^2 - 12x + 2$

Practice Test 2

Part 1

Questions 1–8 of part 1 of this test are multiple choice, and questions 9–18 are student-produced response questions. For questions 9–18, use the grids provided by your instructor. You have 25 minutes to complete this part of the test.

Note to instructor: A blank sheet of grids is located at the end of the answer key.

1. What is the radius of the base of a cylinder whose volume is $5,086.8$ cubic yards and height is 20 yards?

   (A) 9 yards
   (B) 18 yards
   (C) 14 yards
   (D) 6 yards
   (E) 12 yards

2. The coordinates of $\triangle PQR$ are $P(1,-4)$, $Q(3,6)$, and $R(-4,1)$. What are the coordinates of $\triangle P'Q'R'$ after translated by $(-2,4)$?

   (A) $P'(-1,-8)$, $Q'(5,2)$, $R'(-2,-3)$
   (B) $P'(-3,0)$, $Q'(-5,2)$, $R'(-6,-3)$
   (C) $P'(3,-8)$, $Q'(5,10)$, $R'(-2,5)$
   (D) $P'(-1,0)$, $Q'(1,10)$, $R'(-6,5)$
   (E) $P'(3,0)$, $Q'(1,2)$, $R'(-2,3)$

3. Mr. and Mrs. Jones have more than ten children. If you square the number of boys and square the number of girls in the family, then add the results together you get 100. How many children do Mr. and Mrs. Jones have?

   (A) 11
   (B) 12
   (C) 14
   (D) 15
   (E) 16

4. Michelle, Whitney, and Michael each had different lunches. One had soup, one had a sandwich, and one had a salad. Michelle did not have a sandwich. Whitney did not have soup or a sandwich. What did each person have for lunch?

   (A) Michelle, soup; Whitney, sandwich; Michael, salad
   (B) Michelle, salad; Whitney, sandwich; Michael, soup
   (C) Michelle, soup; Whitney, salad; Michael, sandwich
   (D) Michelle, sandwich; Whitney, salad; Michael, soup
   (E) Michelle, salad; Whitney, soup; Michael, sandwich

5. B is between A and C, D is between B and C, and C is between B and E. $\overline{AE} = 28$, $\overline{BC} = 10$, and $\overline{AB} = \overline{DB} = \overline{DC}$. What is the length of $\overline{CE}$?

   (A) 5
   (B) 10
   (C) 12
   (D) 13
   (E) 15

6. Adam was traveling from Atlanta to New Orleans by train. His lunch arrived after one-third of the trip was over. When he finished lunch, he had half the distance traveled before lunch left to go. What fraction of the trip did Adam travel while eating lunch?

(A) $\frac{1}{6}$

(B) $\frac{1}{3}$

(C) $\frac{3}{4}$

(D) $\frac{1}{4}$

(E) $\frac{1}{2}$

7. Which of the following is a solution of $4x^2 - 17x = -13$?

(A) -7
(B) -3
(C) 1
(D) [illegible]
(E) [illegible]

8. Jack bought 2 slices of pizza, two cartons of milk, and one chocolate chip cookie for lunch and spent $2.05. Matt bought a sandwich from home, so he just bought a carton of milk and 3 cookies for $0.75. Lauren bought a slice of pizza and a carton of milk for $0.95. How much each does a slice of pizza, a carton of milk, and a chocolate chip cookie cost?

(A) Pizza, $0.65; Milk, $0.30; Cookies, $0.15
(B) Pizza, $0.75; Milk, $0.35; Cookies, $0.25
(C) Pizza, $1.00; Milk, $0.50; Cookies, $0.50
(D) Pizza, $0.75; Milk, $0.25; Cookies, $0.25
(E) Pizza, $0.75; Milk, $0.50; Cookies, $0.25

9. Suppose two dice are rolled. What is the probability of rolling a sum greater than an eight?

10. The probability of being accepted to University College is $\frac{5}{[illegible]}$. What is the probability of not being accepted?

11. Find the value of x so that the mean of $\{3, 5, 7, 8, x\}$ is 9.

12. If one leg of a right triangle is 17.2 cm and the other leg is 22.5 cm, what is the length of the hypotenuse?

13. If X is the midpoint of $\overline{AB}$, $\overline{AX} = 3x + 8$, and $\overline{XB} = 5x - 6$, find the measure of $\overline{AX}$.

14. If Jimmy's car gets 24 miles for each gallon of gasoline, and she has 8 gallons of gasoline in her tank, how far can she drive before having to stop and get gas?

15. The function rule is $3x(x + 5)$. What does $f(x)$ equal when $x = 2$?

x	$f(x)$
4	108
3	72
2	

16. Solve: $3(x + 2) - 6 = 6(x - 5)$

17. Sean earns a weekly base salary of $150 per week plus a commission of 20% of his total sales for the week. Last week Sean earned a total of $510. What was the total, in dollars, of Sean's sales last week?

18. A bricklayer uses 7 bricks per square foot of surface covered. If he covers an area 2 feet wide and 5 feet high, about how many bricks will he use?

Part 2

Questions 1–20 of part 2 of this test are multiple choice. You have 25 minutes to complete this part of the test.

1. A line p has equation $y = \frac{1}{4}x + 3$. If $q \perp p$ and q passes through $(5, -2)$, what is the equation of q?

   (A) $y = \frac{1}{4}x - 18$
   (B) $y = 4x - 18$
   (C) $y = -\frac{1}{4}x + 18$
   (D) $y = -4 + 18$
   (E) $y = \frac{1}{4}x + 18$

2. Find the area of a triangle with vertices $A = (-3, 6)$, $B = (-7, -4)$, and $C = (6, -4)$.

   (A) 60 square units
   (B) 39 square units
   (C) 78 square units
   (D) [illegible] square units
   (E) 52 square units

3. One integer is 2 less than another integer. Three times the reciprocal of the lesser integer plus five times the reciprocal of the greater integer is $\frac{7}{8}$. What are the two integers?

   (A) 8, 10
   (B) 12, 14
   (C) 6, 12
   (D) 8, 14
   (E) 10, 14

4. If $-2x + 7y = -8$ and $x - 6y = 10$, then $x - y = ?$

   (A) 0
   (B) 6
   (C) $\frac{3}{2}$
   (D) $\frac{5}{3}$
   (E) -2

5. In 1992 you bought a rare baseball card for $500 that you expect to increase in value 12% each year for the next 15 years. Estimate the cards value in 2002.

   (A) $861.47
   (B) $986.91
   (C) $1237.98
   (D) $1557.99
   (E) $2001.76

6. Kim can run 2 miles in 25 minutes, what is the rate she runs in miles per hour?

   (A) 0.08
   (B) 0.8
   (C) 3
   (D) 4.8
   (E) 5

7. The base of an isosceles triangle is 18 centimeters long. The altitude to the base is 12 centimeters long. What is the approximate measure of a base angle of the triangle?

   (A) 33.7°
   (B) 36.9°
   (C) 38.7°
   (D) 53.1°
   (E) 56.3°

8. Simplify $\frac{1 \div \frac{1}{x}}{\frac{1}{x}}$.

   (A) 1
   (B) $\frac{1}{x}$
   (C) x
   (D) x^2
   (E) x^3

9. Of the following numbers, which is the greatest?

(A) $\frac{1}{\sqrt{3}}$

(B) $\frac{1}{3\sqrt{3}}$

(C) $\frac{3}{\sqrt{3}}$

(D) $\frac{1}{3}$

(E) $3\sqrt{3}$

10. The side lengths of a cube are doubled. How many times larger is the surface area of the new cube?

(A) 2 times
(B) 3 times
(C) 4 times
(D) [illegible] times
(E) 16 times

11. If $abc = 8$ and $b = c$ then $a = ?$

(A) b^2

(B) $\frac{8}{b^2}$

(C) $8b^2$

(D) $\frac{1}{b^2}$

(E) $\frac{1}{8b^2}$

12. Angela travels 209.2 km in 2 hours, what is her average speed?

(A) 104.6 h/km
(B) 104.6 km/h
(C) 72 h/km
(D) 20.92 km/h
(E) 20.92 h/km

13. A firefighter needs to use a 14-meter ladder to enter a window which is 13.5 meters above the ground. How far from the base of the building should the ladder be placed?

(A) 3.7 m
(B) 0.5 m
(C) 1 m
(D) 2.5 m
(E) 3 m

14. Harrison steps outside his house to see the hot air balloon pass by. He raises his eyes at a 35° angle to view the balloon. If the balloon is 5,000 feet above the ground, about how far is it from Harrison? Hint: Harrison's eye level is 5.2 feet from the ground.

(A) 6,100 feet
(B) 8,700 feet
(C) 7,100 feet
(D) 2,900 feet
(E) 4,500 feet

15. The Sweet Shoppe has 6 flavors of ice cream, 4 toppings, and 3 kinds of sprinkles. How many different sundaes can be made using one flavor of ice cream, one topping, and one kind of sprinkles?

(A) 13
(B) 19
(C) 36
(D) 72
(E) 24

16. The variable x and y vary inversely. When x is 9, y is 36. If x is 3, what is y?

(A) 3
(B) 12
(C) 30
(D) [illegible]
(E) [illegible]

17. Which is the graph of: $f(x) = \begin{cases} 2 & \text{if } x \leq 3 \\ 3 + x & \text{if } 3 < x < 5 \\ \frac{1}{2}x & \text{if } x \geq 5 \end{cases}$

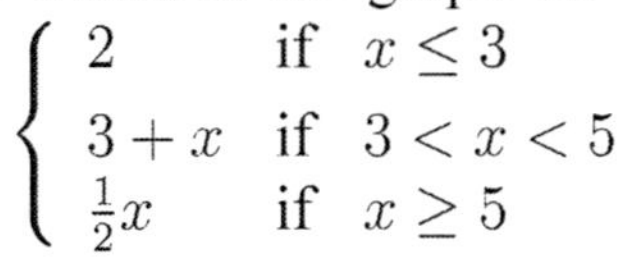

(A)

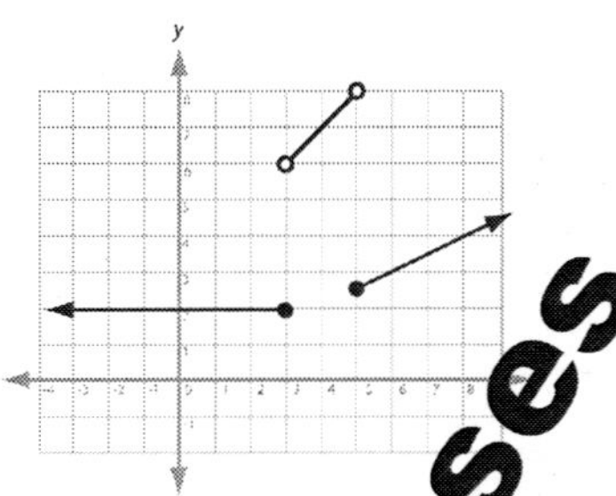

(B)

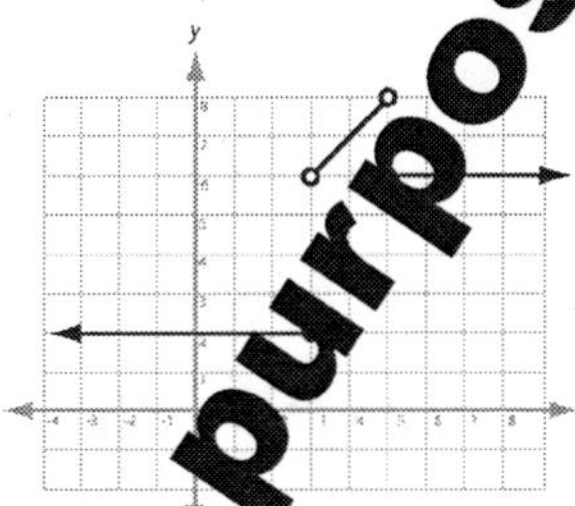

(C)

(D)

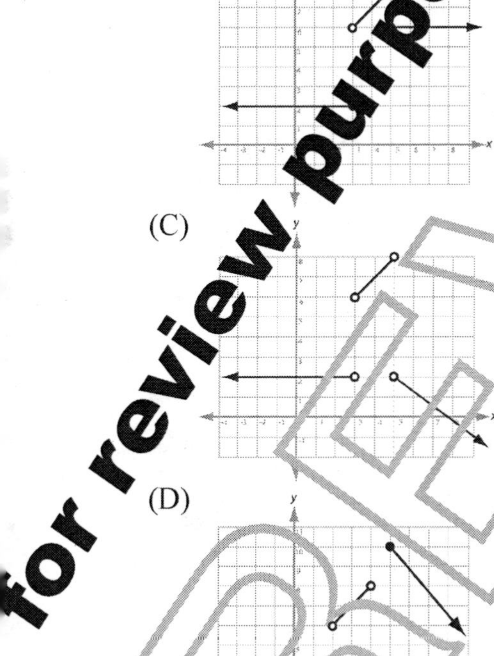

(E)

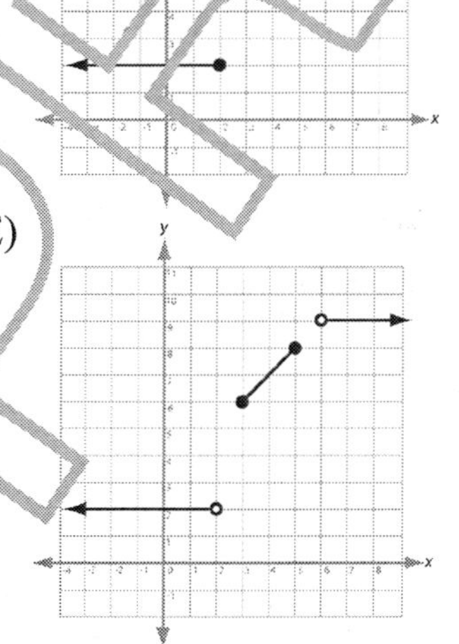

18. A bank charges a $10 fee if the account balance is less than $200. If the balance is in between $200 and $500 there is a $5 fee. If at least $500 is in the account, there is no fee. Graph the fee schedule for different account balances.

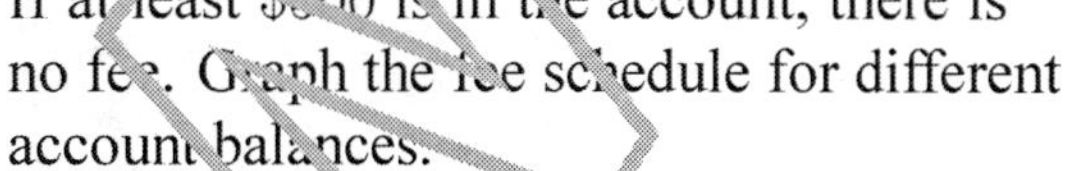

(A)

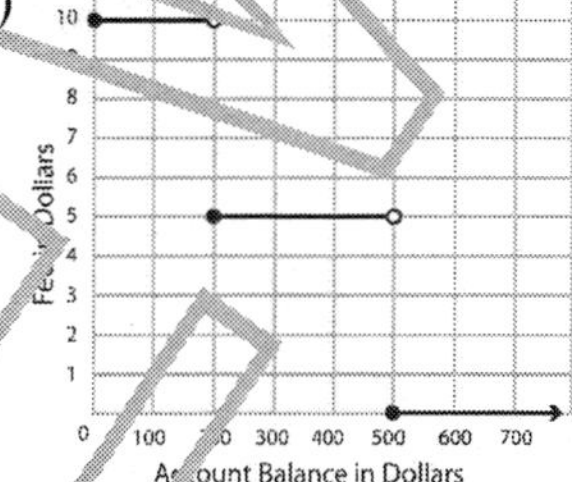

(B)

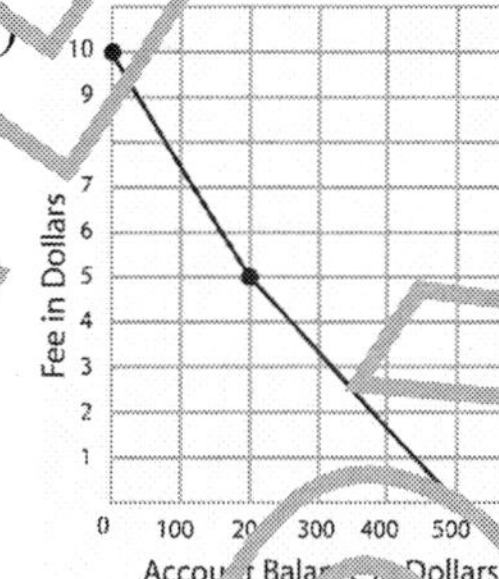

(C)

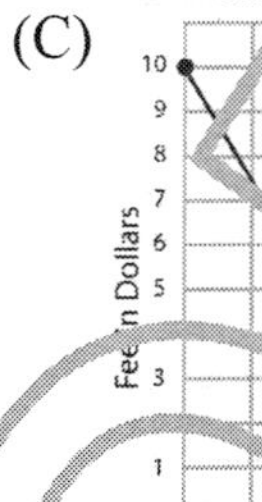

(D)

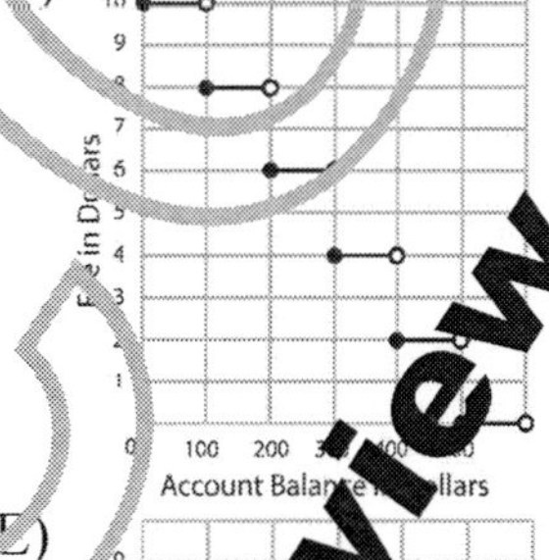

(E)

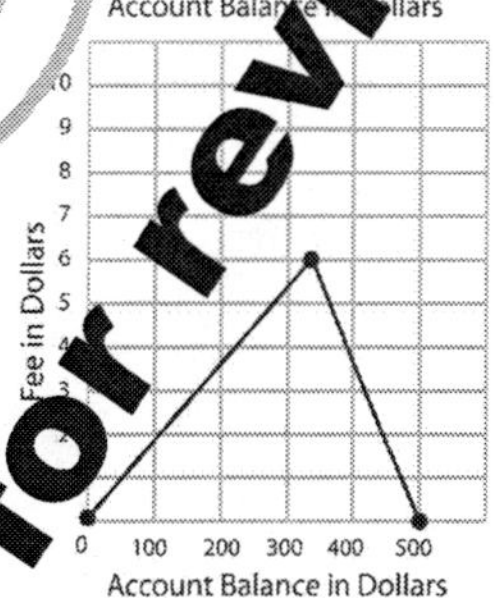

19. Suppose $\triangle DEF$ has vertices $D(-8,-2)$, $E(-5,-2)$, and $F(-8,-7)$. If $\triangle DEF$ is rotated 90° counterclockwise about the origin, what are the coordinates of the vertices of $\triangle D'E'F'$?

(A) $D'(2,-8)$, $E'(2,-5)$, $F'(7,-8)$
(B) $D'(-8,2)$, $E'(-5,2)$, $F'(-8,7)$
(C) $D'(7,-8)$, $E'(2,-8)$, $F'(2,-5)$
(D) $D'(2,-8)$, $E'(2,-5)$, $F'(8,-7)$
(E) $D'(2,-5)$, $E'(2,-8)$, $F'(7,-8)$

20. Which of the following is the graph of the solution set of $|3y+9|>12$?

(A)
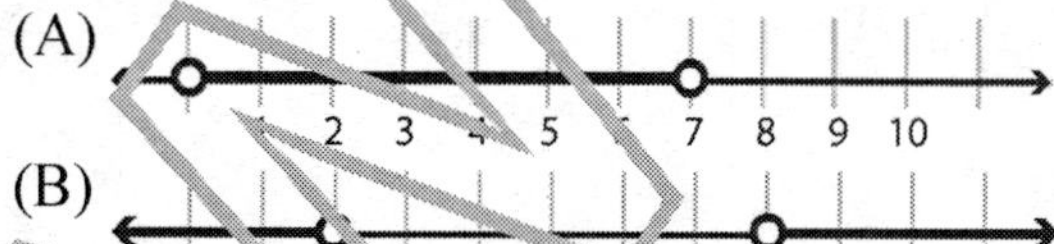

(B)
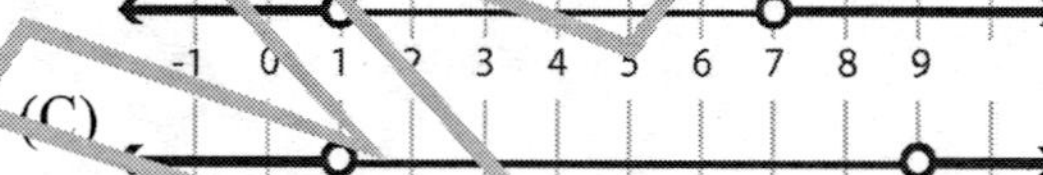

(C)
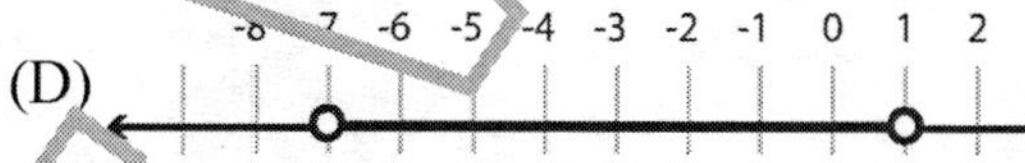

(D)

(E)
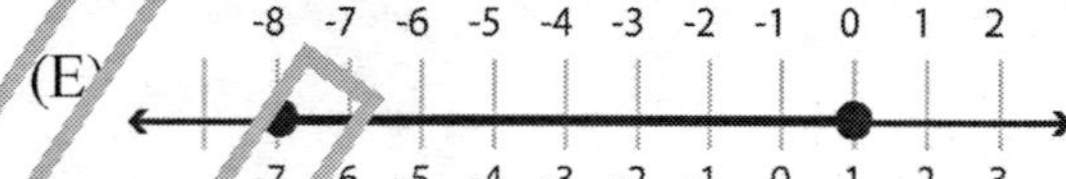

Part 3

Questions 1–16 of part 3 of this test are multiple choice. You have 20 minutes to complete this part of the test.

1. Michael is adding on a garage to his house. What do the planes of the floor and back wall of the house most closely represent?

   (A) coplanar planes
   (B) intersecting planes
   (C) parallel planes
   (D) bisecting planes
   (E) none of the above

2. The table shows the sales of athletic footwear for a recent year. Which measure of central tendency is the greatest?

Type of Shoe	Sales (in billions)
Aerobic	0.3
Baseball/softball	2.0
Basketball	2.3
Cross Training	0.2
Hiking	1.0
Running	2.6
Soccer	0.2
Sports Sandals	0.3
Tennis	0.8
Walking	1.2

   (A) mean
   (B) median
   (C) mode
   (D) None of these
   (E) Not enough information

3. The average of 7, 5, 9, 3, and $2x$ is x. What is the value of x?

   (A) 2
   (B) 2.4
   (C) 4
   (D) 6
   (E) 8

4. Suppose that in a one-hour television show, 24 minutes are used for advertising. If you start watching the show at a random time, what is the probability that you tune in during a commercial?

   (A) 0.24
   (B) 0.4
   (C) 0.6
   (D) 0.[illegible]7
   (E) Cannot be determined

5. The volume of a cube is 512 cubic centimeters. What is the length of one side?

   (A) 175 centimeters
   (B) 170 centimeters
   (C) 8 centimeters
   (D) 256 centimeters
   (E) 24 centimeters

6. Don sells 100 hamburgers per day for \$2 each, so his daily revenue is \$200. He estimates that for every 25 cents he increases the price of a hamburger, he will sell 5 fewer. What range of prices can he charge so that his daily revenue is at least \$225?

   (A) \$3.00 – \$4.00
   (B) \$4.00 – \$4.[illegible]0
   (C) \$[illegible].00 – \$4.[illegible]0
   (D) \$2.25 – \$4.50
   (E) \$2.50 – \$4.50

7. You are assigned an Algebra Project worth 150 points. There is a total of 42 five-point and two-point sections. How many two-point sections are included in the project?

(A) 18
(B) 20
(C) 22
(D) 24
(E) 26

8.

Column A	Column B
The slope of a line perpendicular to $y = \frac{10}{7}x + \frac{1}{7}$	The slope of a line through $(-2, -4)$ and $(8, 3)$.

Choose the statement that is true.

(A) The two quantities are equal.
(B) The quantity in Column A is greater.
(C) The quantity in Column B is greater.
(D) The quantity in column B is undefined.
(E) There is not enough information to determine the relationship.

9. Which of the following pairs of numbers has a mean of 64?

(A) 4 and 6
(B) 2 and 32
(C) 16 and 256
(D) 2 and 1024
(E) 32 and 96

10. Evaluate $p - (p + 3q) - r$ if $p = -7$, $q = 4$, and $r = -16$.

(A) -21
(B) 51
(C) -19
(D) 77
(E) 45

11. In a science experiment, students hung a cup from a spring and measured the length of the spring when candies were added to it. Their data are shown in the table below. Which statement is true?

# of Candies	**Length of Spring**
0	0.0
1	1.3
2	2.5
3	5.1
4	6.7
5	8.9
6	10.8
7	12.7
8	14.0
9	15.6

(A) The relation is a function because for each x value, there is exactly one y value.
(B) The relation is a function because the range values increase.
(C) The relation is not a function because only a line can be a function.
(D) The relation is not a function because there are two y values for some x values.
(E) Not enough information is given.

12. You are given the following information about $\triangle ABC$ and $\triangle XYZ$.

I. $\angle A \cong \angle X$
II. $\angle B \cong \angle Y$
III. $\overline{AB} \cong \overline{XY}$
IV. $\overline{BC} \cong \overline{YZ}$

Which combination cannot be used to prove $\triangle ABC \cong \triangle XYZ$?

(A) I, II, and IV
(B) I, III, and IV
(C) II, III, and IV
(D) I, II, and III
(E) All (I, II, III, and IV)

13. Suppose a rose bush grows one inch every year. If x is how old the rose bush is (in years), and $f(x)$ is how tall the rose bush is (in inches), what type of function is $f(x)$?

(A) identity function
(B) constant function
(C) step function
(D) absolute value function
(E) Not enough information given

14. $-\frac{1}{2} \times -\frac{2}{3} \times -\frac{3}{4} \times \frac{4}{5} =$

(A) $\frac{5}{6}$
(B) $-\frac{5}{6}$
(C) $-\frac{1}{6}$
(D) $-\frac{1}{5}$
(E) $\frac{1}{5}$

15. What is true about a system of equations in three variables that has 1 solution?

(A) planes intersect in the same plane
(B) planes have no points in common
(C) planes intersect in one line
(D) planes intersect in two lines
(E) planes intersect in one point

16. Evaluate the expression, $2b(4a + a^2)$, if $a = -9$ and $b = \frac{2}{3}$.

(A) -156
(B) 138
(C) 60
(D) -351
(E) 156

Index

Absolute value, 13
Acknowledgements, ii
Addition
  of polynomials, 110
Algebra
  consecutive integer problems, 181
  multi-step problems, 8[illegible]
  one-step problems with addition and subtraction, 69
  one-step problems with multiplication and division, [illegible]
  two step problems, 79
    with fractions, 80
  vocabulary, 50
  word problems, 52
    setting up, 54
Algorithms, 212, 213
Alternate Exterior Angles, 272
Alternate Interior Angles, 272
Angles, 264
  acute, 265
  adjacent, 269
  alternate exterior, 272
  alternate interior, 272
  central, 267
  complementary, 271
  corresponding, 272
  obtuse, 265
  right, 265
  straight, 265
  sum of polygon interior, 273
  sum of triangle interior, 279
  supplementary, 271
  vertical, 270
Area
  circle, 302
  squares and rectangles, 298
  triangles, 299
  two-step problems, 30[illegible]
Arguments, 218
  inductive and deductive, 219
Associative Property of Addition, 57
Associative Property of Multiplication, 57

Bar Graphs, 240
Base, 42, 5[illegible]
Binomials, 108, 122
  multiplication
    FOIL method, 115
Box-and-Whisker Plot, 232

Cartesian Plane, 65, 140, 165
Central Angle, Circle, 300
Circle Definitions, 300
Circle Graphs, 242
Circumference, Circle, 301
Coefficient, 50
  leading, 50
Collinear Lines, 172
Combination, 259
Combining like terms, 82
Commutative Property of Addition, 57
Commutative Property of Multiplication, 57
Complement, 1[illegible]
Complementary Angles, 2[illegible]
Completing the Square, 137
Conclusion, 218
Congruent Figures, [illegible], 276
Constant, 50
Contrapositive, 220
Converse, 220
Corresponding Angles, 272
Counterexample, 220

Decimal Word Problems, 29
Decimals
  changing to fractions, 28
  changing to percents, 30

changing with whole numbers to mixed numbers, 29

Deductive Reasoning, 218

Degree, 50

Denominator, 16, 48, 113

Dependent Events, 247

Diagnostic Test, 1

Diameter, Circle, 301

Difference of Two Squares, 134

Digits, 12

Dilation, 323

Direct and Indirect Variation, 104

Discount, Finding the Amount of, 36

Discounted Sale Price, Finding, 37

Distance Formula

$d = \sqrt{(y_2 - y_1)^2 + (x_2 - x_1)^2}$, 143

Distributive Property, 57

Domain, [illegible], 191

Element, 17

Empirical Rule, 251

Empty Set

$\emptyset$, 17

Equations

finding using two points or a point and slope, 151

linear, 140

linear systems

solving by adding or subtracting, 176

solving by substitution, 175

solving systems of equations, 172

of perpendicular lines, 159

solving with absolute values, 90

solving with like terms, 82

writing from data, 161

Equilateral Triangle, 282

Evaluation Chart, 9

Exponential Growth and Decay, 200

Exponents, 42, 108

division with, 44

multiplication using, 43

multiplying polynomials, 111

of polynomials, 109

simplifying binomial expressions, 116

when subtracting polynomials, 110

Extremes, 231

Factoring

by grouping, 121

difference of two squares, 128

of polynomials, 118

quadratic equations, 133

trinomials, 122

trinomials with two variables, 127

FOIL method, 115, 121, 122

for multiplying binomials, *see* Binomials

Fractions

changing to decimals, 27

changing to percents, 31

improper, 62

word problems, 26

Frequency Table, 237

Functions, 191

notation, 192

piecewise, 202

qualitative behavior, 196

real-life, 198

recognizing, 193

Geometric Probability, 253

Geometric Relationships of Plane Figures, 305

Graphing

a line knowing a point and slope, 150

fractional values, 60

horizontal and vertical lines, 142

inequalities, 152

linear data, 162

linear equations, 140

non-linear equations, 165

on a number line, 60

systems of inequalities, 178

Graphs

bar, 240

circle, 242

line, 241

pie, 242

Greatest Common Factor, 24

Histogram, 238
Hypotenuse, 283

Identity Property of Addition, 57
Identity Property of Multiplication, 57
Independent Events, 247
Inductive reasoning, *see* Patterns, 218
Inequalities
  graphing, 74
  graphing systems of, 1[illegible]8
  multi-step, 88
  solution sets, 74
  solving by addition and subtraction, 75
  solving by multiplication and division, 76
  solving with absolute values, 90
  systems of, 172
  word problems, 93
Inequality
  definition, 50
Inscribed Angle, Circle, 300
Integers, 12, 11[illegible]
  even, 12
  odd, 12
Intercepts of a Line, 145
Intersecting Lines, 172, 174
Intersection
  ∩, 19[illegible]
Interval, 202
Inverse, 2[illegible]0
Inverse Property of Addition, 57
Inverse Property of Multiplication, 57
Irrational numbers, [illegible]1
Isosceles Triangle, 282

Least Common Multiple, 25
Line, 296
  segment, 296
Line Graphs, 241
Line of Reflection, 316
Linear Equation, 140, 148
Lines
  collinear
  coinciding, 172
  diagonal, 296
  horizontal, 296
  intersecting, 172, 174, 296
  of best fit, 235
  parallel, 172, 296
  perpendicular, 296
  skew, 296
  vertical, 296
Logic, 21[illegible]

Major Arc, Circle, 300
Mathematical Reasoning, 218
Mean, 224, 225
Measures of Central Tendency, 228
Median, 226, 231
Midpoint of a Line Segment
  $M = \left(\frac{x_1+x_2}{2}, \frac{y_1+y_2}{2}\right)$, 144
Minor Arc, Circle, 300
Mixed Numbers
  changing to decimals, [illegible]8
  changing to percents, [illegible]2
Mode, 227
Monomials, 108
  adding and subtracting, 108
  multiplying, 111
  multiplying by polynomials, 1[illegible]
Multi-Step Algebra Problems, [illegible]

Negative Numbers
  multiplying and dividing with, 72
Normal Distribution, 251
Number line, 60
  vertical, 64
Numerator, 16

One-Step Algebra Problems
  addition and subtraction, 69
  multiplication and division, 70
Order of operations, 15, 16
Ordered Pair, 65, 140, 184
Origin, 65

Parabola, 165

Parallel lines, 172
Parentheses
removing, 84
Parentheses, removing and simplifying
polynomials, 114
Patterns
geometric, 206
inductive reasoning, 214
number, 205
Percent of Increase and Decrease, Finding, 35
Percent of Total, Finding, 34
Percent Word Problems, 33
Percents
changing to decimals, 30
changing to fractions, 31
changing to mixed numbers, 32
Perfect Squares, 128, 136
Perimeter, [illegible]
Permutation, 256
Perpendicular Lines
equations of, 159
π, [illegible]
Point-Slope Form of an Equation
$y - y_1 = m(x - x_1)$, 151
Polygons, 296, 297
sum of interior angles, 273
Polynomial(s), 108, 118
adding, 109
dividing by monomials, 113
factoring, 118
greatest common factor, 118–120
multiplying by monomials, 112
subtracting, 110
Practice Test 1, 328
Practice Test 2, 335
Preface, xi
Premises
of an argument, 218
Prime Numbers, 12
Probability, 245
Probability Distribution, 251
continuous, 251
discrete, 251
Product, 118
Proportions, 101
Proposition, 218
Pythagorean Theorem, 283
applications, 285

Quadratic equation, 132
$ax^2 + bx + c = 0$, 138
Quadratic Equations
finding the vertex, 166
Quadratic formula
$\frac{-b \pm \sqrt{b^2 - 4ac}}{2a}$, 138
Quartiles, 231

Radical Equations, 87
Radius, Circle, 301
Range, 184, 191, 223
Rational numbers, 11
Ratios, 99
Ray, 296
Real numbers, 11
Reflection, 316
line of, 316
Relations, 184
Right Triangle, 282, 283
Rotation, 321

Sales Tax, 38
Scale Factor, 323
Scatter Plot, 233
Secant, Circle, 300
Sentence, 50
Sequences, 205
limit, 208
Series
arithmetic, 2[illegible]
geometric, 210
limit, 208
Set, 17
Similar Figures, 276
Similar Triangles, 280
Slope, 148, 172
changing the slope of a line, 157

$m = \dfrac{y_2 - y_1}{x_2 - x_1}$, 146
Slope Intercept Form of a Line
  $y = mx + b$, 148, 172
Square roots, 45, 48, 62
  adding and subtracting, 46
  dividing, 48
  multiplying, 47
Statistics, 223
Stem-and-Leaf Plot, 229
Subset
  $\subseteq$, 18
Substitution
  numbers for variables, 51
Subtraction
  of polynomials, 110
Subtrahend, 110
Supplementary Angles, 271

Table of Contents, x
Tables
  data, 239
Tally Chart, 237
Tangent, Circle, 300
Term, 50
Translation, 319
Transversal, 272
Tree Diagram, 249
Triangles
  corresponding sides, 280
  equilateral, 282
  isosceles, 282
  proportional sides, 280
  right, 282
  side and angle relationships, 282
  similar, 280
  $30 - 60 - 90$, 287
  $45 - 45 - 90$, 287
  sum of interior angles, 279
Trigonometric Functions
  $\cos A = \dfrac{\text{adj.}}{\text{hyp.}}$, 290
  $\tan A = \dfrac{\text{opp.}}{\text{adj.}}$, 290
  $\sin A = \dfrac{\text{opp.}}{\text{hyp.}}$, 290
Trinomials, 108
  factoring, 122
  factoring with two variables, 127
Two Step Algebra Problems, 79
  with fractions, 80
Two-Step Area Problems, 303

Union
  $\cup$, 20

Variable, 50, 69, 83, 108, 109, 111
  coefficient of negative one, 7[illegible]
Venn Diagram, 19, 21
Volume
  cubes, 309
  cylinders, 310
  rectangular prisms, 308
  rectangular solids, 307

Whole numbers, 12
Word Problems, 14
  algebra, 52
    setting up, 54
  changing to algebraic equations, [illegible]
  decimals, 29
  distance, 98
  fractions, 26
  percent, 3[illegible]
  rate, 96
  ratios and proportions, 102
  systems of equations, 179
  time of travel, 9[illegible]
  two-step problems, 14

x-axis, 65, 67

y-axis, 65, 67
y-intercept, 172
  changing the intercept of a line, 157

SAT

Please fill out the form completely, and return by mail or fax to American Book Company.

Purchase Order #: ____________________ Date: ____________________

Contact Person: ____

School Name (and District, if any): ____

____

Billing Address: Street Address: ☐ same as billing

____________________ ____________________

Attn: ____________________ Attn: ____________________

____________________ ____________________

____________________ ____________________

Phone: ____________________ E-Mail: ____________________

Credit Card #: ____________________ Exp Date: ____________________

Authorized Signature: ____

er Number	Product Title	Pricing* (10 books)	Qty	Pricing (30+ books)	Qty	Pricing (30 e-books)	Qty	Pricing (30 books+e-books)	Qty	Total Cost
AT-M1205	SAT Math Test Preparation Guide	$169.90 (1 set of 10 books)		$329.70 (1 set of 30 books)		$329.70 (1 set of 30 books)		$599.40 (30 books+ 30 e-books)		
AT-R1205	SAT Reading Test Preparation Guide	$169.90 (1 set of 10 books)		$329.70 (1 set of 30 books)		$329.70 (1 set of 30 books)		$599.40 (30 books+ 30 e-books)		
AT-W0705	SAT Writing Test Preparation Guide	$169.90 (1 set of 10 books)		$329.70 (1 set of 30 books)		$329.70 (1 set of 30 books)		$599.40 (30 books+ 30 e-books)		
CT-E0708	ACT English Test Preparation Guide	$129.90 (1 set of 10 books)		$299.70 (1 set of 30 books)		$299.70 (1 set of 30 books)		$539.40 (30 books+ 30 e-books)		
CT-M0708	ACT Mathematics Test Preparation Guide	$139.90 (1 set of 10 books)		$329.70 (1 set of 30 books)		$329.70 (1 set of 30 books)		$599.40 (30 books+ 30 e-books)		
CT-R0708	ACT Reading Test Preparation Guide	$129.90 (1 set of 10 books)		$299.70 (1 set of 30 books)		$299.70 (1 set of 30 books)		$539.40 (30 books+ 30 e-books)		
CT-S0608	ACT Science Test Preparation Guide	$129.90 (1 set of 10 books)		$299.70 (1 set of 30 books)		$299.70 (1 set of 30 books)		$539.40 (30 books+ 30 e-books)		
CT-W0711	ACT Writing Test Preparation Guide	$129.90 (1 set of 10 books)		$299.70 (1 set of 30 books)		$299.70 (1 set of 30 books)		$539.40 (30 books+ 30 e-books)		
GR-1007	Basics Made Easy: Grammar and Usage Review	$119.90 (1 set of 10 books)		$269.70 (1 set of 30 books)		$269.70 (1 set of 30 books)		$479.40 (30 books+ 30 e-books)		
MR-1000	Basics Made Easy: Mathematics Review	$139.90 (1 set of 10 books)		$329.70 (1 set of 30 books)		$329.70 (1 set of 30 books)		$599.40 (30 books+ 30 e-books)		
RR-0609	Basics Made Easy: Reading Review	$129.90 (1 set of 10 books)		$299.70 (1 set of 30 books)		$299.70 (1 set of 30 books)		$539.40 (30 books+ 30 e-books)		
WR-0609	Basics Made Easy: Writing Review	$139.90 (1 set of 10 books)		$329.70 (1 set of 30 books)		$329.70 (1 set of 30 books)		$599.40 (30 books+ 30 e-books)		
S-L0710	Projecting Success! Language Arts Digital Slides	$39.00 (1 digital set)								
ASH-M0106	Mathematics Flash Cards	$69.00 (1 set)								
ASH-0804	High School Science Flash Cards (English Version)	$59.00 (1 set)								
ASH-B0906	High School Science Flash Cards (English/Spanish)	$79.00 (1 set)								
ASH-P0808	High School Periodic Table Flash Cards	$59.00 (1 set)								

1

*Minimum order is 1 set of 10 books of the same subject.

Subtotal ________

Shipping & Handling 12%
Shipping 6% on print and digital packages
Shipping waived on digital resources. ________

Total ________

erican Book Company • PO Box 2638 • Woodstock, GA 30188-1383
l Free Phone: 1-888-264-5877 • Toll-Free Fax: 1-866-827-3240
b Site: www.americanbookcompany.com